UNDERSTANDING THE PHYSICS OF TOYS

Principles, Theory and Exercises

UNDERSTANDING THE PHYSICS OF TOYS

Principles, Theory and Exercises

S Rajasekar
Bharathidasan University, India

R Velusamy
Ayya Nadar Janaki Ammal College, India

Miguel A F Sanjuán
Universidad Rey Juan Carlos, Spain

World Scientific

NEW JERSEY • LONDON • SINGAPORE • BEIJING • SHANGHAI • HONG KONG • TAIPEI • CHENNAI • TOKYO

Published by

World Scientific Publishing Co. Pte. Ltd.
5 Toh Tuck Link, Singapore 596224
USA office: 27 Warren Street, Suite 401-402, Hackensack, NJ 07601
UK office: 57 Shelton Street, Covent Garden, London WC2H 9HE

British Library Cataloguing-in-Publication Data
A catalogue record for this book is available from the British Library.

UNDERSTANDING THE PHYSICS OF TOYS
Principles, Theory and Exercises

ISBN 978-981-126-800-7 (hardcover)
ISBN 978-981-126-801-4 (ebook for institutions)
ISBN 978-981-126-802-1 (ebook for individuals)

For any available supplementary material, please visit
https://www.worldscientific.com/worldscibooks/10.1142/13190#t=suppl

Desk Editor: Rhaimie Wahap

Dedication Page

To Professor M. Lakshmanan
To my students
To my wife, Céline, and my daughters, Alicia and Mónica

Preface

Science toys are generally a source of entertainment, display surprising actions, decorate houses and offices and can be played with for hours. Such toys are fascinating to the children, high school and college students, teachers and scientists. Many great scientists were captivated by toys – Ludwig Prandtl, Wolfgang Pauli, Niels Bohr, Gaspard-Gustave de Coriolis, James Clerk Maxwell, Michael Faraday to just mention a few. Study of toys leads to scientific exploration, engineering applications, technological developments, permanently attract children and students to the excitement of science and paves the way for a promising career in science. Interestingly, most of the science toys are based on the principles of physics.

A difficulty in teaching subjects like mechanics and electricity and magnetism is simulating students to probe and understand the behaviour of objects of everyday life. Science toys are helpful for this. With the use of simple toys, abstract physical concepts and statements can be illustrated and many physical qualitative interconnections can be well established. Moreover, the demonstration of many fundamentals of physics and engineering through the working principles of popular toys is inexpensive, quickly reaches the senses and inspire the students better learning and more discoveries. The systematic way of setting up of theoretical model equations for the toys provides a remarkable experience to students on construction of mathematical model equations to physical and engineering systems. Experiments with a great variety of toys were made in microgravity environment of space by different space shuttle missions for the purpose of testing some basic laws of physics.

A course on *The Physics of Toys* designed based on this book can be taught as a full course at any level and even to the students who have never underwent a physics course. Further, the toys can be used to illustrate various concepts and principles in different branches of physics and engineering including the current research topics. Some toys are recommended to kids to keep them calm, focus, listen and participate by mental health counselors and therapists.

Considering the above and catering to the needs of graduate and master-level programme students in physics and engineering the present book covers more than 40 wide range of popular toys. For each toy the presentation covers various features including history, construction, working principle, theoretical model, a solved problem and $5-10$ exercises. The mechanical toys presented in this book are useful for teaching centre of mass, forces, energy, momentum, conservation of energy and momentum, stability of equilibrium points, wave propagation and types of simple motions. Spinning toys are suitable for addressing the notions of moment of inertia, centripetal force, torque, angular momentum, friction, role of centre of mass, unexpected effects of minute disturbances, various rotational motions and conversion of oscillations into rotations and one form of energy to another form. Flying and throwable toys are helpful to describe the basis of the applied aviation physics, influence of drag force, pressure difference, lift force and Bernoulli's principle. The toys working with the principles of heat and thermodynamics are simple things to illustrate blackbody radiation and absorption, capillary effect, liquid-vapour evaporation, thermal diffusion and energy conversion. The electric and magnetic toys are very thoughtful to demonstrate force versus energy, electromotive force, escape velocity, conversion of magnetic energy to kinetic energy, magnetic dipole interaction, influence of applied voltage, corona discharge, triboelectricity, electrostatic induction etc.

The book contains 8 parts. Part-I reviews the basic concepts required for the understanding of functioning of the various toys. Part-II deals with the mechanical toys such as roly-poly, seesaw, Newton's cradle, slinky, woodpecker and Cartesian diver. Spinning toys are considered in Part-III. Particularly, the fascinating spinning top, tippe top, PhiTOP, Euler's disk, fidget spinner, rattleback, hurricane balls, the notched stick, gyroscope, the buzzer, yo-yo, astrojax and hula hoop are covered. Part-IV presents flying toys, namely, balsa gliders, mangus glider and kite. The throwable toys frisbee, boomerang, skipping stones and bouncing popper are covered in Part-V. The toys drinking bird, putt-putt boat and Crookes radiometer working under the principles of heat and thermodynamics are treated in the next part. Part-VII is devoted to the electric and magnetic toys. The toys Gauss riffle, the Levitron, an electric train, a simple electromagnetic train, fun-fly stick and plasma ball are discussed. The last part, Part-VIII, presents briefly the features of certain other popular toys including balancing toys, gyro-ring, toy guns, floating ping-pong ball, helicopter toy, pinwheel, tornado bottle, magnetic levitation of iron balls, magnetic seal and ball, simple DC motors and kaleidoscope.

We are thankful to K.P.N. Murthy, K. Murali, P. Philominathan and V. Chinnathambi for their suggestions. We extend our thanks to our family members for their great support and cooperation during the course of preparation of this work.

S. Rajasekar
School of Physics
Bharathidasan University
Tiruchirapalli 620024
India

R. Velusamy (Retired)
Department of Physics
Ayya Nadar Janaki Ammal College
Sivakasi 626124
India

Miguel A.F. Sanjuán
Department of Physics
Universidad Rey Juan Carlos
28933 Móstoles, Madrid
Spain

Contents

Part I: Introduction

Chapter 1

Fundamental Concepts

Everything in the universe is in motion. Study of motion is called *mechanics*. There are several laws that explain the motion and causes for the changes of motion of bodies. To understand and study motions one must know many quantities like displacement, velocity, rotational angle, linear momentum, angular momentum, force, torque, linear acceleration, angular acceleration, potential energy, kinetic energy, work done, power, etc. One must become familiar with laws governing the motion of bodies. To study the motion of the objects, one has to learn how to setup the equations of motion and solve them. One must know how to interpret the solutions of the equations of motion. Bernoulli's principle is useful to understand the flight of flying toys.

There are different types of motion, from simple to very complicated ones. Dynamical systems including toys can exhibit translational, rotational, rolling, slipping, oscillatory, spinning, precession, nutation and many more types of motion. As a toy is simply a machine, the working of it can be explained only by the understanding of the motion of the toy. We can say that the physics of a toy is just the study of motion of that toy. Therefore, we give a brief introduction to the fundamental concepts involved in the motion of objects in this chapter.

All motions are described relative to a frame of reference of a coordinate system. Choosing a proper coordinate system will simplify the mathematics of solving the equations of motion of the bodies. We must also know how to relate a laboratory coordinate system with the body centred coordinate system attached with the moving body. All toys interact with their surroundings. Therefore, we must know how does the interaction of a toy with its surrounding affect its working. Though many toys look simple and seem to have simple motions, it is quite difficult to explain their working with simple theories and explanations. The function of certain toys are governed by the various laws of electricity and magnetism. The dynamics of toys is often modelled by appropriate model equations. Usually these model equations are ordinary differential equations, particularly, nonlinear equations. Systematic general methods are not available to construct exact analytical solutions for nonlinear differential equations. Therefore, one may look for steady state solutions (equilibrium points) and analyse their stability under weak perturbations. Stability analysis is useful, for example, in the study of toys such as gyroscope, Levitron, hurricane balls, gee-haw whammy-diddle and hula hoop, just to mention a few. We review these in the present chapter.

1.1 Kinematic Quantities of Motion

Kinematics of motion of a body refers to the description of the position, orientation and their changes with respect to time. Coordinate systems are needed to represent the position of the body and orientation. For example, if the Cartesian coordinate system (x, y, z) along the three mutually perpendicular directions are chosen then the position vector $\mathbf{r}$ of the body with respect to the origin can be given as

$$\mathbf{r} = x\mathbf{i} + y\mathbf{j} + z\mathbf{k}. \tag{1.1}$$

The displacement between two points (x_1, y_1, z_1) and (x_2, y_2, z_2) can be given as

$$\mathbf{r} = \mathbf{r}_2 - \mathbf{r}_1 = (x_2 - x_1)\mathbf{i} + (y_2 - y_1)\mathbf{j} + (z_2 - z_1)\mathbf{k}. \tag{1.2}$$

For a body in motion $\mathbf{r}$ changes with time. The rate of change of displacement is called the *velocity* of the body $\mathbf{v} = \mathrm{d}\mathbf{r}/\mathrm{d}t = \dot{\mathbf{r}}$. The magnitude of $\mathbf{v}$ is called the *speed* of the body. A body is said to be in uniform motion if its velocity is constant with respect to time. For nonuniform motion velocity changes with time. The rate of change of velocity is called *linear acceleration* $\mathbf{a} = \mathrm{d}\mathbf{v}/\mathrm{d}t$. The product of the mass and velocity is called the *linear momentum* of the body denoted as $\mathbf{p} = m\mathbf{v}$. The rate of change of linear momentum $\mathrm{d}\mathbf{p}/\mathrm{d}t$ is the force $\mathbf{F}$ acting on the body.

A rotating body is also a moving body. The rotation angle θ is measured with respect to the direction of a rotating vector with respect to a reference direction attached to a coordinate system. So, just like the linear velocity, we define the angular velocity $\boldsymbol{\omega}$ as the rate of change of angle $\boldsymbol{\theta}$ as $\boldsymbol{\omega} = \mathrm{d}\boldsymbol{\theta}/\mathrm{d}t$. Any rotation has to take place in a plane perpendicular to an axis of rotation. The angular velocity $\boldsymbol{\omega}$ is taken as the direction of the axis of rotation perpendicular to the plane in which the rotation takes place. The rate of change of the angular velocity gives the angular acceleration $\boldsymbol{\alpha} = \mathrm{d}\boldsymbol{\omega}/\mathrm{d}t$.

We get the linear momentum by multiplying the linear velocity by mass m. Mass is the inertial quantity which opposes the linear motion of an object. Similar to mass there is an inertial quantity which opposes the rotational motion of an object. It is called the *moment of inertia* I. The product of I and the angular velocity $\boldsymbol{\omega}$ gives the *angular momentum* $\mathbf{L} = I\boldsymbol{\omega}$. We can prove that the moment of linear momentum is the angular momentum $\mathbf{L} = \mathbf{r} \times \mathbf{p}$. The twisting force of a rotating body is termed as *torque* $\boldsymbol{\tau} = \mathbf{r} \times \mathbf{F}$ where $\mathbf{F}$ is the applied force and $\mathbf{r}$ is the position vector. Its magnitude is $\tau = rF\sin\theta$ where θ is the angle between $\mathbf{r}$ and $\mathbf{F}$. Also, $\boldsymbol{\tau} = \mathrm{d}\mathbf{L}/\mathrm{d}t$.

The dynamics of any object can be described using the kinematic quantities described above.

1.2 Moment of Inertia

Moment of inertia [1] plays the role in rotational motion that mass plays in linear motion. Both are inertial quantities offering resistance of a body to changes in its motion. Moment of inertia is calculated with respect to an axis of rotation. If the axis of rotation changes, then the moment of inertia may change. If there is a point mass m at a perpendicular distance r from the axis of rotation, then the moment of inertia is defined as $I = mr^2$. If we have a system of N particles of masses m_1, m_2, ..., m_N at perpendicular distances $\mathbf{r}_1$, $\mathbf{r}_1$, ..., $\mathbf{r}_N$, respectively, from the axis, then the moment of inertia of the system is given by

$$I = \sum_{i=1}^{N} m_i |\mathbf{r}_i|^2. \tag{1.3}$$

Suppose a solid mass of uniform density is distributed over a region, then the summation in the above equation is replaced with an integral as

$$I = \int r^2 \mathrm{d}m, \tag{1.4}$$

where $\mathrm{d}m$ is the elemental mass $\mathrm{d}m$ at a distance r from the axis of rotation. If the mass distribution is not uniform in the solid then $\mathrm{d}m$ is replaced by $\rho(r)\mathrm{d}V$ where $\rho(r)$ is the density of the solid at position r.

If I_{CoM} is the moment of inertia about an axis passing through the centre of mass (CoM) the moment of inertia about a parallel axis at a distance d from the axis passing through the CoM G is given by the parallel-axis theorem as

$$I_{\text{parallel-axis}} = I_{\text{CoM}} + md^2. \tag{1.5}$$

For a plane laminar body in the $x - y$ plane the moment of inertia about the perpendicular axis I_z is given by the perpendicular-axis theorem as

$$I_z = I_x + I_y, \tag{1.6}$$

where I_x and I_y are the moments of inertia of the laminar body about the x and y-axes, respectively.

For a system of N particles of masses $m_1, m_2, \ldots, m_N$ with position vectors $\mathbf{r}_1, \mathbf{r}_2, \ldots, \mathbf{r}_N$ from the origin, the total angular momentum of the body about the axis passing through the origin is given as

$$\begin{aligned}
\mathbf{L} &= \sum_{i=1}^{N} \mathbf{r}_i \times m_i \dot{\mathbf{r}}_i \\
&= \sum_{i=1}^{N} m_i \mathbf{r}_i \times (\boldsymbol{\omega} \times \mathbf{r}_i) \\
&= \sum_{i=1}^{N} m_i \left[|\mathbf{r}_i|^2 \boldsymbol{\omega} - (\mathbf{r}_i \cdot \boldsymbol{\omega}) \mathbf{r}_i \right] \\
&= \sum_{i=1}^{N} m_i \left[\left(x_i^2 + y_i^2 + z_i^2\right) \left(\omega_x \mathbf{i} + \omega_y \mathbf{j} + \omega_z \mathbf{k}\right) \right. \\
&\qquad \left. - \left(x_i \omega_x + y_i \omega_y + z_i \omega_z\right) \left(x_i \mathbf{i} + y_i \mathbf{j} + z_i \mathbf{k}\right) \right].
\end{aligned} \tag{1.7}$$

This equation can be written in tensor form as

$$\begin{bmatrix} L_x \\ L_y \\ L_z \end{bmatrix} = \begin{bmatrix} I_{xx} & I_{xy} & I_{xz} \\ I_{yx} & I_{yy} & I_{yz} \\ I_{zx} & I_{zy} & I_{zz} \end{bmatrix} \begin{bmatrix} \omega_x \\ \omega_y \\ \omega_z \end{bmatrix}, \tag{1.8}$$

where

$$I_{xx} = \sum_{i=1}^{N} m_i \left(y_i^2 + z_i^2\right), \quad I_{yy} = \sum_{i=1}^{N} m_i \left(z_i^2 + x_i^2\right), \tag{1.9a}$$

$$I_{zz} = \sum_{i=1}^{N} m_i \left(x_i^2 + y_i^2\right), \quad I_{xy} = I_{yx} = -\sum_{i=1}^{N} m_i x_i y_i, \tag{1.9b}$$

$$I_{yz} = I_{zy} = -\sum_{i=1}^{N} m_i y_i z_i, \quad I_{zx} = I_{xz} = -\sum_{i=1}^{N} m_i z_i x_i. \tag{1.9c}$$

For a solid body with uniform density the tensor components can be written by replacing the summation into integration as

$$I_{xx} = \int \left(y^2 + z^2\right) \mathrm{d}m, \quad I_{yy} = \int \left(z^2 + x^2\right) \mathrm{d}m, \tag{1.10a}$$

$$I_{zz} = \int \left(x^2 + y^2\right) \mathrm{d}m, \quad I_{xy} = I_{yx} = -\int xy\mathrm{d}m, \tag{1.10b}$$

$$I_{yz} = I_{zy} = -\int yz\mathrm{d}m, \quad I_{zx} = I_{xz} = -\int zx\mathrm{d}m. \tag{1.10c}$$

Equation (1.8) can be written as $\mathbf{L} = I\boldsymbol{\omega}$ where I is the symmetric tensor called *moment of inertia tensor*.

It is always possible to find three mutually perpendicular axes (principal axes) in which the moment of inertia tensor is diagonal with three principal moments of inertia I_1, I_2 and I_3 as

$$I = \begin{bmatrix} I_1 & 0 & 0 \\ 0 & I_2 & 0 \\ 0 & 0 & I_3 \end{bmatrix}. \tag{1.11}$$

Moment of inertia tensor can be found in any coordinate system. Then finding the principal moments of inertia and the principal axes become just finding the eigenvalues and eigenvectors of the moment of inertia determinant. Solving the determinant

$$\begin{vmatrix} I_{xx} - I & I_{xy} & I_{xz} \\ I_{yx} & I_{yy} - I & I_{yz} \\ I_{zx} & I_{zy} & I_{zz} - I \end{vmatrix} = 0 \tag{1.12}$$

will give a cubic equation in I. The three roots of the equation will give the principal moments of inertia I_1, I_2 and I_3. Corresponding to these three eigenvalues, one can find three eigenvectors which are the principal axes for the moment of inertia of the given solid. In most problems these three axes would be the most preferred coordinate system to setup the equations of motion.

Solved Problem 1:

Consider a hollow rod of length L and mass M. Two beads of mass m each with negligible size are kept at the centre initially. The rod rotates with respect to the centre vertical axis with angular velocity ω_0 when the beads are at the centre of the rod. What will be the angular velocity of the rod when the two beads move to the two ends of the rod? The moment of inertia of the rod of length L and mass M through the centre vertical axis is $ML^2/12$.

As no external torque acts the angular momentum is conserved. Therefore,

$$L_{\text{i}} = I_0\omega_0 = \frac{1}{12}ML^2\omega_0 \tag{1.13}$$

and

$$L_{\text{f}} = I\omega. \tag{1.14}$$

Using the parallel-axis theorem we obtain

$$I = \frac{1}{12}ML^2 + m\left(\frac{L}{2}\right)^2 + m\left(\frac{L}{2}\right)^2 = \left(\frac{M}{12} + \frac{m}{2}\right)L^2. \tag{1.15}$$

As the angular momentum is conserved $L_{\text{i}} = L_{\text{f}}$. That is

$$\frac{1}{12}ML^2\omega_0 = \left(\frac{M}{12} + \frac{m}{2}\right)L^2\omega \tag{1.16}$$

which gives

$$\omega = \frac{M\omega_0}{M + 6m}. \tag{1.17}$$

1.3 Dynamics: Equations of Motion and Conservation Laws

Equations of motion for dynamics are mathematical equations relating the kinetic quantities like displacement, velocity, linear momentum, linear acceleration, angle of rotation, angular velocity and angular acceleration with dynamic quantities like force, torque, energy, work and power.

The first law of Newton states that every body continues in a state of rest or in uniform motion along a straight-line unless compelled by an external force acts on that body to change that state. So, force is a dynamical quantity which changes the state of rest or uniform motion of a body along a straight-line. The force is defined by the second law of motion which states that the rate of change of linear momentum is proportional to the force acting on the body, $\mathrm{d}\mathbf{p}/\mathrm{d}t = k\mathbf{F}$. The units of $\mathbf{p}$ and $\mathbf{F}$ are chosen in such a way that k is made unity. Therefore, the Newton's second law gives $\mathrm{d}\mathbf{p}/\mathrm{d}t = \mathbf{F}$. For a system of N particles we write

$$\frac{\mathrm{d}\mathbf{p}_i}{\mathrm{d}t} = \mathbf{F}_i, \quad i = 1, 2, \ldots, N. \tag{1.18}$$

A *couple* is defined as two equal and opposite forces acting at a distance separation $\mathbf{r}$. The *torque* is defined as the moment of the force $\boldsymbol{\tau} = \mathbf{r} \times \mathbf{F}$. The rate of change of angular momentum of a rotating body is given by $\mathrm{d}\mathbf{L}/\mathrm{d}t = \boldsymbol{\tau}$. $\mathrm{d}\mathbf{p}/\mathrm{d}t = \mathbf{F}$ is the equation of motion of a body for its linear motion and $d\mathbf{L}/dt = \boldsymbol{\tau}$ is the equation of motion for the rotational motion. If we know the forces acting on the body, then $d\mathbf{p}/dt = \mathbf{F}$ can be solved to find all the kinetic quantities of the particle like position and momentum as a function of time. Thus, the force acting on the system must be known to study the dynamics of the body.

Another important quantity associated with the motion of a body is the energy of the body. Energy is a scalar quantity associated with the motion of the body and its position. The *kinetic energy* is the energy associated with the motion of the body. It is given by K.E. $= \frac{1}{2}mv^2$ for translational motion and K.E. $= \frac{1}{2}I\omega^2$ for rotational motion of a body. The energy associated with a body due to its position is called *potential energy*. Energy can be given to a body or extracted from it by doing work by an external force. *Work* is defined as $W = \mathbf{F} \cdot \mathbf{r}$ or $W = \boldsymbol{\tau} \cdot \boldsymbol{\theta}$. If energy is transferred to the body then W is positive and if energy is extracted from the body then W is negative.

The work done by a force on a body is stored as potential energy, for example, if a body of mass m is lifted to a height h against the gravitational force mg then the work done mgh is the potential energy of a body. Total energy of a body is given by $E =$ K.E. + P.E.. For a system of many particles the total energy is the sum of the potential and kinetic energies of each particle. The rate of doing of work is called *power* $P = dW/dt$. If the total work done on a body over a closed path in a force field is zero then the force is called *conservative force*. Gravitational force, electric and magnetic forces are all conservative forces. Any conservative force can be given as a negative gradient of a scalar function as $\mathbf{F} = -\nabla V$ where V is the potential energy at the point.

If a system does not interact with its environment and remains isolated then certain mechanical properties of the system do not change with time. Such quantities are called *constants of motion* and they are said to be *conserved*. In mechanics the total energy, total linear momentum and total angular momentum are all conserved for an isolated system. Interaction within particles of the system will not change these quantities.

Energy is defined as the capacity to do work. It exists in various forms. But one type of energy can be converted to another form. As long as no energy is given to the system or extracted from the system, the total energy of all the particles of the system will be always constant. Just like the vector sum of the total linear momentum is conserved, the total vector sum of angular momentum is also conserved. These conservation laws help to make certain predictions about the motion of the particles of a system.

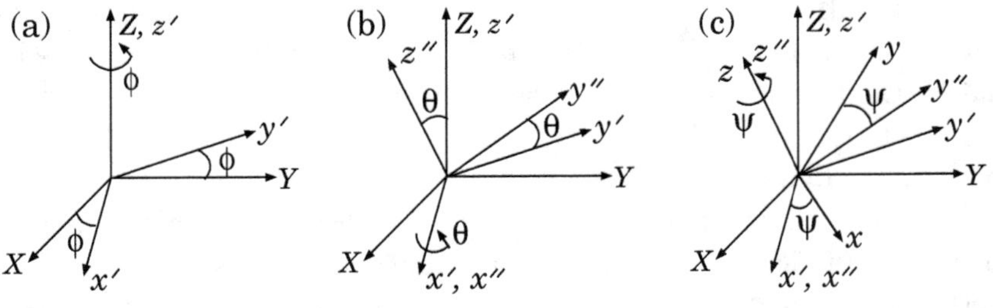

Fig. 1.1: Three transformations of the coordinate system (X, Y, Z) to the system (x, y, z). (a) ϕ-rotation. (b) θ-rotation. (c) ψ-rotation.

1.4 Euler Angles

Euler angles are very useful to describe the motion of a rigid body that rotates about a single point [2-4]. They are particularly useful to study the motions of a gyroscope or a top. They can also be used to explain the motion of an aircraft which rotates about its CoM. To specify the angular orientation of a rigid body, we need three independent angles. These three angles, called *Euler angles*, are obtained by three transformations of coordinate axes, from a fixed coordinate system (X, Y, Z) to a three mutually perpendicular body centred coordinate system (x, y, z). The body centred axes are the principal axes of the moment of inertia tensor of the body.

Consider the three fixed axes (X, Y, Z) system rotated about an angle ϕ with respect to the Z-axis to get new axes (x', y', z') as shown in Fig. 1.1a. The transformation between (x', y', z') and (X, Y, Z) can be given as

$$\begin{bmatrix} x' \\ y' \\ z' \end{bmatrix} = \begin{bmatrix} \cos\phi & \sin\phi & 0 \\ -\sin\phi & \cos\phi & 0 \\ 0 & 0 & 1 \end{bmatrix} \begin{bmatrix} X \\ Y \\ Z \end{bmatrix} = [T_\phi] \begin{bmatrix} X \\ Y \\ Z \end{bmatrix}. \tag{1.19}$$

Next, we rotate about the x' axis into the (x'', y'', z'') system as shown in Fig. 1.1b. The transformation equation for θ-rotation can be given as

$$\begin{bmatrix} x'' \\ y'' \\ z'' \end{bmatrix} = \begin{bmatrix} 1 & 0 & 0 \\ 0 & \cos\theta & \sin\theta \\ 0 & -\sin\theta & \cos\theta \end{bmatrix} \begin{bmatrix} x' \\ y' \\ z' \end{bmatrix} = [T_\theta] \begin{bmatrix} x' \\ y' \\ z' \end{bmatrix}. \tag{1.20}$$

The final rotation of angle ψ about the z'' axis will bring the (x'', y'', z'') system into the body fixed axis system (x, y, z) as shown in Fig. 1.1c. The transformation equation for ψ rotation is given as

$$\begin{bmatrix} x \\ y \\ z \end{bmatrix} = \begin{bmatrix} \cos\psi & \sin\psi & 0 \\ -\sin\psi & \cos\psi & 0 \\ 0 & 0 & 1 \end{bmatrix} \begin{bmatrix} x'' \\ y'' \\ z'' \end{bmatrix} = [T_\psi] \begin{bmatrix} x'' \\ y'' \\ z'' \end{bmatrix}. \tag{1.21}$$

The final transformation from the laboratory fixed coordinate system (X, Y, Z) to the body fixed rotating system (x, y, z) is obtained using the Eqs. (1.19)-(1.21) as

$$\begin{bmatrix} x \\ y \\ z \end{bmatrix} = [T_\psi][T_\theta][T_\phi] \begin{bmatrix} X \\ Y \\ Z \end{bmatrix}. \tag{1.22}$$

$T = [T_\psi][T_\theta][T_\phi]$ is found by multiplying the three matrices as

$$T = \begin{bmatrix} T_{11} & T_{12} & T_{13} \\ T_{21} & T_{22} & T_{23} \\ T_{31} & T_{32} & T_{33} \end{bmatrix}, \tag{1.23a}$$

where

$$T_{11} = \cos\psi\cos\phi - \cos\theta\sin\phi\sin\psi, \tag{1.23b}$$

$$T_{12} = \cos\psi\sin\phi + \cos\theta\cos\phi\sin\psi, \quad T_{13} = \sin\theta\sin\psi, \tag{1.23c}$$

$$T_{21} = -\sin\psi\cos\phi - \cos\theta\sin\phi\cos\psi, \tag{1.23d}$$

$$T_{22} = -\sin\psi\sin\phi + \cos\theta\cos\phi\cos\psi, \quad T_{23} = \sin\theta\cos\psi, \tag{1.23e}$$

$$T_{31} = \sin\phi\sin\theta, \quad T_{32} = \cos\phi\sin\theta, \quad T_{33} = \cos\theta. \tag{1.23f}$$

This is the final body fixed coordinate system with angular velocities ω_x, ω_y and ω_z which will be used for describing the rotational motion of the rigid body.

1.5 Types of Motion

In this section we review some of the simple motions.

1.5.1 *Linear Motion*

The simplest type of motion is linear motion which is also called as *rectilinear motion*. It is a one-dimensional motion along a straight line. Hence, it can be described using just one coordinate. If there is no force acting on the body, then the body will move with a uniform velocity along a straight line. In a nonuniform linear motion, a force acts to change the velocity of the body. For a body with constant mass the Newton's equation of motion for linear motion is given by $m\ddot{x} = F$. For uniform motion $F = 0$. Thus, we get $\ddot{x} = 0$ and $x = At + B$ where A and B are constants to be determined by the initial conditions. If x_0 is the initial position and v_0 is the initial velocity then

$$x = v_0 t + x_0. \tag{1.24}$$

If the force is a constant then the acceleration a is constant. Then we get $\ddot{x} = a$ and $x = \frac{1}{2}at^2 + Ct + D$. For the initial conditions $x(t = 0) = x_0$ and $v(t = 0) = v_0$ we find $D = x_0$ and $C = v_0$. Then

$$x = a_0 + v_0 t + \frac{1}{2}at^2. \tag{1.25}$$

The kinetic energy for linear motion is given by K.E. $= \frac{1}{2}mv^2$.

1.5.2 *Rotational Motion*

A pure rotation refers to a special case of rotational motion. In a pure rotational motion every particle of a body moves along a circle perpendicular to the axis of rotation. The equations of kinematics of pure rotational motion will be similar to that of a linear motion given by Eqs. (1.24) and (1.25). If θ is the angle of rotation, ω the angular velocity and α the angular acceleration then

$$\omega(t) = \omega_0 + \alpha t, \tag{1.26}$$

$$\theta(t) = \theta_0 + \omega_0 t + \frac{1}{2}\alpha t^2, \tag{1.27}$$

where θ_0 is the initial angle and ω_0 is the initial angular velocity. The kinetic energy due to rotational motion is given by K.E. $= \frac{1}{2}I\omega^2$ where I is the moment of inertia of the body about the axis of rotation.

1.5.3 *Rolling Motion*

Rolling motion is a type of motion which has both rotational and translational motions. The rotational axis and the CoM of the body are not stationary. In pure rolling an object rolls without of slip along a flat surface with its axis parallel to the surface. The point of contact of the rolling body at the surface on which it rolls does not slip. The point of contact is instantaneously at rest. If the friction acting at the point of contact is large enough, then slipping will not occur. Consider a wheel pulled across a horizontal surface by a force F as shown in Fig. 1.2a. Figure 1.2b gives the velocity and acceleration of the body rolling without slipping. We see that the point of contact A is at rest with respect to the surface momentarily as it does not slip. The wheel has angular velocity ω. It rotates about the axis perpendicular to x and parallel to the plane of the horizontal surface passing through the contact point A. R is the radius of the wheel. If v_{CoM} is the velocity of the CoM and $\mathbf{v}_{\text{A}}$ is the velocity of the point A with respect to the surface, we can write

$$\mathbf{v}_{\text{A}} = -R\omega\mathbf{i} + v_{\text{CoM}}\mathbf{i}. \tag{1.28}$$

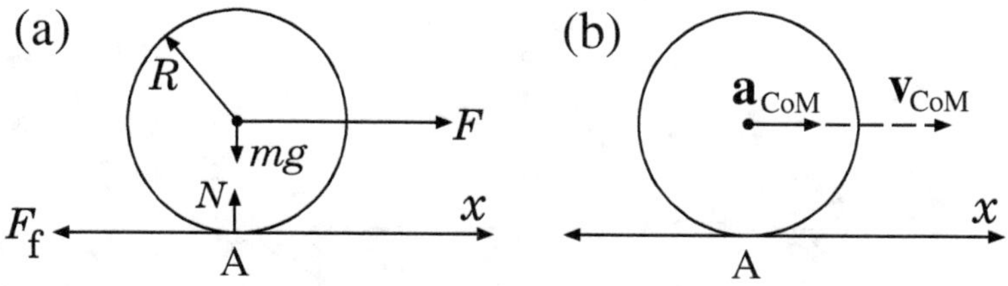

Fig. 1.2: (a) A force F pulls the wheel and the friction F_f is acting opposite to the direction of motion x. (b) Velocity and acceleration of a rolling wheel without slipping.

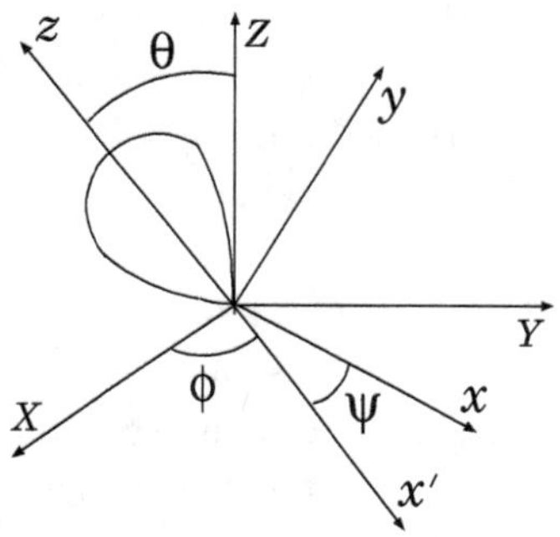

Fig. 1.3: Representation of the angles ϕ, θ and ψ. Also, see Fig. 1.1.

For rolling without slipping $\mathbf{v}_\text{A} = 0$ and $\mathbf{v}_\text{CoM} = R\omega\mathbf{i}$. The acceleration of the CoM is given by

$$\mathbf{a}_\text{CoM} = \dot{\mathbf{v}}_\text{CoM} = R\dot{\omega}\mathbf{i} = R\,\alpha\,\mathbf{i}. \tag{1.29}$$

When the static friction is absent, then the point A will not be at rest with respect to the surface. So, only kinetic friction will act on the body. As $v_\text{A} \neq 0$ we have $\omega \neq v_\text{CoM}/R$ and $\alpha \neq a_\text{CoM}/R$.

1.5.4 *Spin, Precession and Nutation*

There are other types of motion associated with the spinning of a rigid body like a top. (X, Y, Z) is the fixed coordinate system and (x, y, z) is the body fixed coordinate system. The Euler angles (ϕ, θ, ψ) which define the orientation of the body are shown in Fig. 1.3. Suppose a torque is applied to the body about the symmetry axis z. Then, it will *spin* about

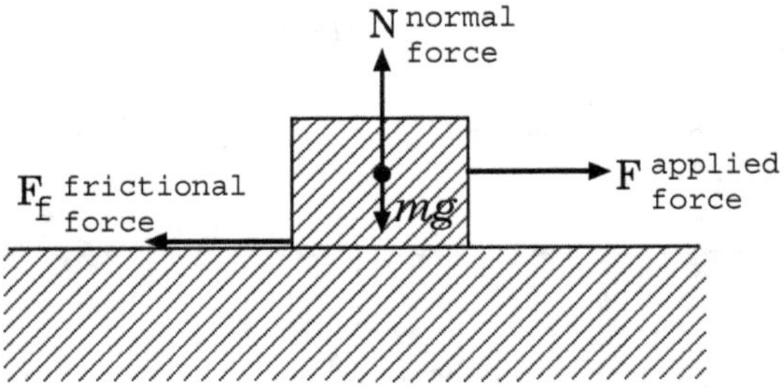

Fig. 1.4: Forces on a moving body over a surface.

that axis. If the symmetric solid body is kept with a sharp tip resting on the surface vertically without spinning about the vertical axis, then it will fall immediately. Now, if it is made to spin by giving a torque along the vertical symmetry axis, then the body will start spinning. It will keep spinning for a longer time. The frictional forces due to air and the surface on which it spins will decrease its energy. Thus, it will soon stop spinning and fall down due to gravity. A spinning top carries angular momentum. As angular momentum is conserved, the symmetric body will keep on spinning until an external torque acts on it and changes the angular momentum.

If a top is spinning about its symmetric axis at an angle θ with respect to the vertical axis, then its CoM and its points of contact on the surface will not be in vertical line. The force of gravity and the reaction force will generate a couple and change the angular momentum. Due to this change of angular momentum the top will have a *precession motion* about the vertical axis Z. So, ϕ is called the *precession angle*.

In addition to this precession motion the top will have different types of oscillatory motion due to the variation of the angle θ between two limits. This motion in θ is called *nutation*. The types of nutation motion are dependent on the initial conditions.

1.6 Friction

Friction is a force between two surfaces that are sliding or trying to slide across one another. It acts always in the direction opposite to the direction of the motion of the body. Friction dissipates energy and hence slows down the motion of the moving body. The directions of the forces acting on a moving body over a surface is given in Fig. 1.4. The normal force $\mathbf{N}$ is equal to the gravitational force mg but acts in the opposite direction: $\mathbf{N} = -m\mathbf{g}$.

There are in general five types of friction: *static friction, kinetic friction, rolling friction, sliding friction and fluid friction.* Static friction is the friction which keeps a body at rest. Only when an external applied force exceeds this static frictional force F_s the body will move. The ratio of the static frictional force F_s to the normal force N gives the *coefficient of static friction* $\mu_s = F_s/N$. The maximum force of static friction does not depend on the area of contact. The limiting friction is the highest value of the static friction which comes into play when the object is about to slide over the surface on which it is at rest. It depends on the nature of the material, its eveness and its interaction with the material of the surface.

Static friction keeps the body at rest. As long as the external applied force is less than the maximum static frictional force the body will not move. Once the force of static friction is overcome by the external force, the body starts moving. But a moving body in the absence of the applied force comes into rest soon.

A frictional force acts on a moving body also. The force of kinetic friction is what slows down the moving body. The ratio of kinetic frictional force F_k to the normal force gives the coefficient of kinetic friction $\mu_k = F_k/N$. Generally, the kinetic frictional force will be less than the static frictional force.

There are two types of kinetic friction. The kinetic friction acting between the object with a sliding motion and the surface is called *sliding friction*. Another type of kinetic friction is the *rolling friction* which arises due to rotational motion of a disc-like or any other circular object on a surface. Usually the rolling frictional force is much less than the sliding frictional force. *Fluid friction* occurs if there is a relative velocity difference between the motions of two layers of liquid or a surface.

1.7 Lagrangian Dynamics

If we have a system with N particles there will be $3N$ degrees of freedom as each particle can move in three independent directions x, y, z. So, the configuration space in which the system evolves has $3N$ independent coordinates. Any point on the configuration space at a time t represents the position of the N particles at the time t. The point moves in a trajectory in the $3N$ dimensional configuration space in a definite curve under the influences of the forces acting on the particles of the system. For some of the systems the motion will be constrained to move satisfying certain conditions expressed as functional equations of coordinates. These equations are called *constraint equations*. The constraint equations may be of the form

$$f_i(r_1, r_2, \ldots, r_N, t) = 0 \tag{1.30}$$

or of the type

$$f_i(r_1, r_2, \ldots, r_N, t) \neq 0, \tag{1.31}$$

where $r_i = (x_i, y_i, z_i)$. Constraint equations of type given by Eq. (1.30) are called *holonomic constraints* and equations of the type of Eq. (1.31) are called *nonholonomic constraints*. If time appears explicitly in the constraint equations, then they are called *rehonomic* and if time is not explicitly present in the constraint equations they are called *scleronomic.*

Suppose a particle is constrained to move in a plane then the constraint equation will be

$$ax + by + cz + d = 0. \tag{1.32}$$

The above constraint equation makes one coordinate dependent on other coordinates. Therefore, all the coordinates are not independent. Hence, each constraint equation reduces the degrees of freedom of the system by one. If there are k constraint equations, then the system will have only $n = 3N - k$ degrees of freedom. It is possible to study the motion of such a system with just $n = 3N - k$ coordinates. These coordinates are called the *generalized coordinates* of the system and denoted as $(q_1, q_2, \ldots, q_n)$. Moreover, the forces responsible for the constraint motion are not known in advance. Consequently, these forces cannot be included in Newton's equations of motion. So, Lagrangian dynamics was developed to treat the problems having constraint equations [2,3].

In Lagrangian dynamics $n = 3N - k$ generalized coordinates $(q_1, q_2, \ldots, q_n)$ for the system are chosen. They are all independent corresponding to $n = 3N - k$ degrees of freedom. The generalized coordinates are written in terms of coordinates of the particles

$$q_i = q_i(r_1, r_2, \ldots r_N, t), \quad i = 1, 2, \ldots, n. \tag{1.33}$$

The inverse transformations are given by

$$r_i = r_i(q_1, q_2, \ldots, q_n, t), \quad i = 1, 2, \ldots, N. \tag{1.34}$$

All q_i are independent. But all r_i are not independent due to k-constraint equations. Lagrangian dynamics considers the motion of the system in the configuration space $(q_1, q_2, \ldots, q_n)$. The system traverses a path in this configuration space as a function of time.

Newtonian mechanics uses the forces to setup the equations of motion. In Lagrangian dynamics energy functions are used. The total kinetic energy is

$$T = \frac{1}{2}\sum_{i=1}^{N} m_i|\mathbf{r}_i|^2. \tag{1.35}$$

The potential energy is given as a function of coordinates as

$$V = V(r_1, r_2, \ldots, r_N). \tag{1.36}$$

Then, a function called *Lagrangian* is defined as $L = T - V$. Using the transformation Eqs. (1.34) L can be converted into functions of generalized coordinates q_i and generalized velocities $\dot{q}_i$:

$$L = L(q_1, q_2, \ldots, q_n, \dot{q}_1, \dot{q}_2, \ldots, \dot{q}_n). \tag{1.37}$$

One can derive the equations of motion using Hamilton's principle of least action. It states that the system moves along a curve in configuration space in which the action $S = \int L\mathrm{d}t$ is an extremum. Hence,

$$\delta S = \delta \int L\mathrm{d}t = 0. \tag{1.38}$$

Using L given in Eq. (1.37) we obtain the Lagrangian equations of motion using Eq. (1.38). The Lagrangian equations of motion are found to be

$$\frac{\mathrm{d}}{\mathrm{d}t}\left(\frac{\partial L}{\partial \dot{q}_i}\right) - \frac{\partial L}{\partial q_i} = 0, \quad i = 1, 2, \ldots, n. \tag{1.39}$$

Solving the n-equations given by Eqs. (1.39), we can find n-generalized coordinates $q_i(t)$ as a function of time. Then, using the transformation Eqs. (1.34) and the k-constraint equations the coordinates of the particle $r_1, r_2, \ldots, r_n$ can be found as a function of time.

The generalized force for the coordinates q_i is defined as

$$Q_i = \sum_{j=1}^{N} \mathbf{F}_j \cdot \frac{\partial \mathbf{r}_j}{\partial q_i}, \quad i = 1, 2, \ldots, n. \tag{1.40}$$

Therefore, for both conservative force and nonconservative force the generalized Lagrangian equations are given as

$$\frac{\mathrm{d}}{\mathrm{d}t}\left(\frac{\partial T}{\partial \dot{q}_i}\right) - \frac{\partial T}{\partial q_i} = Q_i, \quad i = 1, 2, \ldots, n. \tag{1.41}$$

If $Q_i = -c\dot{q}_i$ it leads to energy dissipation corresponding to a system with frictional forces.

1.8 Equilibrium Points and Their Stability Analysis

For nonlinear systems [5] general methods are not available to construct exact analytical solutions. However, we can easily determine all the possible *equilibrium points*, also called ***stationary states*** and analyse their stability without solving the equation of motion of a system which does not have explicit time-dependent terms. We can always rewrite such an n-th order ordinary differential equation into a system of n first-order equations.

1.8.1 *Equilibrium Points*

Consider a system of n first-order equations as

$$\dot{\mathbf{X}} = \mathbf{F}(\mathbf{X}), \tag{1.42a}$$

where

$$\mathbf{X} = (x_1, x_2, \ldots, x_n)^{\mathrm{T}} \in \mathrm{R}^n, \quad \mathbf{F} = (F_1, F_2, \ldots, F_n). \tag{1.42b}$$

Equations (1.42a) have no explicit time dependence. An *equilibrium point* is a state which does not change in time, that is,

$$\dot{\mathbf{X}} = 0 = \mathbf{F}(\mathbf{X}). \tag{1.43}$$

The roots of the equation $\mathbf{F}(\mathbf{X}) = 0$ are the equilibrium points. We denote an equilibrium point as $\mathbf{X}^*$. When the trajectories starting in the neighbourhood of $\mathbf{X}^*$ approach it asymptotically as $t \to \infty$, then $\mathbf{X}^*$ is said to be *stable*. If the neighbouring trajectories move away from $\mathbf{X}^*$ as $t \to \infty$, then it is said to be *unstable*.

Conditions for stability can be obtained by linear stability analysis. A system can be subjected to a weak perturbation or a small disturbance near an equilibrium point and one can identify the conditions for the perturbation to die out. In the next subsection, for simplicity, we outline the linear stability analysis for a two-dimensional dynamical system and then extend it to higher-dimensional systems.

1.8.2 *Stability Criteria*

Consider a two-dimensional system of the form

$$\dot{x} = P(x, y), \quad \dot{y} = Q(x, y). \tag{1.44}$$

Denote an equilibrium point of the system as $\mathbf{X}^* = (x^*, y^*)$ so that $P(x^*, y^*) = Q(x^*, y^*) = 0$. To analyse the stability of $\mathbf{X}^*$ we disturb it as

$$x = x^* + \xi(t), \quad y = y^* + \eta(t), \quad |\xi|, |\eta| \ll 1. \tag{1.45}$$

Substitution of (1.45) in (1.44) gives

$$\dot{x}^* + \dot{\xi} = P(x^*, y^*) + P_x|_{x^*,y^*}\xi + P_y|_{x^*,y^*}\eta$$
$$+ \text{ higher order terms in } (\xi, \eta), \tag{1.46a}$$
$$\dot{y}^* + \dot{\eta} = Q(x^*, y^*) + Q_x|_{x^*,y^*}\xi + Q_y|_{x^*,y^*}\eta$$
$$+ \text{ higher order terms in } (\xi, \eta). \tag{1.46b}$$

Substituting $\dot{x}^* = P(x^*, y^*)$, $\dot{y}^* = Q(x^*, y^*)$ and neglecting the higher-order terms, we obtain

$$\dot{\xi} = a\xi + b\eta, \quad \dot{\eta} = c\xi + d\eta, \tag{1.47a}$$

where

$$a = P_x|_{x^*,y^*}, \quad b = P_y|_{x^*,y^*}, \quad c = Q_x|_{x^*,y^*}, \quad d = Q_y|_{x^*,y^*}. \tag{1.47b}$$

From Eqs. (1.47) we obain

$$\ddot{\xi} - (a+d)\dot{\xi} + (ad - bc)\xi = 0. \tag{1.48}$$

The solution of Eq. (1.48) is

$$\xi(t) = A\mathrm{e}^{\lambda_+ t} + B\mathrm{e}^{\lambda_- t}, \tag{1.49a}$$

where

$$\lambda_\pm = \frac{1}{2}\left[(a+d) \pm \sqrt{(a+d)^2 - 4(ad-bc)}\right], \quad ad - bc \neq 0. \tag{1.49b}$$

Note that $\lambda_\pm$ are the eigenvalues of the matrix $M = \begin{pmatrix} a & b \\ c & d \end{pmatrix}$, $\det M \neq 0$ where a, b, c and d are given by Eq. (1.47b). Use of the obtained solution ξ in $\dot{\xi} = a\xi + b\eta$ gives

$$\eta(t) = C\mathrm{e}^{\lambda_+ t} + D\mathrm{e}^{\lambda_- t}, \quad C = \frac{A(\lambda_+ - a)}{b}, \quad D = \frac{B(\lambda_- - a)}{b}. \tag{1.50}$$

An equilibrium point is asymptotically stable if $\xi(t)$ and $\eta(t) \to 0$ as $t \to \infty$. From Eqs. (1.49) and (1.50) we observe that the decay of ξ and η depends on the sign of $\lambda_\pm$. For $\mathbf{X}^*$ to be stable the real parts of both the eigenvalues have to be negative. If at least one eigenvalue has positive real part, then $\mathbf{X}^*$ becomes unstable.

What about the stability of equilibrium points of first-order systems? For a first-order system $\dot{x} = P(x)$ we find

$$\dot{\xi} = \lambda\xi, \quad \lambda = \partial P/\partial x|_{x=x^*}. \tag{1.51}$$

The solution of this equation is $\xi(t) = \mathrm{e}^{\lambda t}\xi(0)$. If $\lambda < 0$, then $\xi(t) \to 0$ as $t \to \infty$ and hence x^* is stable. For $\lambda > 0$, $|\xi(t)| \to \infty$ as $t \to \infty$ and x^* is unstable.

For a broad classification of equilibrium points into stable node/star, unstable node/star, etc. based on the nature of the eigenvalues one may refer to [5]. For a higher-dimensional system an equilibrium point is stable if all the eigenvalues have negative real parts.

Solved Problem 2:

Determine the stability of the equilibrium points of the quadratic oscillator

$$\ddot{x} + d\dot{x} - \alpha x + \beta x^2 = 0, \quad d,\, \alpha,\, \beta > 0. \tag{1.52}$$

We rewrite the given system as

$$\dot{x} = y = P, \;\; \dot{y} = -dy + \alpha x - \beta x^2 = Q. \tag{1.53}$$

The equilibrium points are obtained by substituting $\dot{x} = \dot{y} = 0$. This gives

$$(x^*, y^*) = (0, 0), \;\; (\alpha/\beta, 0). \tag{1.54}$$

The stability determining eigenvalues are obtained from

$$\det(M - \lambda I) = \begin{vmatrix} \partial P/\partial x - \lambda & \partial P/\partial y \\ \partial Q/\partial x & \partial Q/\partial y - \lambda \end{vmatrix}_{(x^*, y^*)} = 0. \tag{1.55}$$

For the given system, Eq. (1.55) becomes

$$\begin{vmatrix} -\lambda & 1 \\ \alpha - 2\beta x^* & -d - \lambda \end{vmatrix} = \lambda^2 + d\lambda - (\alpha - 2\beta x^*) = 0. \tag{1.56}$$

For $(x^*, y^*) = (0, 0)$ we obtain

$$\lambda_{\pm} = \frac{1}{2}\left[-d \pm \sqrt{d^2 + 4\alpha}\right]. \tag{1.57}$$

That is, $\lambda_+ > 0$ while $\lambda_- < 0$. As one of the eigenvalues has positive real part the equilibrium point $(x^*, y^*) = (0, 0)$ is always unstable.

Next, for $(x^*, y^*) = (\alpha/\beta, 0)$ we obtain

$$\lambda_{\pm} = \frac{1}{2}\left[-d \pm \sqrt{d^2 - 4\alpha}\right]. \tag{1.58}$$

As both d and α are > 0 the quantity $d^2 - 4\alpha$ is always $< d^2$. Therefore, the signs of real parts of both λ_+ and λ_- are always negative and hence the equilibrium point is always stable.

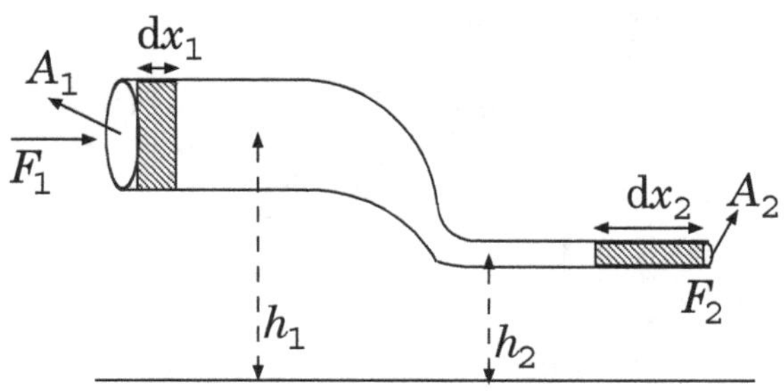

Fig. 1.5: A pipe with two different cross-sections at two different heights.

1.9 Bernoulli's Principle

If the fluid motion is frictionless, then the conservation of energy leads to Bernoulli's principle. It relates the pressure of an incompressible fluid at a point to its elevation and its speed. The Bernoulli's principle states that for an incompressible, non-viscous fluid, the sum of the gravitational potential energy of elevation, the energy associated with the fluid pressure and the kinetic energy of the fluid motion remains constant.

Consider a pipe as shown in Fig. 1.5 with two different areas of cross-sections A_1 and A_2 and at two different heights h_1 and h_2 through which an incompressible fluid is flowing. The pressure of the fluid and velocity at A_1 area of cross-section are P_1 and v_1, respectively. At A_2 area of cross-section, the pressure and velocity are P_2 and v_2, respectively. Consider two sections of same volume $\mathrm{d}V$ of the fluid at both ends: $\mathrm{d}V = A_1\mathrm{d}x_1 = A_2\mathrm{d}x_2$. The sides of the tube are at elevations h_1 and h_2. Since the force is given by $F = \text{pressure} \times \text{area}$, the work done on volume $\mathrm{d}V$ of the fluid is

$$\begin{aligned} \mathrm{d}W &= F_1\mathrm{d}x_1 - F_2\mathrm{d}x_2 \\ &= P_1A_1\mathrm{d}x_1 - P_2A_2\mathrm{d}x_2 \\ &= (P_1 - P_2)\,\mathrm{d}V\,. \end{aligned} \tag{1.59}$$

The change in the kinetic energy on the same volume is

$$\mathrm{d}K = \frac{1}{2}m_2v_2^2 - \frac{1}{2}m_1v_1^2 = \frac{1}{2}\rho\mathrm{d}V\left(v_2^2 - v_1^2\right)\,. \tag{1.60}$$

The change in the potential energy for the same volume $\mathrm{d}V$ is

$$\mathrm{d}U = m_2gh_2 - m_1gh_1 = \rho\mathrm{d}Vg\,(h_2 - h_1)\,. \tag{1.61}$$

From conservation of energy, we get $\mathrm{d}W = \mathrm{d}K + \mathrm{d}U$, that is,

$$(P_1 - P_2)\,\mathrm{d}V = \frac{1}{2}\rho\mathrm{d}V\left(v_2^2 - v_1^2\right) + \rho\mathrm{d}Vg\,(h_2 - h_1)\,. \tag{1.62}$$

This equation can be rewritten as

$$P_1 + \frac{1}{2}\rho v_1^2 + \rho g h_1 = P_2 + \frac{1}{2}\rho v_2^2 + \rho g h_2 \,. \tag{1.63}$$

This is the Bernoulli's equation which can be given as

$$P + \frac{1}{2}\rho v^2 + \rho g h = \text{a constant} \,. \tag{1.64}$$

When the elevation h is constant, then an increase in velocity v will reduce the pressure at that point.

1.10 Laws of Electromagnetism

The operation of toys such as an electric train, Levitron, Gauss rifle and simple DC motors are based on the laws of electromagnetism [6,7]. Therefore, in this section we discuss the electromagnetic laws such as Faraday's law of induction, Lenz's law, Biot-Savart law, eddy current and Lorentz force.

Faraday's law of induction is a fundamental law of nature discovered by Michael Faraday in 1831. It is one of the four Maxwell's equations which describes all electromagnetic phenomena of nature. Faraday's *law of induction* states that any change in magnetic flux linked to a closed coil of wire induces an voltage in the coil. If the magnetic flux in the coil changes by $\mathrm{d}\phi$ in time $\mathrm{d}t$, then the induced emf E is given by $E = -\mathrm{d}\phi/\mathrm{d}t$. The negative sign in the Faraday's equation comes due to Lenz's law. *Lenz's law* states that the polarity of the induced emf gnerated by a change of magnetic flux is such that it generates a current whose magnetic field is in a direction that opposes the change that the original magnetic field produced.

Another consequence of Faraday's law of induction is the eddy current generated in a conductor due to changing magnetic field. These currents circulate like a swirling eddies in a stream on the conductors. They flow in closed loops perpendicular to the plane of the magnetic field. This eddy current will produce its own magnetic field. According to Lenz's law, this magnetic field will oppose the change of magnetic field which created it. So, the eddy currents create resistance which are used in eddy current breaking of many devices like rotating power tools and roller coasters. In the toy electromagnetic train also it opposes the motion of the train.

Current is the source of magnetic field. *Biot-Savart law* gives the magnetic field $\mathbf{B}$ produced by a current carrying circuit. Suppose a steady current I flows through an elemental length $\mathrm{d}\mathbf{l}$ of a circuit. The magnetic field due to this current element $I\mathrm{d}\mathbf{l}$ is given by Biot-Savart law as

$$\mathrm{d}\mathbf{B}(\mathbf{r}) = \frac{\mu_0 I \mathrm{d}\mathbf{l} \times \mathbf{r}_1}{|\mathbf{r}|^2}, \tag{1.65}$$

where $\mathbf{r}_1 = \mathbf{r}/|\mathbf{r}|$ is the unit vector of the vector $\mathbf{r}$ connecting the current element and the observation point of $\mathbf{B}$. The total field due to the whole circuit is obtained by integrating over the whole wire length of the circuit as

$$\mathrm{d}\mathbf{B} = \frac{\mu_0}{4\pi} \int_{\mathrm{wire}} \frac{I\mathrm{d}\mathbf{l} \times \mathbf{r}_1}{|\mathbf{r}|^2}. \tag{1.66}$$

Solved Problem 3:

A circular loop of radius R carries a current I. Find the magnetic field at its centre.

We have

$$\mathrm{d}\mathbf{B} = \frac{\mu_0}{4\pi} \frac{I\mathrm{d}\mathbf{l} \times \mathbf{R}}{R^3}. \tag{1.67}$$

As $\mathrm{d}\mathbf{l}$ and $\mathbf{R}$ are perpendicular and $\mathrm{d}l = R\mathrm{d}\theta$ we write

$$\mathrm{d}\mathbf{B} = \frac{\mu_0}{4\pi} \frac{IR^2\mathrm{d}\theta}{R^3}\mathbf{z} = \frac{\mu_0}{4\pi} \frac{I\mathrm{d}\theta}{R}\mathbf{z}, \tag{1.68}$$

where $\mathbf{z}$ is the direction perpendicular to the plane of the loop containing $\mathrm{d}\mathbf{l}$ and $\mathbf{R}$ vectors. Then,

$$\mathbf{B} = \frac{\mu_0}{4\pi} \frac{I}{R}\mathbf{z} \int_0^{2\pi} \mathrm{d}\theta = \frac{\mu_0 I}{2R}\mathbf{z}. \tag{1.69}$$

Lorentz force is the force exerted on a charged particle of charge q moving with velocity $\mathbf{v}$ through an electric field $\mathbf{E}$ and magnetic field $\mathbf{B}$. This electromagentic force is given by

$$\mathbf{F} = q\mathbf{E} + q\mathbf{v} \times \mathbf{B}. \tag{1.70}$$

The magnetic force acts only on moving charges. Since a current represents a movement of charges in the wire the magnetic force will act on these charges. If a charge $\mathrm{d}q$ flows through the wire in time $\mathrm{d}t$ and if the velocity of the charges is $\mathrm{d}\mathbf{X}'/\mathrm{d}t$, then the magnetic force given by Lorentz force on a circuit carrying element $\mathrm{d}\mathbf{X}'$ can be given as

$$\mathrm{d}\mathbf{F} = \mathrm{d}q\mathbf{v} \times \mathbf{B} = \mathrm{d}q\frac{\mathrm{d}\mathbf{X}'}{\mathrm{d}t} \times \mathbf{B} = \frac{\mathrm{d}q}{\mathrm{d}t}\mathrm{d}\mathbf{X}' \times \mathbf{B} = I\mathrm{d}\mathbf{X}' \times \mathbf{B}. \tag{1.71}$$

The force $\mathrm{d}\mathbf{F}$ acts perpendicular to both $\mathrm{d}\mathbf{X}'$ and $\mathbf{B}$. The total force on the wire carrying current can be found by integrating Eq. (1.71) over the length of the wire as

$$\mathbf{F} = \int_{\mathrm{wire}} I\mathrm{d}\mathbf{X}' \times \mathbf{B}. \tag{1.72}$$

1.11 Conclusion

Only the basic concepts connected with the study of working of toys considered in this book have been discussed here. The study of motion of objects requires very advanced mathematical techniques in setting of the equations of motion and solving them. The study of motion of a system with many bodies will be still more complicated. The mathematical techniques needed to solve such systems are quite advanced and still only a very limited number of problems can be exactly solved.

Solving many mechanical problems need many different approximation methods and numerical techniques. Only a few fundamental types of motion has been discussed in this chapter. But there are many types of complicated motions like orbital motion for central forces, periodic motions, chaotic motions, irregular but nonchaotic motion etc. [5,8,9]. Mechanics deals with all types of motion. One has to learn various advanced mathematical techniques, approximation methods and computational techniques for solving the mechanical problems. Interpreting the solutions is also not a simple task.

1.12 Bibliography

[1] R. Feynman, R. Leighton and M. Sands, *The Feynman Lectures on Physics, Vol. I* (California Institute of Technology, California, 2013) (online edition).

[2] H. Goldstein, C. Poole and J. Safko, *Classical Mechanics* (Addison Wesley, New York, 2001) 3rd edition.

[3] J.R. Taylor, *Classical Mechanics* (University Science Books, California, 2004).

[4] P. Kim, *Rigid Body Dynamics for Beginners* (CreateSpace Independent Publsihing Platform, California, 2013).

[5] M. Lakshmanan and S. Rajasekar, *Nonlinear Dynamics: Integrability, Chaos and Patterns* (Springer, New York, 2002).

[6] D.J. Griffiths, *Introduction to Electrodynamics* (Cambridge University Press, Cambridge, 2017) IVth edition.

[7] J.D. Jackson, *Classical Electrodynamics* (John Wiley, New York, 1998) 3rd edition.

[8] S.H. Strogatz, *Nonlinear Dynamics and Chaos: With Applications to Physics, Biology, Chemistry, and Engineering* (CRC Press, Boca Raton, 2014) 2nd edition.

[9] K.T. Alligood, T.D. Sauer and J.A. Yorke, *Chaos: An Introduction to Dynamical Systems* (Springer, New York, 1996).

1.13 Exercises

1.1 Suppose a body of mass m is at a perpendicular distance r to an axis of rotation, then its moment of inertia of the body about the axis of rotation is mr^2. Obtain $\mathbf{L} = \mathbf{r} \times \mathbf{p}$ from $\mathbf{L} = I\omega$.

1.2 Assume that a particle of mass m is at a distance $\mathbf{r}$ in a plane perpendicular to an axis which rotates the particle with angular velocity $\boldsymbol{\omega}$. Show that $\mathrm{d}\mathbf{r}/\mathrm{d}t = \boldsymbol{\omega} \times \mathbf{r}$.

1.3 Prove the parallel-axis theorem of moment of inertia.

1.4 Prove the perpendicular-axis theorem of moment of inertia.

1.5 With respect to a coordinate system (x, y, z) three masses of 1 kg each are located at the points $(1, 0, 0)$, $(0, 1, 2)$ and $(0, 2, 1)$. Determine the moment of inertia tensor of the system.

1.6 Find the principal moments of inertia and principal axes for the moment of inertia of the previous exercise.

1.7 Apply Stokes theorem to prove that the conservative force can be given as a gradiant of a scalar function.

1.8 Show that the kinetic energy of a rolling body without slipping is given by the kinetic energy of the CoM plus the rotational energy of the body about its axis passing through the CoM parallel to the surface.

1.9 Obtain Newton's equations of motion from Lagrange's equation of motion.

1.10 Find the magnetic dipole moment m of a spherical magnet of radius 0.95 cm and residual magnetism $B_{\mathrm{r}} = 1.24\,\mathrm{T}$.

Part II: Mechanical Toys

Chapter 1

Roly-Poly

Balancing class of toys get upright themselves when pushed over. A popular such type of toys is the roly-poly (Fig. 1.1), which is considered as one of the oldest toys in history. It is also called *tilting doll*, *tumbler* and *wobbly man*. Roly-poly toys are used to symbolize the ability to have success, recover from misfortune and overcome adversity [1]. In a class room, it can be used to describe to-and-fro oscillation about a stable equilibrium point/position. Michael Faraday used a roly-poly to explain the concepts of centre of gravity and equilibrium in the "*A Course of Six Lectures on the Various Forces of Matter*" delivered during the Christmas holidays of 1859-60 before a Juvenile Auditory of the Royal Institution of Great Britain [2]. Another popular toy that also tends to make it upright position is a *weeble*, typically egg-shaped. A popular advertising slogan for weeble-like toys is "*Weebles wobble, but they don't fall down*" [3].

Fig. 1.1: A roly-poly toy.

A typical roly-poly has its bottom with a shape round, roughly a hemisphere as is the case of the toy shown in Fig. 1.1. Inner part is usually hollow with a weight at the bottom. When gravity acts on a body, every particle of which it is composed, is attracted towards the earth. The resultant force is the weight of the body. If M is the total mass of the body and g, the acceleration due to gravity, then the weight Mg is the resultant of the force of gravity acting on every particle of the body. The point through which this resultant force acts is called *centre of gravity* (CoG). The CoG of the roly-poly is made vertically below the centre of the hemisphere. Because of this, when the toy is tilted the CoG raises. When the toy is gently released after tilting, it wobbles for a few moments and returns to the upright position with the lowest CoG height. Toy manufacturers produce a variety of looking roly-poly toys.

1.1 Type-I and II Balancing Toys

The variants of roly-poly includes roly-poly minnie mouse with extra weights, balancing horse and rider, Thanjavur doll (if made of terracotta, generates a dance-like movement with a slow oscillation), balancing frog, balancing bicycle man on a rope, and balancing circus man. All of them have a fixed pivot point or a fixed axis of rotation. They are termed as *type-I* balancing toys. On the other hand, the roly-poly and weebles have a varying pivot point as they roll over a plane surface and are termed *type-II* balancing toys. Table 1.1 presents the differences between *type-I* and *type-II* balancing toys [4].

1.2 Centre of Gravity

If g is the same on all the particles of a body, then the CoG will be the centre of mass (CoM) of the body. The CoG can be defined as an imaginary point at which all the weight of the body is considered to be concentrated.

Because the resultant gravitation force acts through the CoG, the moment of the forces of the particles of the body must be equal to the moment of the total force through the CoG. If $\mathbf{r}_1, \mathbf{r}_2, \ldots, \mathbf{r}_N$ are the position vectors of particles of masses $m_1, m_2, \ldots, m_N$, respectively, and $\mathbf{r}_\mathrm{c}$ is the position vector of the CoG with respect to an origin O, then $Mg\mathbf{r}_\mathrm{c} = \sum_{i=1}^{N} m_i g\mathbf{r}_i$. For any small object, g is uniform, so that the CoG is given by $\mathbf{r}_\mathrm{c} = (1/M)\sum_{i=1}^{N} m_i\mathbf{r}_i$. If we consider the body to be a continuous

Table 1.1: Differences between type-I and type-II balancing toys.

No.	Type-I	Type-II
1.	Examples: Balancing frog and balancing bird.	Examples: Roly-poly and weebles.
2.	The pivot point or the rotation axis is fixed.	The pivot point or the rotation axis is instantaneous.
3.	The pivot can be a point or a line.	The point of contact with the plane surface is the instantaneous pivot point for a spherical toy, whereas for a cylindrical toy the instantaneous axis of rotation is the line of contact.
4.	The CoG is vertically below the fixed pivot.	In the upright position the CoG of a spherical (cylindrical) toy is vertically below the centre of the sphere (axis of the cylinder) and above the pivot point (axis of rotation).
5.	Tipping the toy raises the CoG.	Tilting raises the CoG.
6.	The path of the CoG is circular.	The path of the CoG is a cycloid.

system, then the above equation can be written as

$$\mathbf{r}_{\rm c} = \frac{1}{M}\int \mathbf{r} {\rm d}m\,. \tag{1.1}$$

Consequently, the components of $\mathbf{r}_{\rm c}$ are given by

$$x_{\rm c} = \frac{1}{M}\int x\,{\rm d}m, \quad y_{\rm c} = \frac{1}{M}\int y\,{\rm d}m, \quad z_{\rm c} = \frac{1}{M}\int z\,{\rm d}m. \tag{1.2}$$

If the body with a volume V has a uniform mass density ρ, then the Eqs. (1.2) can be written as

$$x_{\rm c} = \frac{\rho}{M}\int x\,{\rm d}V, \quad y_{\rm c} = \frac{\rho}{M}\int y\,{\rm d}V, \quad z_{\rm c} = \frac{\rho}{M}\int z\,{\rm d}V. \tag{1.3}$$

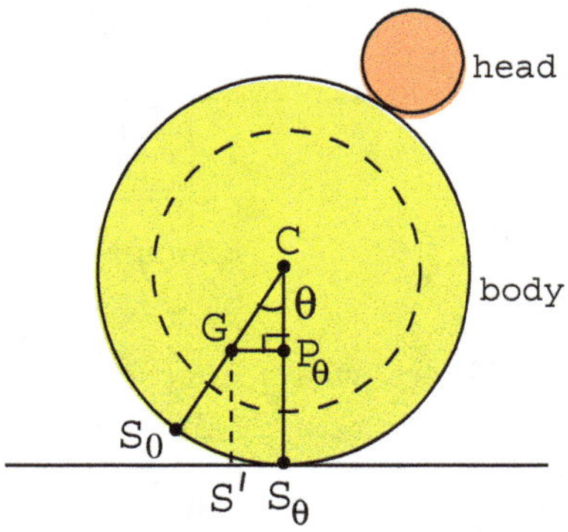

Fig. 1.2: A spherical roly-poly rotated by an angle θ from its upright position.

1.3 Analysis of a Model of the Roly-Poly

Consider a spherical roly-poly with a total mass M as shown in Fig. 1.2 where C is the centre of the larger sphere with radius r and G is the CoG [4]. The point of support is S_θ and P_θ represents the projection of G on to the vertical line from S_θ. G is vertically below C and above S_0. At an angle θ, the height $h(\theta)$ of the CoG from the surface is $\overline{S_\theta P_\theta}$. In the following we obtain expressions for the torque $\tau(\theta)$ and the height $h(\theta)$ and then we analyze the stability of the equilibrium points of the toy [4].

We have

$$\overline{S_0G} = h_0,\quad \overline{CG} = r - h_0,\quad \cos\theta = \frac{\overline{CP_\theta}}{\overline{CG}},\quad \sin\theta = \frac{\overline{GP_\theta}}{\overline{CG}}. \tag{1.4}$$

Then, the height $h(\theta)$ is given by

$$h(\theta) = \overline{S_\theta P_\theta} = \overline{CS_\theta} - \overline{CP_\theta} = r - \overline{CG}\cos\theta = r - (r - h_0)\cos\theta. \tag{1.5}$$

The torque is

$$\tau(\theta) = Mg\overline{GP_\theta} = Mg\overline{CG}\sin\theta = Mg(r - h_0)\sin\theta\,, \tag{1.6}$$

where g is the gravitational acceleration.

Next, from Eq. (1.5) we write $h'(\theta) = \mathrm{d}h(\theta)/\mathrm{d}\theta = (r - h_0)\sin\theta$. Denote θ^* as an equilibrium position of the toy. It is a root of the equation $h'(\theta) = \mathrm{d}h(\theta)/\mathrm{d}\theta = 0$. Equation (1.5) gives $\sin\theta^* = 0$. That is, $\theta^* = 0, \pi$. If θ^* is locally a minimum of $h(\theta)$, then it is stable and is unstable when it becomes locally a maximum. The condition for a point θ to be a minimum of $h(\theta)$ is

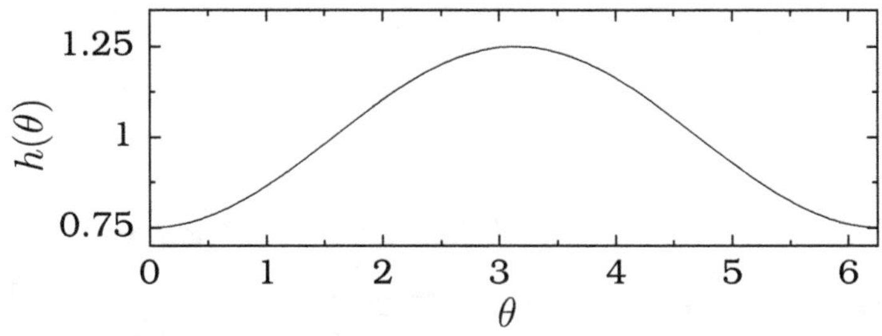

Fig. 1.3: Variation of the height $h(\theta)$ with the angle θ with respect to its upright position.

$\mathrm{d}h/\mathrm{d}\theta = 0$ and $\mathrm{d}^2h/\mathrm{d}\theta^2 > 0$. Therefore, an equilibrium position is stable if $\mathrm{d}^2h/\mathrm{d}\theta^2|_{\theta=\theta^*} > 0$. When $\mathrm{d}^2h/\mathrm{d}\theta^2|_{\theta=\theta^*} < 0$ the equilibrium point becomes unstable.

From Eq. (1.5), we obtain

$$h''(\theta^*) = (r - h_0)\cos\theta^* \,. \tag{1.7}$$

For $\theta^* = 0$

$$h''(\theta^* = 0) = r - h_0 \,. \tag{1.8}$$

The requirement for the upright position to be stable is $h_0 < r$, that is, the CoG must be below the centre of the large sphere. Figure 1.3 shows the plot of $h(\theta)$ versus θ with $r = 1$ and $h_0 = 0.75$. As θ increases from zero, $h(\theta)$ increases from h_0 and reaches the maximum value $2r - h_0 = 5/4$ at $\theta = \pi$. Further increase in θ from $\theta = \pi$, makes it decrease $h(\theta)$ from $2r - h_0$ becoming h_0 at $\theta = 2\pi$. The curve $h(\theta)$ has a minimum at $\theta = 0$ and hence the upright position of the toy is stable. When the toy is released from a slightly tilted position, it undergoes a damped to-and-fro oscillation. The toy loses its initial energy slowly due to air resistance and friction with the surface of the floor and finally returns to its upright position.

Notice that for $\theta^* = \pi$ we have

$$h''(\theta^* = \pi) = -(r - h_0) \,. \tag{1.9}$$

Therefore, as long as the height of the CoG h_0 at $\theta = 0$ is vertically below the centre of the large sphere, then $h''(\theta^* = \pi) < 0$ and $\theta^* = \pi$ corresponds to an unstable position. *When does the roly-poly fall down?* (see the exercise 1.2 at the end of the present chapter).

Why an egg does not tend to stand upright on its less pointed end, while a roly-poly toy does [5]? For a roly-poly at its upright position, the CoG

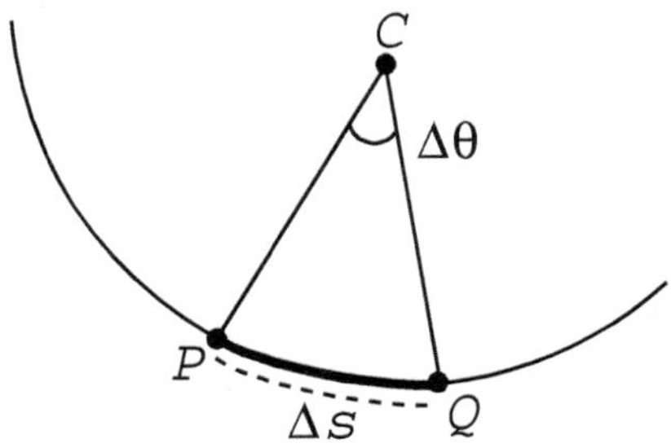

Fig. 1.4: PC (QC) is a straight-line normal to the curve at the point P (Q). C is the centre of curvature at the point P.

is below the centre of curvature of its vertex, thereby making the upright position ($\theta = 0$) a stable equilibrium. That is, for a roly-poly toy with hemispherical bottom, the height of the CoG in its upright position from the surface of the floor is smaller than the radius of the of the hemispherical bottom. For an egg, the CoG is vertically above the centre of curvature of its vertex, and hence its upright position is unstable.

The above analysis is for a spherical or cylindrical roly-poly. One can extend the above treatment to an ellipsoidal or elliptic-cylindrical roly-poly with an elliptic rolling cross-section [5]. Suppose we consider a solid of revolution that is resting on its vertex over a horizontal surface. We introduce the notion of curvature, which plays an important role on the stability of the roly-poly toy.

Consider the curve shown in Fig. 1.4. $\Delta\theta$ be the angle between CQ and CP, and the points P and Q be close to each other. Δs is the arc length $\widehat{PQ}$ along the curve. At the point P the radius of the curvature is defined as

$$r = \lim_{\Delta\theta \to 0} \frac{\Delta s}{\Delta\theta} = \frac{\mathrm{d}s}{\mathrm{d}\theta}\,. \tag{1.10}$$

The centre of curvature C at the point P on the curve is the position of the intersection point as the arc length $\widehat{PQ} \to 0$. r can be considered as the radius of the circle approximating the curve about the point of interest.

Suppose we consider an equilibrium position of, say, a solid revolution, and assume that it is on its vertex over a horizontal plane. *When does the equilibrium point stable?* The equilibrium point becomes stable provided the radius of curvature of the contour of the solid at its vertex is larger than the vertical height of the CoG from the plane [6].

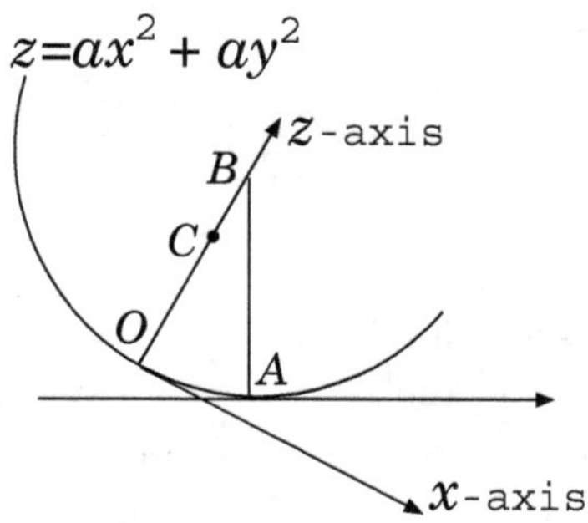

Fig. 1.5: The projection of the lower part of a roly-poly with the paraboloid shape on the $x - z$ plane.

Solved Problem 1:

Consider a roly-poly toy with the lower part having the shape of a paraboloid of revolution described by the equation $z = ax^2 + ay^2$ in the coordinate system as shown in Fig. 1.5 projected on the $x - z$ plane for a tilted position. $O(0, 0)$ is the equilibrium resting point, $C(0, c)$ is the CoG at the equilibrium point (with no tilt), $A(x_0, z_0)$ is the resting point at the tilted position and $B(0, z)$ is the new CoG. Show that this roly-poly will restore its balance as long as the condition $c < \frac{1}{2a} + ax_0^2$ is satisfied. Hence, at the point O, the above condition for restoring the balance becomes $0 \leq c \leq 1/(2a)$ [7].

We have the paraboloid equation $z = ax^2 + ay^2$. Projecting in the x-z plane gives the parabola $z = ax^2$. The slope of the tangent at $A(x_0, y_0)$ is found to be $\left.\frac{\mathrm{d}z}{\mathrm{d}x}\right|_A = 2ax_0$. Since AB is perpendicular to the tangent, the slope of AB is $-1/(2ax_0)$. Hence, the equation of the line AB is obtained as

$$z - z_0 = -\frac{(x - x_0)}{2ax_0} = -\frac{x}{2ax_0} + \frac{1}{2a}. \tag{1.11}$$

Substituting $z_0 = ax_0^2$ we write

$$z = -\frac{x}{2ax_0} + \frac{1}{2a} + ax_0^2. \tag{1.12}$$

Since B lies on the line AB and z-axis, $x = 0$, so $z = 1/(2a) + ax_0^2$. The coordinates of B are $B(0, 1/(2a) + ax_0^2)$. Then, $OB = 1/(2a) + ax_0^2$. As $OC = c$ and as $OC < OB$ for the roly-poly toy to restore its balance, we get the condition $c < 1/(2a) + ax_0^2$. Therefore, at $O(0,0)$, the equilibrium position, we get $0 < c < 1/(2a)$.

1.4 Conclusion

Though roly-poly toys are simple, children can learn a lot from them. They can understand the stable and unstable equilibrium points. They can also learn that the toy can execute oscillations in its stable equilibrium position. They can be taught how the stability of an equilibrium position is determined by the CoG position and the radius of curvature of the surface in contact with the horizontal plane. Suitably designed roly-poly toys can be useful for developing motor skills for small children as a child can bat at it without it is rolling away. As it is a simple toy to make, the students can be asked to design roly-poly toys of different shape. For advanced students a method has been proposed to design an arbitrary roly-poly toy [7] which theoretically guarantee the stability of the toy.

1.5 Bibliography

[1] J.A. Kyburz, Asian Folk Stud. **53**, 1 (1994).
[2] R. Turner, The Phys. Teach. **30**, 542 (1992).
[3] S. De, Math. Mag. **90**, 99 (2017).
[4] S.I. Hong, Phys. Edu. **51**, 013003 (2016).
[5] S.I. Hong, Eur. J. Phys. **37**, 062001 (2016).
[6] S. De, Math. Mag. **90**, 99 (2017).
[7] H. Zhao, C. Hong, J. Lin, X. Jin and W. Xu, Comp. Aided Geomet. Design **43**, 226 (2016).

1.6 Exercises

1.1 Show that the CoG is independent of the origin of a coordinate system.

1.2 For the roly-poly shown in Fig 1.2 find the condition for it to fall down when pushed.

1.3 Consider an elliptic toy with coordinates fixed on it as shown in Fig. 1.6. C is the geometrical centre with $(x, y) = (0, b)$ and G is the CoG at $(x, y) = (0, g)$. If the toy is tilted to touch the horizontal plane surface at a point S, then AB represents the surface of the floor and SP is normal to the tangent line at the point S. The point P is

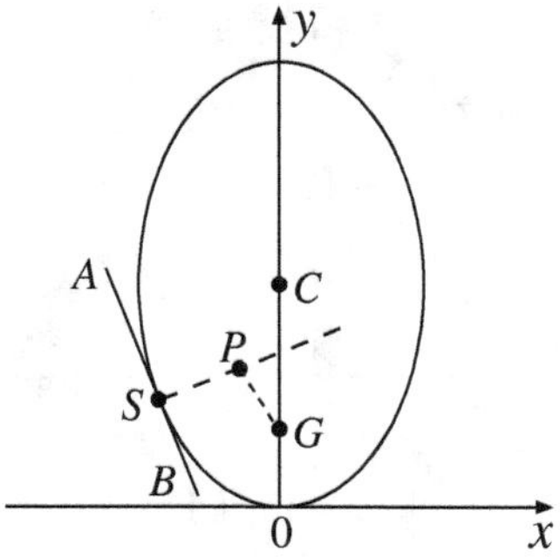

Fig. 1.6: An upright elliptic roly-poly toy.

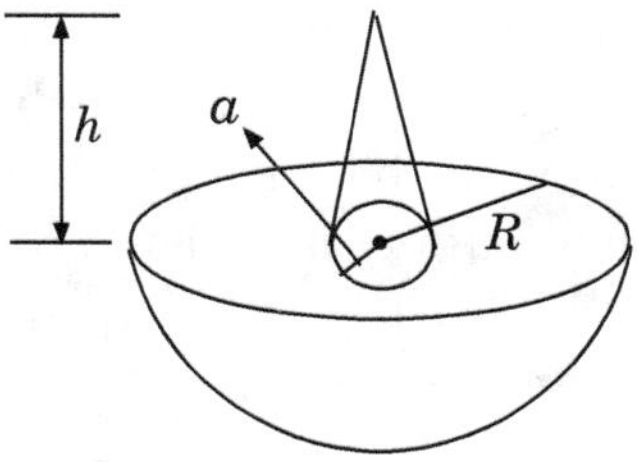

Fig. 1.7: A roly-poly toy.

the orthogonal projection of G to the normal line. The height SP is the CoG at the point S from the surface of the floor. For the elliptic cross-section of the upright toy

$$\left(\frac{x}{a}\right)^2 + \left(\frac{y-b}{b}\right)^2 = 1\,, \quad a > 0,\ b > 0,\ \rho = \frac{b}{a}\,. \tag{1.13}$$

Denote the contact point of the toy as (α, β) and (α^*, β^*) as the stable equilibrium point. Determine the stable equilibrium point (α^*, β^*) [5] for
(a) $\rho = 1$, (b) $0 < \rho < 1$ and (c) $\rho > 1$.

1.4 A roly-poly is made of a solid hemisphere as the lower part and a solid cone as the upper part as shown in Fig. 1.7 with uniform mass density ρ. Show that its CoG is given by

$$y_0 = \frac{1}{M_1 + M_2}\left[\frac{3R}{8}M_1 + \left(R + \frac{h}{4}\right)M_2\right]\,, \tag{1.14}$$

where $M_1 = 2\pi R^3\rho/3$ and $M_2 = \pi a^2 h\rho/3$.

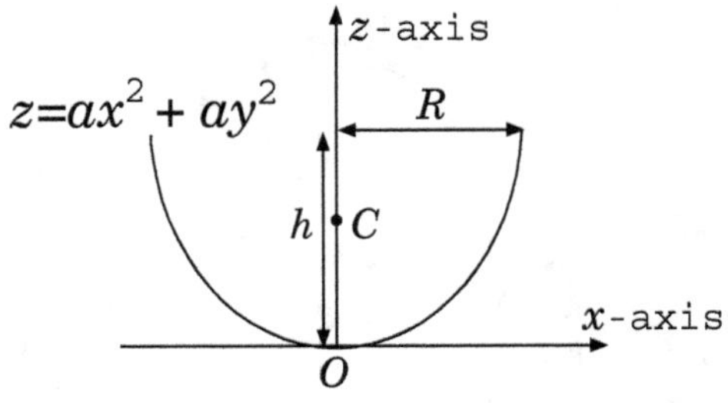

Fig. 1.8: A roly-poly with the lower part having the shape of a paraboloid.

1.5 Consider a roly-poly with the lower part having the shape of a paraboloid described by the function $z = ax^2 + ay^2$. Let the lower part be filled up to a height h with uniform mass density ρ as shown in Fig. 1.8 and the upper part has a negligible mass. Show that the distance of the CoG from the bottom O is given by $C = 2aR^2/3$. Next, assume that the condition for the balance of the toy is $C \leq 1/(2a)$ (refer to the solved problem 1 in this chapter). Then, find the condition on aR for the balance of the toy.

Chapter 2

Seesaw

A seesaw is a playing equipment in a playground. It is believed to be invented in 17th century by Korean girls who were not permitted to cross their courtyard walls. They used the seesaw to loft them high enough to see the outside world. A typical seesaw is a long, narrow board supported usually at its centre by a single pivot point (also called *fulcrum*). The pivot prevents the falling of seesaw and allows rotational motion of the board and not translational motion. Further, the total weight of the seesaw and riders on the board are supported by the pivot.

When one end of the board of a seesaw moves down the other end moves up. In Fig. 2.1 when the person A is trying to lift the person B, then the downward force is the weight of the person B. Essentially, when the pivot is at the centre, on one side the *effort* is applied whereas on the other side *load* is present.

Seesaws are also called *teeter-totter*, *teeter board*, *tilt*, *tilting board*, *teedle board*, *dandle* and *dandle board*. The up and down movement of a seesaw resembles like the forward and backward movement of a *saw* and hence the name seesaw. Its origin is *scie*-the French word for saw and the

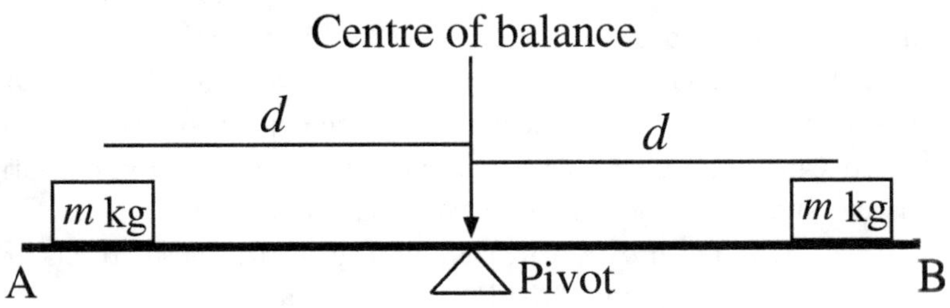

Fig. 2.1: A seesaw.

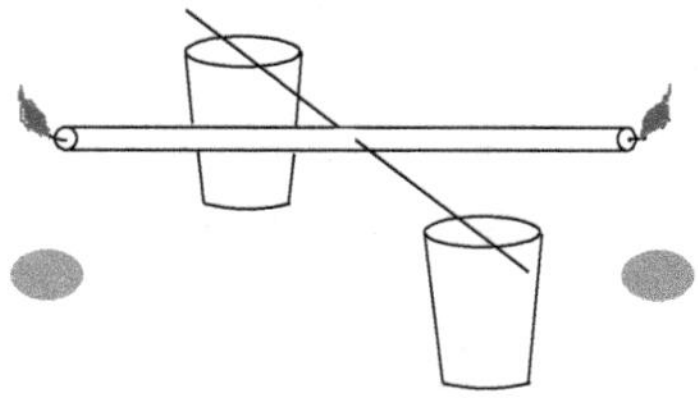

Fig. 2.2: A candle seesaw.

Anglo-Saxon *saw*. *Scie-saw* then became *seesaw*. Seesaws are of different shapes and designs like aeroplanes, helicopters and animals.

Interestingly, a candle burning at the two ends can oscillate about horizontal, forming a beautiful *candle seesaw* (Fig. 2.2). The candle seesaw is one of the popular physical experiments presented by Arthur Good (under the pseudonym Tom Tit) in the French magazine *L'Illustration*, titled *La Science Amusante* towards the end of 19th century [1-3]. Let us lit both ends of the candle. In a few seconds the candle begins to oscillate like a seesaw. Tom Tit described it as [2,4]: One drop of stearin drops on one of the saucers placed underneath. The equilibrium of the candle is broken, and its other end descends, making the end that just lost a drop of stearin ascends. But this motion makes more drops fall from the end that just descended, which now in turn becomes the lighter one. Thus, this end ascends once more, while the other descends, leading to oscillations.

Tjossem, Case and Bass developed a theoretical model for predicting the maximum growth rate of oscillation when the dripping rate and the candle seesaw frequency match each other [5]. Theodorakis and Paridi [4] setup a quantitative model for the candle seesaw oscillation. The results of their model are very in close agreement with the experimental results.

For an experimental study of a seesaw, a reader may refer to [6]. Seesaw-like dynamics has been found applications in many fields of science. We are going to mention a few of them. A seesaw-like movement of the relay region has been simulated in silico upon the recovery stage of myosin [7]. The seesaw mechanism has been utilized in the study of stability of the generalized Chaplygin gas in the dynamics of active and sterile neutrinos [8], massive and mixed neutrinos [9] and alar vertical discrepancy [10], a type of reconstructive problem in the surgery on the nose to change its shape or improve its function. In a RF MEMS switch design a seesaw-type movable part has been proposed to implement a capacitive shunt across a coplanar waveguide transmission line [11].

Because a seesaw is a simple machine-lever, we first present about simple machines. Then, we bring out certain fundamentals of a seesaw. We analyse the motion of a seesaw through a theoretical model. Expressions for amplitude and time period of oscillation of a seesaw will be obtained.

2.1 Machines

A *machine* is a mechanical device which receives an input amount of work and converts that into an output amount of work. In an ideal machine with no frictional losses, the output work will be same as the input. Generally, machines will be complex. But a simple machine is a mechanical device that changes the magnitude or direction of the force. Any machine is a complex construction of many simple machines. There are six simple machines, namely, lever, pully, inclined plane, screw, wedge and wheel and axle.

Simple machines are used as they offer mechanical advantage. Mechanical advantage is a measurement of the force amplification of a machine defined for an ideal machine as

$$A_{\mathrm{IM}} = \frac{\text{resistance (load) force}}{\text{effort (applied) force}} \,. \tag{2.1}$$

As an ideal machine has no friction, then from the conservation of energy

$$\text{input work} = \text{output work}\,, \tag{2.2}$$

that is,

$$\text{effort force} \times \text{effort distance} = \text{resistance force} \times \text{resistance distance}\,. \tag{2.3}$$

Then,

$$A_{\mathrm{IM}} = \frac{\text{effort distance}}{\text{resistance distance}} \,. \tag{2.4}$$

Since a compound machine is formed by combining many simple machines, the product of each simple machine's A_{IM} gives the compound machine's A_{IM}.

Seesaw is an example of a simple machine-lever. A lever consists of an inflexible rod placed over a fulcrum. There are three classes of levers according to where the load and effort are located with respect to the fulcrum. A class-1 lever has the fulcrum placed between the effort and load as shown in Fig. 2.3a. The effort in a class-1 lever is in one direction and the load moves in opposite direction. Its mechanical advantage is

$$A_{\mathrm{IM}} = \frac{\text{effort distance}}{\text{load distance}} \,. \tag{2.5}$$

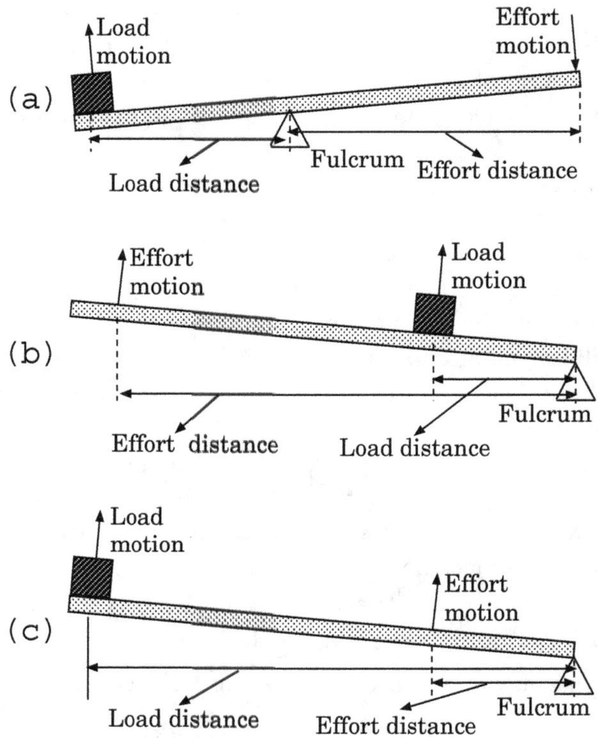

Fig. 2.3: (a) Class-1 lever, (b) class-2 lever and (c) class-3 lever.

In a seesaw, the fulcrum is placed at the centre. So, its $A_{\mathrm{IM}} = 1$.

In a class-2 lever the load is placed between the effort and the fulcrum as shown in Fig. 2.3b. In the class-2 lever, the movement of the load is in the same direction as that of the effort. In a class-3 lever, the effort is applied between the load and the fulcrum as shown in Fig. 2.3c. In class-3 lever, the load motion and effort motion are all in the same direction.

2.2 Some Basics of Seesaw

In this section we present some basics of seesaw.

2.2.1 *A Simple Seesaw Model*

It is easy to make a seesaw model with a pencil and a ruler [12]. The model is shown in Fig. 2.4. Tape a pencil to a table using two pieces of tape.

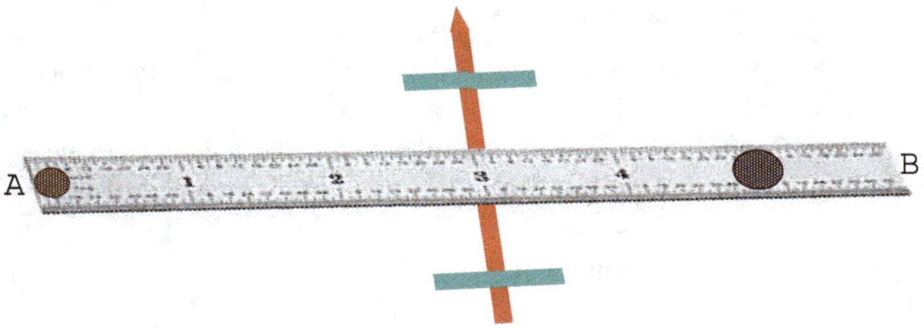

Fig. 2.4: A simple seesaw made of a ruler, a pencil and two pieces of tape.

Place a ruler across the pencil and adjust its position suitably so that it is balanced. The balance point on the ruler can be marked by a marker. Take two coins of different weight. Place the lesser weight coin at the edge of one side of the ruler, say side A. Then, initially place the other coin at the edge of the other side of the ruler (side B). The side A would be in 'up' position while the side B would be in 'down' position. With a small step size, move the position of the heavier weight coin towards the pencil. At one particular position of the coin in side B the ruler would be balanced. For a balance the heavier rider should be closer to the pivot point while the lighter rider should be farther from the pivot.

2.2.2 *Seesaw as a Big Balance*

It is interesting to notice that a seesaw appears as a big balance [13]. The weight of a rider on one side of it is like an applied force to that end of a lever. As a result the rider on the other end can be lifted by the lever. More and more weight can be lifted by the rider by moving his position (that is, the applied force) further away from the lever. In this way, the lever can be utilized to manipulate the relation between force and distance.

In 2018 four retired soldiers who had took part in the missions on the dangerous Sichuan-Tibet highway, demonstrated the balancing a giant seesaw of 26 m length and 2.3 m width over a duration by driving four cars on the seesaw in the Impossible Challenge show of China Global Television Network. This video record of this amazing event is available in the YouTube [14]. For fascinating seesaws with kinetic art and infinite oscillations see [15,16].

2.2.3 *Balance Equation*

On a seesaw, the torque is produced by a rider's weight. Counterclockwise torque (weight×lever arm) is produced by the left-side rider. Clockwise torque is produced by the right-side rider. Balance is realized when the sum of these torques becomes zero. *What about the angular velocity of a balanced seesaw?* For a seesaw with L persons on the left and R persons on the right the balance equation is

$$\sum_{i=1}^{L} m_i d_i = \sum_{j=1}^{R} m_j d_j \,, \tag{2.6}$$

where m_p is the weight of the person p and d_p is the distance of the person p from the centre of the seesaw.

Solved Problem 1:

A seesaw is of 6 m length with its pivot at its centre. A 10 kg child and a 15 kg child sit at opposite ends. In order to balance the seesaw where should an another 12 kg child sit?

The 10 kg child and a 15 kg child sit at opposite ends, that is, 3 m away from the centre. The downward force on the side of 15 kg child is relatively larger. Therefore, the third child has to sit on the side of the 10 kg child in order to balance the seesaw. At balance the requirement is

$$m_1 d_1 + m_3 d_3 = m_2 d_2 \,, \tag{2.7}$$

where $m_1 = 10\,\text{kg}$, $m_2 = 15\,\text{kg}$, $d_1 = 3\,\text{m}$, $d_2 = 3\,\text{m}$ and $m_3 = 12\,\text{kg}$. Then, we obtain

$$d_3 = (m_2 d_2 - m_1 d_1)/m_3 = (15 \times 3 - 10 \times 3)/12 = 1.25\,\text{m}\,. \tag{2.8}$$

2.3 A Theoretical Model of Seesaw

A theoretical model has been developed by Pecori and Torzo [6] to study the motion of a seesaw. They considered two cases: (i) A seesaw with a rectangular pivot and (ii) a seesaw with a cylindrical pivot. In this section, we discuss the case (i). For the case (ii) see the exercise 2.4.

Consider a flat top board of length L balanced on the edge of another vertical board (pivot) of width d. When the top board oscillates the support is changed alternatively between the left and right edges of the vertical board as shown in Fig. 2.5.

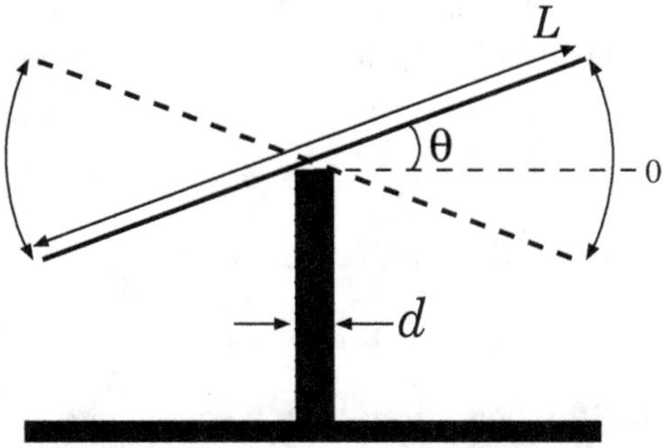

Fig. 2.5: A seesaw with a top board supported alternately by two edges, left and right edges, of a vertical board as the top board rotates counter-clockwise and then clockwise.

Assume that the oscillation amplitude θ is small and further $d \ll L$. For a small amplitude oscillation, it is possible to construct an analytical solution for the motion of the top board. The associated torque τ is due to the horizontal displacement between the centre of mass (CoM) of the top board and the edge of the vertical board. This torque causes the motion of the board. For small amplitude motion the torque is approximately constant and the angular acceleration (α) is also a constant. The motion of the top board is governed by

$$\tau = I\alpha \,, \tag{2.9}$$

where I is the moment of inertia given by (for the derivation see the exercise 2.10 at the end of this chapter)

$$I = \frac{1}{3}m\left(\frac{L}{2}\right)^2 + m\left(\frac{d}{2}\right)^2 = \frac{1}{3}m\left(\frac{L}{2}\right)^2\left(1 + 3\frac{d^2}{L^2}\right) . \tag{2.10}$$

For $d \ll L$ we have $I = mL^2/12$.

We find τ in Eq. (2.9). The torque is rF where r is the distance between the CoM of the top board and the point at which the force acts. Here r is $d/2$ for small θ. The gravitational force F acting at the centre of the board is simply $mg\cos\theta \approx mg$ for small θ. Then $\tau = Fd/2 = mgd/2$.

Next, the angular acceleration α is

$$\alpha = \frac{\tau}{I} = \frac{mgd}{2}\frac{12}{mL^2} = \frac{6gd}{L^2} = C \,. \tag{2.11}$$

On the other hand, denoting the angular acceleration as $\mathrm{d}^2\theta/\mathrm{d}t^2$, Eq. (2.11) becomes $\mathrm{d}^2\theta/\mathrm{d}t^2 = C$. Its solution is

$$\theta(t) = \frac{1}{2}Ct^2 + \theta'(0)t + \theta(0) \,. \tag{2.12}$$

Suppose at $t = 0$ the top board is at its maximum amplitude, say, $\theta_{\max}$ with zero initial velocity [17]. That is, $\theta(0) = \theta_{\max}$ and $\theta'(0) = 0$. Then, the solution is

$$\theta(t) = \frac{1}{2}Ct^2 + \theta_{\max} . \tag{2.13}$$

Since $\theta(t) \leq \theta_{\max}$ the solution can be written as

$$|\theta_{\max} - \theta(t)| = \frac{1}{2}Ct^2 . \tag{2.14}$$

The oscillation of the top board is anharmonic. Denote Δt is the time taken by the board from $\theta_{\max}$ to $\theta = 0$. Equation (2.14) gives

$$\theta_{\max} = \frac{1}{2}C(\Delta t)^2 \quad \text{or} \quad \Delta t = \sqrt{\frac{2\theta_{\max}}{C}} . \tag{2.15}$$

The time period of oscillation T_2 for the top board to go from $\theta_{\max} \to \theta = 0 \to -\theta_{\max} \to \theta = 0 \to \theta_{\max}$ is $4\Delta t$. We obtain

$$T = 4\Delta t = 4\sqrt{\frac{2}{C}}\sqrt{\theta_{\max}} = k\sqrt{\theta_{\max}} , \quad k = 4\sqrt{\frac{2}{C}} . \tag{2.16}$$

If we denote A as the maximum vertical displacement of the top board from its horizontal position, then from Fig. 2.5, we write

$$\sin\theta_{\max} = \frac{A}{(L+d)/2} . \tag{2.17}$$

For small amplitude of oscillation $\sin\theta_{\max} \approx \theta_{\max}$ and neglecting d in $L+d$ since $d \ll L$ we get

$$\theta_{\max} = \frac{A}{L/2} . \tag{2.18}$$

Now,

$$T = k\sqrt{2/L}\sqrt{A} = 4\sqrt{\frac{2}{C} \times \frac{2}{L}}\sqrt{A} = 8\sqrt{\frac{1}{CL}}\sqrt{A} = 8\sqrt{\frac{I}{L\tau}}\sqrt{A} . \tag{2.19}$$

When $I = mL^2/12$ and $\tau = mgd/2$ we obtain

$$T = 4\sqrt{A}\sqrt{\frac{2L}{3gd}} . \tag{2.20}$$

There are two important results: (i) T is independent of the mass of the board and (ii) T is amplitude (A) dependent. In order to confirm this theoretical result (i) Pecori and Torzo [6] performed experiments with a top board with different masses. Their results matched with the theoretical prediction.

Note that because of the air resistance and friction of the top board with the edges of the vertical board, a loss of energy takes place continuously, the top board exhibits damped oscillations and finally comes to a halt at its horizontal rest state.

Suppose that we add to each end of the top board a mass $M \gg m/2$. In this case (see the exercise 2.3 at the end of the present chapter)

$$T = 2\sqrt{A}\sqrt{\frac{2L}{gd}}\,. \tag{2.21}$$

T is independent of M.

2.4 Conclusion

Seesaw is a simple toy. It is available in every children's playground. It can be constructed for demonstration to the school children using just a pencil and a ruler. By balancing the ruler on the pencil, the students can find that the CoM is at the centre of the ruler. By placing different weights from a weight box on the ruler on both sides of the pivot and balancing it, the students can be taught the principle of moments. By shifting the pivot from the centre, and balancing the ruler with different masses at both ends of the ruler, the students can be taught to calculate the mechanical advantage of the simple lever. The students will also enjoy the candle seesaw when demonstrated by the teacher. By constructing a simple seesaw and keeping a small weight in one end, and dropping a heavy weight on the other end, the students can be taught how seesaw acrobatics are done in circus. Seesaw is a simple example for a system exhibiting oscillatory motion where the period depends on the amplitude of the oscillation.

2.5 Bibliography

[1] Tom Tit (Arthur Good), *La Science Amusante* (Librairie Larousse, Paris, 1890).

[2] Tom Tit (Arthur Good), *La Science Amusante, 3e Serie:* 100 *Nouvelles Experiences* (Libraire Larouse, Paris, 1906) pp.11-12.

[3] A.M. Low, *Popular Scientific Recreations* (Ward-Lock, London, 1933) p.145.

[4] S. Theodorakis and K. Paridi, Am. J. Phys. **77**, 1049 (2009).

[5] P.J.H. Tjossem, W.B. Case and R.M. Bass, Am. J. Phys. **87**, 370 (2019).

[6] B. Pecori and G. Torzo, The Phys. Teach. **39**, 491 (2001).
[7] B. Kinstses, Z. Yang and A.M. Csizmadia, J. Biochem. **283**, 34121 (2008).
[8] A.E. Bernardine and O. Bertolami, Phys. Rev. D **81**, 123013 (2010).
[9] S. Bilenky, *Introduction to the Physics of Massive and Mixed Neutrinos* (Springer, Berlin, 2010).
[10] S.M. Hyun, G.S. Medikeri and D.H. Jung, Plastic & Recons. Surgery **136**, 488 (2015).
[11] J.M. Cabral and A.S. Holmes, IEEE MELECON Proceedings (May 16, 2006) pp.288.
[12] https://www.acs.org/content/dam/acsorg/education/resources/k-8/-science-activities/motionenergy/motion/just-weight-and-seesaw.pdf.
[13] https://www.acs.org/content/dam/acsorg/education/resources/k-8/-science-activities/motionenergy/motion/just-weight-and-seesaw.pdf.
[14] https://www.youtube.com/watch?v=2D893rrAxEo.
[15] https://www.youtube.com/watch?v=dglSqRTgFv8.
[16] https://www.youtube.com/watch?v=Bft7Wd8QElw.
[17] R. Akridge, The Phys. Teach. **36**, 507 (1998).

2.6 Exercises

2.1 Compare the principle of a seesaw with that of the claw of a hammer used to remove a nail.

2.2 Mary is 21 kg and is 2.5 m away from the centre of the seesaw. Revathi, to balance the seesaw, adjusted her position to be at 3 m from the centre of the seesaw in the other side. Find the weight of Revathi.

2.3 Consider a seesaw shown in Fig. 2.5. Add to both ends of the top board a mass $M \gg m/2$. For this case show that $I \approx ML^2/2$ and T is given by Eq. (2.20).

2.4 A seesaw with a cylindrical pivot of radius R is shown in Fig. 2.6 [6]. Show that $I = mL^2/12$ and the oscillation of the top board is harmonic with time period of oscillation as $T = L\pi\sqrt{1/(3Rg)}$.

2.5 For a loaded seesaw (mass $M \gg m$ (mass of the top board of the seesaw) is added to each end of the top board of the seesaw) with cylindrical pivot of radius R, determine the time period of oscillation.

2.6 Consider a steel ruler of length 30 cm weighing 20 g lieing flat on a table. Now, nudge the ruler from the zero marking side to fall over the edge of the table. What will be the reading of the ruler at the edge of the table before it falls-off?

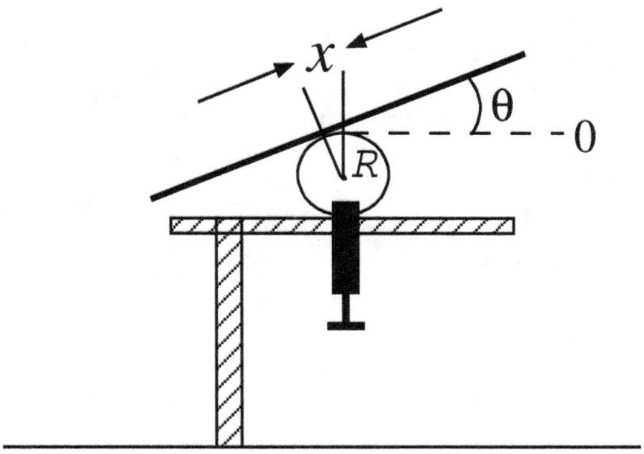

Fig. 2.6: A seesaw with a cylindrical pivot of radius R.

2.7 Consider the previous exercise. Keep the 30 cm end just outside the edge of the table and tie a twine carrying a mass of 5 g at the 30 cm end. At what reading on the ruler will it fall when it is nudged over the edge of the table?

2.8 A seesaw of length 4 m is having its fulcrum at the centre. At one end, a child of mass 10 kg is sitting and at the other end, another child of mass 15 kg is sitting. How much the CoM shifts from the furcrum towards the heavier child if the seesaw mass is negligible?

2.9 Consider the previous exercise. How much the centre of mass (CoM) shift from the furcrum towards the heavier child if the seesaw mass is 10 kg?

2.10 Consider a rectangular plate of length L, thickness t, width ω and mass m oscillating about the edge of a vertical board of thickness d as shown in Fig. 2.5. Show that the moment of inertia about the edge of the vertical board is given by Eq. (2.10).

Chapter 3

Newton's Cradle

A typical Newton's cradle shown in Fig. 3.1 consists of a line of just barely touching identically sized metal balls with each being suspended by two inelastic strings from a metal frame. At the stationary state the neighboring balls almost touch one another. A widely available commercial Newton's cradle has five balls. *What is fascinating about this toy?* We are familiar with the elastic collision of a point mass moving with velocity v_1 with a stationary point mass. After the collision both move in accordance with the conservation of momentum and conservation of kinetic energy. The law of conservation of momentum states that the total momentum of an isolated system remains constant or the total momentum of an isolated system before collision is always equal to the total momentum after collision.

We say that any quantity which does not change with time is said to be *conserved*. It has the same value both before and after an event. We have many conserved quantities in physics. In mechanics, if the system is isolated and frictionless, then the three fundamental quantities, namely, linear momentum, angular momentum and total energy are conserved.

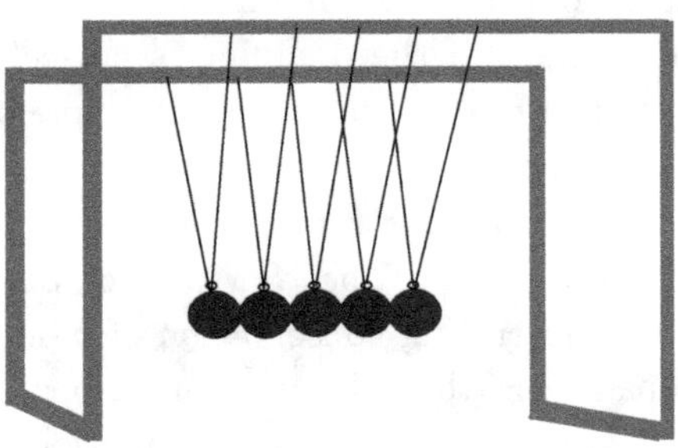

Fig. 3.1: A Newton's cradle with five identical balls.

Conservation of linear momentum is mostly used for describing collisions between objects. If the collisions are elastic (no friction involved), then the total energy will also be conserved. A system is said to be isolated if no external force acts on the particles of the system. If $\mathbf{P}_{1\mathrm{i}}$, $\mathbf{P}_{2\mathrm{i}}$, $\mathbf{P}_{3\mathrm{i}}$, ... are the linear momenta of the particles before an event and $\mathbf{P}_{1\mathrm{f}}$, $\mathbf{P}_{2\mathrm{f}}$, $\mathbf{P}_{3\mathrm{f}}$, ... are the linear momenta after the event then for an isolated system

$$\mathbf{P}_{1\mathrm{i}} + \mathbf{P}_{2\mathrm{i}} + \mathbf{P}_{3\mathrm{i}} + \cdots = \mathbf{P}_{1\mathrm{f}} + \mathbf{P}_{2\mathrm{f}} + \mathbf{P}_{3\mathrm{f}} + \cdots . \tag{3.1}$$

Conservation of linear momentum arises due to Newton's third law which states that for every action there is an equal and opposite reaction. If $\mathbf{F}_{12}$ is the force on particle 1 due to particle 2 and $\mathbf{F}_{21}$ is the force on particle 2 due to particle 1, then according to Newton's third law $\mathbf{F}_{12} = -\mathbf{F}_{21}$. If the force acts between the two particles for a duration $\triangle t$, then the impulses experienced by particle 1 and particle 2 must be equal and opposite in direction:

$$\mathbf{F}_{12}\triangle t = -\mathbf{F}_{21}\triangle t . \tag{3.2}$$

Since the impulse is the change of linear momentum, we get $\triangle\mathbf{P}_1 = -\triangle\mathbf{P}_2$. This implies that the total linear momentum $\mathbf{P}_1 + \mathbf{P}_2$ is a constant.

For an isolated system the total energy is conserved. The energy associated with a system may be in the form of kinetic energy, potential energy, heat energy, chemical energy, etc. The energies of the system may change from one form to another (say kinetic energy to potential energy), but the total energy will remain the same.

In the case of the Newton's cradle, if the left end ball is displaced and released to strike the line of balls, then the ball at the right end alone fly-off and strike with the line of balls causing the left end ball alone to fly-off (Fig. 3.2) making a long-lasting cycle of the above dynamics. When one end of the ball is detached from the other and in motion, then all the remaining balls are almost motionless. If the damping is neglected and there is no dissipation of energy, then the motion can in principle be described by the laws of conservation of energy and momentum.

The Newton's cradle was first introduced four centuries ago by the French physicist Abbé Mariotte. The Newton's cradle is considered as a promising toy tool for illustrating conservation of momentum and energy, studying and demonstrating the role of idealizations and explanation of physical phenomena through simplified approximations.

In this chapter we present the features of the Newton's cradle and the physics behind them.

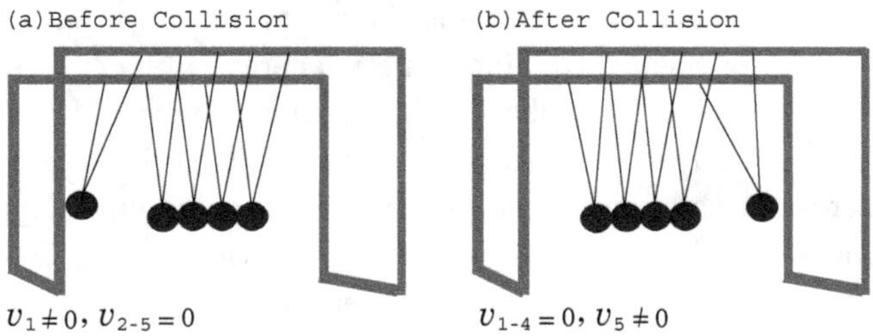

Fig. 3.2: A Newton's cradle with five identical balls before collision and after collision of the left end ball with the remaining line of balls. v_i denotes the velocity of the ith ball.

3.1 Arrangement of a Cradle

A typical Newton's cradle [1,2] shown in Fig. 3.1 consists of a series of balls usually of identical balls. The setup is made with a heavy wooden or plastic base in order to ensure stability. A small ring is attached with the head of each ball. The balls are suspended by thin nylon fishing lines as shown in the Fig. 3.1. The nylon lines hung over two threaded rods. In order to make the balls to swing without the nylon line catching on the threading, the rods are made to have flat surface and milled on them. Lateral adjustment of the positions of the balls can be easily performed with the use of the threaded rod. The height of the balls can also be achieved by changing the length of the nylon line. The balls are arranged on a same plane, they just touch the adjacent balls.

3.2 Working Principle of the Newton's Cradle

Now, we proceed to explain the working of the Newton's cradle [3]. Consider the Newton's cradle with five identical balls as shown in Fig. 3.1 and neglect the air resistance. Initially all the balls are at rest with zero initial velocity, zero kinetic energy and also zero initial potential energy.

Suppose the left end ball (say ball-1) is pulled away to a height, say, h_{top} from the other balls and released with say, zero initial velocity ($v_{\text{top}} = 0$ at time $t = t_0 = 0\,\text{sec}$). At this instant of time its kinetic energy is zero but the potential energy is not zero due to the fact that gravity can make the ball to fall. *What is the potential energy of the ball*-1*?* The ball-1 moves towards right side with increasing velocity. This is because of the work done by gravity resulting in decreasing the potential energy and the change in potential energy is converted into kinetic energy ($mv^2/2$).

At time t_1 sec the ball-1 strikes the ball-2 when it reaches its equilibrium (rest) position with a velocity v_{bottom}. At this position the potential energy becomes zero and the kinetic energy attains its maximum. The first ball stops its motion and its both potential and kinetic energies become zero. However, its energy is not lost but completely transferred to the second ball which is compressed under the force of impact. In a similar manner, the second ball transfers its energy to the third ball and then the fourth. The fourth ball then transfers its energy to the fifth (last) ball. Note that there is no ball right to the fifth ball to compress it.

As a consequence of the conservation of energy the last ball acquires an energy that is the same energy of the first ball just before the collision. The last ball alone now begins to move away in the right with the velocity of the first ball. In the course of its motion, its potential energy increases and the kinetic energy decreases, therefore its velocity gets reduced continuously. At a point of position the kinetic energy becomes zero, the velocity reaches zero and the total energy is simply the potential energy. The gravity then starts pulling the last ball downwards. During the motion the end balls reach the same height (h_{top}). When the last ball returns back, strikes the fourth ball and halts. Then, the first ball begins to move with the same velocity as that of the last ball. The above process continues if there is no loss of energy due to friction with the air, heat generated in the balls and sound generated from the clank of the balls, otherwise the motion of the end balls slow down and the balls finally halt.

Now, one may ask: *When the ball-1 strikes the ball-2, why does not (ball-1) reverse its direction of motion?* Here it comes the conservation of momentum. When the ball-1 collides with the second ball, the momentum of the ball-1 is in the direction of the left to the right. Because the momentum is a vector quantity, reversing the direction of motion of the ball-1 can happen only if there is an outside force. There is no such force. Also, see Sec. 3.5 in the present chapter. In the ideal situation the balls of the Newton's cradle are acted by energy, momentum and gravity. If the collisions are perfectly elastic, the balls swing forever.

Solved Problem 1:

In a Newton's cradle with two identical balls (with mass m) show that if one ball is raised and released, then it will come to rest after transferring all its velocity to the other ball which is initially at rest.

Let v_1 be the velocity of the left-side ball (ball-1) at collision. As the velocity of the other ball (right-side ball, ball-2) v_2 is 0 the total momentum along the x-direction is $P_\mathrm{i} = mv_1$. The total energy $E_\mathrm{i} = mv_1^2/2$. Denote the velocities of the balls-1 and 2 after collision as v_1' and v_2', respectively. Then, the total momentum along the x-direction is

$$P_\mathrm{f} = mv_1' + mv_2' . \tag{3.3}$$

The total energy is

$$E_\mathrm{f} = \frac{1}{2}mv_1'^2 + \frac{1}{2}mv_2'^2 . \tag{3.4}$$

As $P_\mathrm{i} = P_\mathrm{f}$ from the expressions of P_i and P_f we get

$$v_1 = v_1' + v_2' \tag{3.5}$$

and as $E_\mathrm{i} = E_\mathrm{f}$ we obtain

$$v_1^2 = v_1'^2 + v_2'^2 . \tag{3.6}$$

Squaring Eq. (3.5) and equating with Eq. (3.6), we get $2v_1'v_2' = 0$. As ball-2 is not rigid, $v_2' \neq 0$. Therefore $v_1' = 0$, so that Eq. (3.5) gives $v_2' = v_1$.

3.3 Number of Balls Ejected From the Cradle

Consider the Newton's cradle with N balls. When n ($n < N/2$) balls are pulled back and allowed to strike the other balls, for our surprise n balls are ejected from the other end and the remaining balls are almost stationary. *How do the balls know that initially n balls are moving and that exactly n balls should be ejected from the other end?* In the following we find the answer for this question.

Let us consider a simple case of first two balls dropped with the remaining $N-2$ balls initially at rest. Assume that when the first two balls strike the remaining lines of balls with the velocity $v_{\mathrm{bottom}}^{\mathrm{incident}}$, then n balls from the right end move away together with the initial velocity $v_{\mathrm{bottom}}^{\mathrm{ejected}}$. Denote $2m$ and nm are the total masses of the incident first two balls and n number of ejected balls. Conservation of momentum gives

$$2mv_{\mathrm{bottom}}^{\mathrm{incident}} = nmv_{\mathrm{bottom}}^{\mathrm{ejected}} \tag{3.7}$$

which gives

$$v_{\mathrm{bottom}}^{\mathrm{ejected}} = \frac{2}{n}v_{\mathrm{bottom}}^{\mathrm{incident}} . \tag{3.8}$$

For the two unknowns n and $v_{\text{bottom}}^{\text{ejected}}$, we have only one equation. We need another equation. The second equation is, in view of elastic collision of the balls, the equation of conservation of energy. At the bottom (at the time of collision) the conservation of energy gives

$$(\text{P.E.} + \text{K.E.})_2 = (\text{P.E.} + \text{K.E.})_n \,, \tag{3.9}$$

where P.E. and K.E. are potential energy and kinetic energy, respectively, and the subscripts 2 and n refer to incident two balls and ejected n balls, respectively. The heights of the incident two balls and ejected n balls at the time of the incident balls strike the line of balls which are at the bottom are denoted $h_{\text{bottom}}^{\text{incident}}$ and $h_{\text{bottom}}^{\text{ejected}}$, respectively. Then, Eq. (3.9) becomes

$$2mgh_{\text{bottom}}^{\text{incident}} + \frac{1}{2}2m\left(v_{\text{bottom}}^{\text{incident}}\right)^2 = nmgh_{\text{bottom}}^{\text{ejected}} + \frac{1}{2}nm\left(v_{\text{bottom}}^{\text{ejected}}\right)^2 . \tag{3.10}$$

Since $h_{\text{bottom}}^{\text{incident}} = h_{\text{bottom}}^{\text{ejected}} = 0$ we get

$$\left(v_{\text{bottom}}^{\text{incident}}\right)^2 = \frac{n}{2}\left(v_{\text{bottom}}^{\text{ejected}}\right)^2 = \frac{n}{2}\frac{4}{n^2}\left(v_{\text{bottom}}^{\text{incident}}\right)^2 . \tag{3.11}$$

The above equation gives $n = 2$ and in turn, from Eq. (3.8) we obtain $v_{\text{bottom}}^{\text{ejected}} = v_{\text{bottom}}^{\text{incident}}$. The conclusion is that *whenever n balls are dropped on one side, the same number of balls go up on the other side to the same height* [4].

3.4 Duration of Collision

In the Newton's cradle when two balls with the same radii R and same mass m are allowed to collide, then they deform and a finite portion of their surfaces come in contact. Consideration of pressure distribution over the contact region leads to the realization of Hertz's law. Denote s as the change of separation of the centres of the balls. If F is the total force between the balls, then according to Hertz $s \propto F^{2/3}$. The time duration τ of collision is obtained by integrating from $s = 0$ to s_{max}. Here s_{max} is the value of s corresponding to the maximum approach of the centres of the balls. Then, τ is worked out to be [1]

$$\tau = 3.29\left(1 - \sigma^2\right)^{2/5}\left(\frac{m^2}{RE^2 v}\right)^{1/5} . \tag{3.12}$$

In the above equation, v is the velocity of the incident ball, σ is the Poisson's ratio and E is the Young's modulus of the balls.

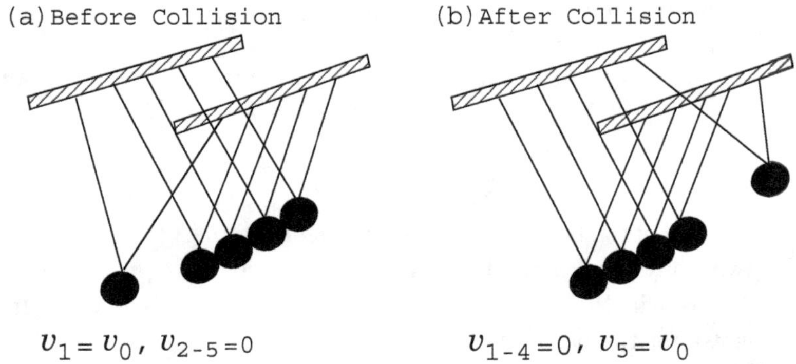

Fig. 3.3: A Newton's cradle with five identical balls before collision and after collision of the left end ball with the remaining line of balls as per the two-body model.

3.5 A Two-Body Model

Now, we consider a simple two-body elastic collision in one dimension [5] to the problem of the left-most ball striking remaining line of balls in the Newton's cradle. In this model ball-1 is one body and the remaining set of $N-1$ balls form another body. That is, we consider the N balls as essentially two masses colliding elastically. The masses of all the balls are equal and denoted as m. Ball-1 is pulled from its equilibrium position and released. The velocity with which the ball-1 collides with the body-2 is, say, $v_{1,\mathrm{b}}$. The velocity of the balls-2-N before collision is $v_{2-N,\mathrm{b}} = 0$. This is shown in Fig. 3.3a for the case of the cradle with 5 balls.

What are the velocities of the two bodies after collision? We can determine them by making use of the conservation of momentum and energy. Denote $v_{1,\mathrm{a}}$ and $v_{2-N,\mathrm{a}}$ as the initial velocities of the body-1 and body-2 after collision, respectively. The laws of conservation of momentum and energy give

$$mv_{1,\mathrm{b}} = mv_{1,\mathrm{a}} + (N-1)mv_{2-N,\mathrm{a}}\,, \tag{3.13}$$

$$\frac{1}{2}mv_{1,\mathrm{b}}^2 = \frac{1}{2}mv_{1,\mathrm{a}}^2 + \frac{1}{2}(N-1)mv_{2-N,\mathrm{a}}^2\,. \tag{3.14}$$

That is,

$$v_{1,\mathrm{b}} = v_{1,\mathrm{a}} + (N-1)v_{2-N,\mathrm{a}}\,, \tag{3.15}$$

$$v_{1,\mathrm{b}}^2 = v_{1,\mathrm{a}}^2 + (N-1)v_{2-N,\mathrm{a}}^2\,. \tag{3.16}$$

Solving these two equations for $v_{1,\mathrm{a}}$ and $v_{2-N,\mathrm{a}}$ gives

$$v_{1,\mathrm{a}} = -\frac{N-2}{N} v_{1,\mathrm{b}}, \quad v_{2-N,\mathrm{a}} = \frac{2}{N} v_{1,\mathrm{b}} \,. \tag{3.17}$$

For the case of $N = 5$

$$v_{1,\mathrm{a}} = -\frac{3}{5} v_{1,\mathrm{b}}, \quad v_{2-5,\mathrm{a}} = \frac{2}{5} v_{1,\mathrm{b}} \tag{3.18}$$

and the motion of the balls after collision is depicted in Fig. 3.3b. According to the two-body model all the balls are in motion after collision. This is not observed in the real experiment with the Newton's cradle. In the experiment after the ball-1 collided with the ball-2 in the line of balls, only the ball-5 is found in motion until colliding with the ball-4.

The model considered here is unsuitable even though it conserves momentum and energy. The two-body model has neglected the following. The balls are finite in size and moreover have elastic properties. Consequently, the momentum and energy of the incident ball propagates through the other balls when collision takes place. The propagation occurs over a finite interval of time. Details of teaching about the nature of models and physics through the Newton's cradle is discussed in [6,7]. For more details about collision of balls in the Newton's cradle one may refer to the Ref. [8]. The coefficient of restitution of the balls can be measured by considering the multiple collisions of two steel balls in the Newton's cradle [9].

3.6 Conclusion

Newton's cradle is a fascinating toy with simple dynamics and can serve as a good experimental tool to explain the laws of linear momentum and mechanical energy. The simple dynamics is correct only if the collisions are totally elastic with no heat loss due to friction and also the collisions take place only between two balls. What we have described in an ideal Newton's cradle is just a special case where the laws of linear momentum and energy is conserved. But there are many other ways the balls could move afterwards while obeying these two conservation laws. It has been found using experimental observations and computer simulation of collision of several balls suspended in a horizontal row that after collision that the individual balls are actually no longer at rest [10].

In demonstrating the Newton's cradle for the conservation of laws of linear momentum and energy to the students, they must be informed that there are many types of motions which conserve both the linear momentum and energy. They must also be informed that the elasticity of the balls, speed of the sound traveling in the lines of balls, gaps between the balls and the rotation of the balls determine the nature of motions of the balls. It has been shown by a simulation of Newton's cradle that the physics involved in it is far from trivial and that the standard textbook explanation is only a first approximation [11].

3.7 Bibliography

[1] D.R. Lovett, K.M. Moulding and S. Anketell-Jones, Eur. J. Phys. **9**, 323 (1988).

[2] R. Ehrlich, The Phys. Teach. **34**, 181 (1996).

[3] C. Schulz, *How Newton's Cradles Work*; http://science.howstuffworks.com/innovation/innovations/newtons-cradle.html.

[4] https://www.phys.vt.edu/outreach/projects-and-demos/demonstrations-wiki/mechanics/newtons-cradle.html.

[5] J.D. Gavenda and J.R. Edgington, The Phys. Teach. **35**, 411 (1997).

[6] C.F. Gauld, Sci. & Edu. **15**, 597 (2006).

[7] C. Gauld and R. Cross, J. Phys. Edu. **56**, 025001 (2021).

[8] R. Cross and C. Gauld, J. Phys. Edu. **56**, 025002 (2021).

[9] R. Cross, Eur. J. Phys. **39**, 025001 (2018).

[10] F. Herrmann and M. Seitz, Am. J. Phys. **50**, 977 (1982).

[11] S. Hutzler, G. Delaney, D. Weaire and F. MacLeon, Am. J. Phys. **72**, 1508 (2004).

3.8 Exercises

3.1 List out a few examples where we can observe effects similar to the one in Newton's cradle, that is, the last ball ejected with a velocity almost equal to the first ball.

3.2 For a five balls cradle prove that when the ball-1 strikes the ball-2, the balls-2 – 5 as a whole cannot move.

3.3 (a) In a N-balls Newton's cradle the left most ball (ball-1) with mass m is brought to a height h_{top} and released with the velocity $v_{\text{top}} = 0$. Calculate the velocity v_{bottom} of the ball-1 when it reaches its rest state (bottom).

(b) In a N-balls Netwon's cradle, with each ball having a mass m, the left most ball (ball-1) is brought to a height h' and released with zero initial velocity. Calculate its velocity at the height $h'/2$.

3.4 Find the interaction time for the collision of two identical steel balls of diameter 0.0254 m and the impact velocity of 0.07 m/sec. The density of steel is 7800 kg/m^3, Young's modulus is 215 GPa and its Poisson's ratio is 0.28.

3.5 Consider the Newton's cradle with five identical balls. Experimentally observe the behavior of the cradle in the following two cases and explain the observation. (i) Middle ball is initially at rest. The left-most two balls and the right-most two balls are brought to the same height and released with zero initial velocity. (ii) Left-most first ball and the right-most two balls are brought to the same height and released with zero initial velocity.

Chapter 4

Falling Slinky

Richard T. James, a naval mechanical engineer, invented the slinky in 1943 and demonstrated its feature, namely, walking over stairs, in 1945 at Gimbels Department Store in Philadelphia. A slinky in a vertically hanging position is depicted in Fig. 4.1. The name slinky means *sleek and graceful*. Initially, Richard James believed that slinky-like springs would be utilized to support sensitive instruments in ships and make them stable at sea during storms. One of the springs was accidentally knocked by James. The spring stepped in a series of arcs to stacks of books, to a tabletop and on to the floor. Finally, it recoiled itself and stood upright.

There is another fascinating display by the slinky. Let us hang a slinky vertically under the force of gravity from its top. Release it gently. The bottom of the slinky remains stationary in mid air while the top of it moves downwards. The top section collapses turn by turn. In a fraction of second the collapsing top section collides with the bottom section. Then, the whole part moves downwards. For a beautiful demonstration of the above dynamics see the YouTube video [1]. The fascinating physics associated with the dynamical features of the slinky in its vertically falling, walk down stairs, modes of oscillation when hanged vertically, waves propagation along it have been well studied. The restoring force of a hanging slinky is due to the coils between the centre of mass (CoM) of the slinky and the support. This is experimentally confirmed [2].

The slinky is used to explore different scientific concepts. In particular, the slinky has been used as a teaching tool in the classroom to simulate properties of waves. From the year of its invention the slinky mesmerized children by its variety of magical display. It is able to walk over stairs. Laser gun-like sound effect can be created using a slinky. In 1999 a slinky postage stamp was issued by the postal service of United States. In 2002 it was

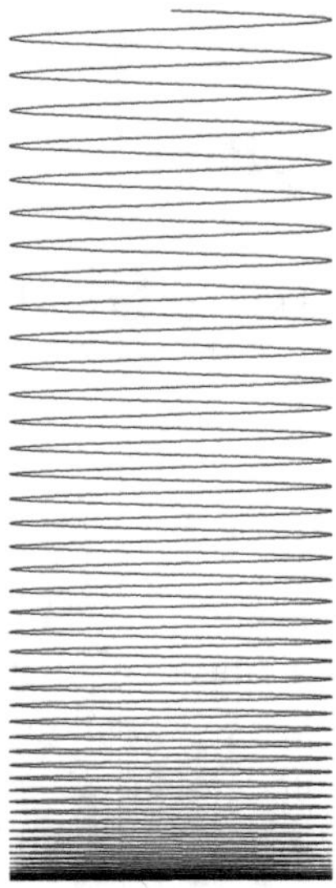

Fig. 4.1: A slinky in a hanging position.

inducted into the National Toy Hall of Fame at The Strong in Rochester, New York. Slinky had appeared on a US postage stamp commemorating the slinky craze of 1945 [3].

To manufacture the slinky and many other related toys James and his wife Betly setup the James Industries in Clifton Heights, Pennsylvania. Nowadays, metal and plastic slinkies are available with different colors mostly with the colors in rainbow order.

Slinkies are remarkable examples of tension springs, that is, they can be under tension as per Hooke's law, but not compression. Compression springs usually possess a finite unstretched length within which the turns are supposed not in contact and the tension is zero. However, such a spring can be compressed to a length within which the turns are in contact and during the compression obey Hooke's law.

The present chapter deals the dynamics of the falling slinky. The walking dynamics and whistling of slinkies will be considered in the next chapter.

4.1 Mathematical Model of a Slinky

In a slinky in its unstretched state the turns are touching and this state is termed as *collapsed*. To separate the turns from touching, a finite tension is needed. To stretch the turns, an external force is to be applied. Suppose the slinky is suspended from its top end stretched under gravity as shown in Fig. 4.1. *What happens if it is released?* Our common sense tells that the whole part of the slinky would fall downward under the action of gravity. But this is not happening. As mentioned in the beginning of this chapter, one can notice a strange physics phenomenon. When the stretched slinky is released, its lower end does not start moving downwards or upwards, but will remain suspended in air for a short time until the top section collapses with the bottom.

What is the physical explanation of the intriguing display of the hanging slinky? Why does the bottom section not move immediately after the release of the top end? The tension pulling up and the gravity pulling down forces are equal and opposite [4]. Further, the collapse of the tension in the slinky takes place from the top down. Finite amount of time is required to transmit this information (release of the top) from the top of the slinky to its bottom. Mathematical models have been developed to describe a falling slinky and a falling spring [5-16]. Here we present the model of Cross and Wheatland [15].

4.1.1 *Equation of Motion*

Consider the slinky as a spring with a total mass m and a spring constant k. Describe the mass distribution along the spring by the dimensionless coordinate ξ with $0 \leq \xi \leq 1$. For an increment $\Delta\xi$ $(= \mathrm{d}\xi)$ in ξ the increment in mass is $\Delta m = m\Delta\xi$. For a spring having N turns the end point of the ith turn has $\xi_i = i/N$ and at time t we have $x_i = x(\xi_i, t)$. The tension of the spring is $f(\xi, t)$.

To describe the dynamical behavior of a falling spring, we must obtain the equation of motion for mass elements along the slinky by taking into account the gravity and the local spring forces at that point. Using Newton's law of motion for a mass element $m\Delta\xi$ at a location ξ_i on the slinky, we get

$$\begin{aligned} m\Delta\xi \left.\frac{\partial^2 x}{\partial t^2}\right|_{\xi_i,t} &= f\left(\xi_i + \Delta\xi, t\right) - f\left(\xi_i, t\right) + m\Delta\xi g \\ &= \Delta\xi \left.\frac{\partial f}{\partial \xi}\right|_{\xi_i,t} + m\Delta\xi g\,. \end{aligned} \tag{4.1}$$

The tension f is given by the Hooke's law as [17]

$$f(\xi, t) = k\left(\frac{\partial x}{\partial \xi} - l_0\right), \tag{4.2}$$

where $\partial x/\partial \xi$ describes the local extension of the slinky and l_0 corresponds to a slinky length at which the tension would be zero (totally collapsed state). From Eq. (4.2) we get

$$\frac{\partial f}{\partial \xi} = k\frac{\partial^2 x}{\partial \xi^2}\,. \tag{4.3}$$

Substituting (4.3) in (4.1), we get the wave equation for a falling or suspended compression spring as

$$m\frac{\partial^2 x}{\partial t^2} = k\frac{\partial^2 x}{\partial \xi^2} + mg\,. \tag{4.4}$$

4.1.2 *The Hanging Slinky*

In a hanging stationary position, the top of the slinky has stretched turns while a small section of length l_1 at the bottom contains collapsed turns, that is, turns are in touch. Denote ξ_1 and $1-\xi_1$ as the mass fractions of stretched and collapsed sections, respectively.

At $t = 0$, the equation for $x_0 = x_0(\xi)$ along the stretched section is obtained from Eq. (4.4) as

$$\frac{\partial^2 x_0}{\partial \xi^2} + \frac{mg}{k} = 0 \tag{4.5}$$

with the boundary conditions

$$\text{at the top}: x_0(\xi = 0) = 0, \quad \text{at the bottom}: \left.\frac{\partial x_0}{\partial \xi}\right|_{\xi=\xi_1} = l_1\,. \tag{4.6}$$

The solution of Eq. (4.5) is

$$x_0(\xi) = -\frac{mg}{2k}\xi^2 + C_1\xi + C_2\,, \tag{4.7}$$

where C_1 and C_2 are constants to be determined by the boundary conditions.

Applying the boundary conditions, we get for $0 \le \xi < \xi_1$

$$x_0(\xi) = l_1\xi + \frac{mg}{k}\xi\left(\xi_1 - \frac{1}{2}\xi\right), \quad 0 \le \xi \le \xi_1\,. \tag{4.8}$$

We need to find $x_0(\xi)$ in the collapsed section ($\xi_1 \le \xi \le 1$) at the bottom of the slinky. In this case

$$\frac{\partial x_0}{\partial \xi} = l_1\,. \tag{4.9}$$

Its solution is given by

$$x_0(\xi) = l_1\xi + C_3\,, \tag{4.10}$$

where C_3 can be obtained by matching the solutions (4.8) and (4.10) at $\xi = \xi_1$. The result is $C_3 = mg\xi_1^2/(2k)$ and

$$x_0(\xi) = l_1\xi + \frac{mg}{2k}\xi_1^2\,, \quad \xi_1 \le \xi \le 1\,. \tag{4.11}$$

Therefore,

$$x_0(\xi) = \begin{cases} l_1\xi + \frac{mg}{k}\xi\left(\xi_1 - \frac{1}{2}\xi\right), & 0 \le \xi \le \xi_1 \\ l_1\xi + \frac{mg}{2k}\xi_1^2, & \xi_1 \le \xi \le 1. \end{cases} \tag{4.12}$$

The point of the centre of mass (CoM) is given by

$$x_{\rm CoM}(0) = \int_0^1 x_0(\xi)\mathrm{d}\xi = \frac{1}{2}l_1 + \frac{mg}{2k}\xi_1^2\left(1 - \frac{1}{3}\xi_1\right). \tag{4.13}$$

Solved Problem 1:

Verify Eq. (4.13).

We write

$$x_{\rm CoM}(0) = \int_0^1 x_0(\xi)\,\mathrm{d}\xi\,. \tag{4.14}$$

Substitution of Eq. (4.12) for x_0 and carrying out the integration gives

$$\begin{aligned} x_{\rm CoM}(0) &= \int_0^{\xi_1}\left[l_1\xi + \frac{mg}{k}\xi\left(\xi_1 - \frac{1}{2}\xi\right)\right]\mathrm{d}\xi + \int_{\xi_1}^1\left[l_1\xi + \frac{mg}{2k}\xi_1^2\right]\mathrm{d}\xi \\ &= \left[\frac{l_1\xi^2}{2}\right]_0^{\xi_1} + \left[\frac{mg\xi_1\xi^2}{2k}\right]_0^{\xi_1} - \left[\frac{mg\xi^3}{6k}\right]_0^{\xi_1}\left[\frac{l_1\xi^2}{2}\right]_{\xi_1}^1 + \left[\frac{mg\xi_1^2\xi}{2k}\right]_{\xi_1}^1 \\ &= \frac{1}{2}l_1 + \frac{mg}{2k}\xi_1^2\left(1 - \frac{1}{3}\xi_1\right). \end{aligned} \tag{4.15}$$

4.1.3 *The Falling Slinky with Instant Collapse of Turns*

What does it happen when we release the top of the slinky at $t = 0$? The slinky starts collapsing from the top to down one turn after another successively. Suppose at $t = t_c$, called *total collapse time*, the collapse reach the bottom. For $0 < t \le t_c$ the slinky essentially consists of the following three sections.

1. The top collapsed part, $0 \le \xi \le \xi_c(t)$.
2. The middle part, $\xi_c < \xi \le \xi_1$, which is unaffected, that is, still in the initial state.
3. The bottom part, $\xi_1 \le \xi \le 1$ which is also unaffected.

The point is that the section $\xi_c < \xi \le 1$ is in the initial state. $x(\xi, t)$ for this section is simply given by Eq. (4.12), that is,

$$x(\xi,t) = \begin{cases} l_1\xi + \frac{mg}{k}\xi\left(\xi_1 - \frac{1}{2}\xi\right), & \xi_c \le \xi \le \xi_1 \\ l_1\xi + \frac{mg}{2k}\xi_1^2, & \xi_1 \le \xi \le 1. \end{cases} \tag{4.16}$$

Next, we proceed to find $x(\xi, t)$ for the top collapsing section. This is obtained by solving Eq. (4.9) and matching its solution with (4.16) at $\xi = \xi_c(t)$. The result is

$$x(\xi,t) = l_1\xi + \frac{mg}{k}\xi_c\left(\xi_1 - \frac{1}{2}\xi_c\right), \quad 0 \le \xi \le \xi_c. \tag{4.17}$$

Thus, we have

$$x(\xi,t) = \begin{cases} l_1\xi + \frac{mg}{k}\xi_c\left(\xi_1 - \frac{1}{2}\xi_c\right), & 0 \le \xi \le \xi_c \\ l_1\xi + \frac{mg}{k}\xi\left(\xi_1 - \frac{1}{2}\xi\right), & \xi_c \le \xi \le \xi_1 \\ l_1\xi + \frac{mg}{2k}\xi_1^2, & \xi_1 \le \xi \le 1. \end{cases} \tag{4.18}$$

The time dependence of $\xi_c(t)$ can be obtained by applying Newton's second law. The top section of the slinky is with $0 \le \xi \le \xi_c$. The mass of this section is $m\xi_c$, while its velocity $\partial x/\partial t$ is given by

$$v = \frac{\partial x}{\partial t} = \frac{mg}{k}\left(\xi_1 - \xi_c\right)\frac{\mathrm{d}\xi_c}{\mathrm{d}t}. \tag{4.19}$$

The total momentum of the slinky is $m\xi_c v$, since the middle and the bottom sections are stationary. At time t the net impulse on the slinky due to gravity is mgt and is equal to $m\xi_c v$. This gives

$$\xi_c\left(\xi_1 - \xi_c\right)\frac{\mathrm{d}\xi_c}{\mathrm{d}t} = \frac{k}{m}t. \tag{4.20}$$

Integrating the above equation with $\xi_c(0) = 0$ gives

$$\xi_c^3 - \frac{3}{2}\xi_1\xi_c^2 + \frac{3k}{2m}t^2 = 0. \tag{4.21}$$

Analytical expressions for all the roots of a cubic equation are presented in Ref. [18]. The first positive root of the above cubic equation is the desired ξ_c. As mentioned earlier, t_c is the time at which the top section of the slinky collapses with the bottom section. When this happens, $\xi_c(t) = \xi_1$ and then Eq. (4.21) gives

$$t_c = \sqrt{\frac{m}{3k}\xi_1^3}. \tag{4.22}$$

Since the CoM falls with acceleration g from rest ($x_{\mathrm{CoM}}(0)$), we write

$$x_{\mathrm{CoM}}(t) = x_{\mathrm{CoM}}(0) + \frac{1}{2}gt^2. \tag{4.23}$$

4.1.4 *What Does Happen After the Time t_c?*

At $t = t_c$ adjacent turns touch with each other, $\xi_c = 1$, $v(t_c) = -gt_c$ and the bottom of the slinky also starts moving downwards. That is, for $t > t_c$ the whole slinky falls downwards. From Eq. (4.18), we can compute the values of vertical positions of top of the slinky ($x_t(t_c)$), bottom of the slinky ($x_b(t_c)$) and $x_{CoM}(t_c)$. Knowing these values at $t = t_c$, we can easily determine the values of the above quantities for $t_c < t < t'$ where t' is the time at which the slinky lands on the floor (see the exercise 4.4 at the end of the present chapter). *How do you determine the time t'?*

4.2 Illustration of Falling of the Slinky

Now, we illustrate the falling of the slinky through the model described above. The chosen values of the parameters are: mass $m = 0.2$ kg, number of turns$= 50$, radius of the slinky $r = 0.03$ m, length in a vertically hanging position$= 1$ m, collapsed length$= 0.06$ m, $\xi = 0.9$ and the spring constant $k = 0.84$ N/m.

Figure 4.2 shows the solution, that is, the vertical position ($-x$) versus the horizontal position at some selected values of t. The solid circle indicates the location of the CoM. Equations (4.12) and (4.18) give the values of the vertical positions of end of the turns only at time $t = 0$ and $0 < t < t_c$. The vertical positions of various mass points between the starting and end points of each turn are needed to draw the slinky. For this purpose assume that ξ_j and ξ_{j+1} are the starting and end values of a turn. Then, the locations of ξ_j and ξ_{j+1} are $x(\xi_j)$ and $x(\xi_{j+1})$, respectively. The turn (coil) between $x(\xi_j)$ and $x(\xi_{j+1})$ can be drawn as follows. Divide the interval $[x(\xi_j), x(\xi_{j+1})]$ into, say, 101 point locations and denote them as

$$X_i = x(\xi_j) + \frac{i}{100}[x(\xi_{j+1}) - x(\xi_j)], \quad i = 0, 1, \ldots, 100. \tag{4.24}$$

Next, define

$$\theta_i = \frac{X_i - x(\xi_i)}{x(\xi_{j+1}) - x(\xi_j)}\,. \tag{4.25}$$

Then, the vertical positions Y_i's corresponding to X_i's are given by

$$Y_i = r\sin(2\pi\theta_i)\,. \tag{4.26}$$

For $t = 0$, the solution is given by Eq. (4.12), while $x_{CoM}(0)$ is given by Eq. (4.13). The value of t_c calculated from Eq. (4.22) is 0.240535 s.

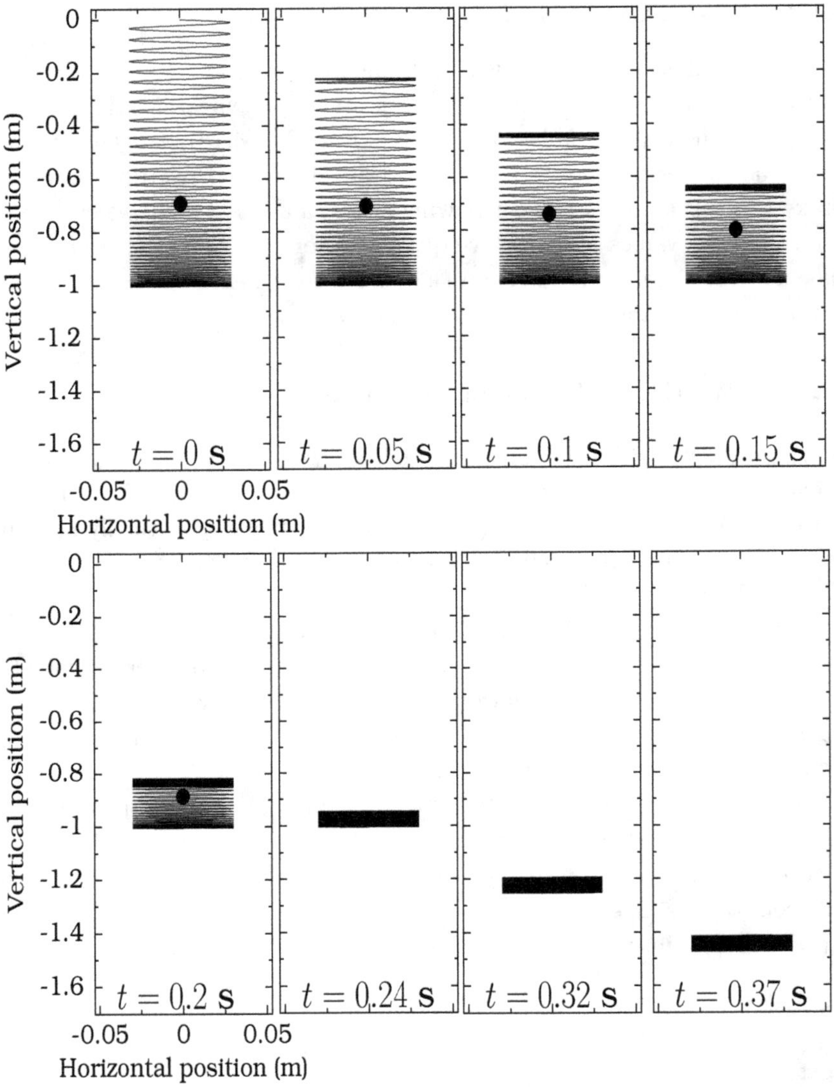

Fig. 4.2: Position of a vertically falling slinky at various times. The solid circle represents the position of the CoM of the slinky. Here $t_c = 0.240535\,\text{s}$.

For $0 < t \leq t_c$ the solution is plotted using Eq. (4.18) and (4.21). We can clearly notice that at $t = 0.0$, 0.05, 0.1, 0.15, 0.2 and $0.24 < t_c$ the bottom section of the slinky is stationary while the top section moves downward. In Fig. 4.2 at $t = t_c \approx 0.24$ the adjacent turns are in touch and for $t > t_c$

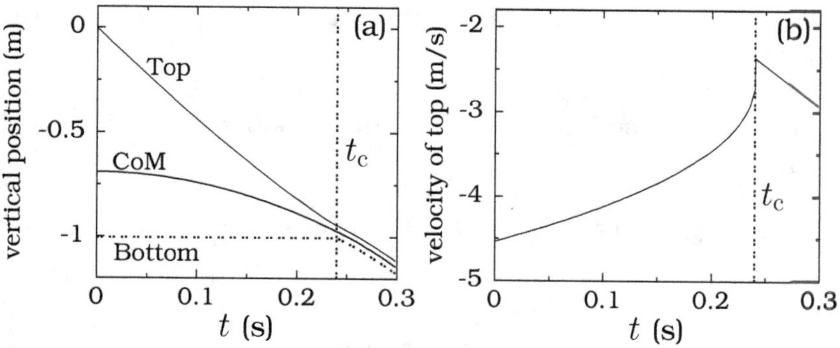

Fig. 4.3: (a) Time development of vertical positions of top, bottom and CoM of the slinky computed from the model. (b) The variation of velocity of the top of the slinky with time.

the whole slinky falls. That is, only when the tension wave front reaches the bottom, it comes into feeling that the top has been released and begins to fall.

Figure 4.3a depicts the variation of vertical positions of the top, bottom and CoM in the slinky. Velocity of the top versus time is shown in Fig. 4.3b. At $t = 0$, $v(0) = -g(m\xi_1/k)^{1/2} = -4.5\,\mathrm{m/s}$ (see the exercise 4.3 at the end of the present chapter). For $t < t_c$ as t increases from 0 the velocity of the top decreases from $v(0)$. At $t = t_c$ there is a sharp rise in the value of v. Then, v increases linearly from $v(t_c)$ for $t > t_c$. For $0 \leq t \leq t_c$ the location of CoM in the slinky varies but always lies between its top and bottom. For $t > t_c$, the location of the CoM (not the vertical position from the origin) remains the same.

4.3 Falling Elastic Bars

Just like the falling slinky, for a falling elastic bar also, the lower end of the bar does not start falling until the deformation change reaches the bottom [13]. Let x be the distance of the bar from one end to a point P, when it is not strained (kept at horizontal position). When it is suspended from that end, the position of P changes into $x + u(x, t)$ where $u(x, t)$ is the deformation field. If $\tau(x)$ is the stress and $\partial u/\partial x$ is the strain, then the Young's modulus $E = \tau(x)/(\partial u/\partial x)$. By applying Newton's equation of motion we can get a similar equation of motion like the slinky. The equation changes into a homogeneous wave equation in the coordinate system moving with the CoM. We can find the velocity of the wave as shown in the following [13].

Let ρ be the linear mass density of the elastic bar. Applying the Newton's equation of motion for a segment of thickness Δl gives the deformation field $u(x, t)$,

$$\rho\Delta l\frac{\partial^2 u}{\partial t^2} = \tau(x + \Delta l, t) - \tau(x, t) + \rho\Delta l g = \Delta l\frac{\partial \tau}{\partial x} + \rho\Delta l g\,. \tag{4.27}$$

That is,

$$\rho\frac{\partial^2 u}{\partial t^2} = \frac{\partial \tau}{\partial x} + \rho g\,. \tag{4.28}$$

As $\tau = E\partial u/\partial x$,

$$\frac{\partial \tau}{\partial x} = E\frac{\partial^2 u}{\partial x^2}\,. \tag{4.29}$$

Substitution of Eq. (4.29) in Eq. (4.28) gives

$$\rho\frac{\partial^2 u}{\partial t^2} = E\frac{\partial^2 u}{\partial x^2} + \rho g\,. \tag{4.30}$$

This is the equation of motion similar to that of a falling slinky. In a coordinate system moving with the CoM

$$u' = u - \frac{1}{2}gt^2, \quad \frac{\partial^2 u'}{\partial t^2} = \frac{\partial^2 u}{\partial x^2} - g\,. \tag{4.31}$$

Substitution of Eq. (4.31) in Eq. (4.30) leads to the homogeneous equation

$$\frac{\partial^2 u}{\partial x^2} = \frac{\rho}{E}\frac{\partial^2 u'}{\partial t^2} = \frac{1}{C^2}\frac{\partial^2 u'}{\partial t^2}\,, \tag{4.32}$$

where $C = \sqrt{E/\rho}$ is the speed of sound in the bar.

4.4 Conclusion

In this chapter, the governing physical principles of a falling slinky have been explained. At first sight, as the bottom of the falling slinky remains stationary for a small time, it appears to defy the laws of physics. When a slinky is hanging, the downward gravitational force is balanced by the upward tension of the slinky, thus it is in equilibrium. When the slinky is released from a vertical hanging position, the tension at the top is released, and hence the slinky starts falling down. But the bottom of the slinky remains utterly motionless until the whole slinky collapses to the bottom. *There after the collapsed slinky falls down under gravitational force alone.* As the tension is released at the top, that information does not reach the bottom instantaneously. A tension wave takes that information to the bottom after a finite time. As the tension wave moves with a finite velocity, ahead of the tension wave front, the turns are still stretched due to tension, and hence they are stationary. Once the wave front strikes the bottom, no tension exists anywhere to balance the gravity. So, the bottom falls along with the slinky due to the gravitational force.

This simple toy can explain to the students that action at a distance cannot take place without a disturbance reaching that point. For example, in the case of two separate charges interacting among themselves, the disturbance travels with the speed of light. Though slinky is a simple toy, it has many practical applications in physics and engineering activities [5]. Using a digital camera with provision for seeing light speed motions, a falling slinky can be used as an undergraduate physics lecture demonstration. It can be used to explain the physics of a tension spring and wave propagation in a spring. The slinky can be used to construct a simple coupled oscillator with a soup can [19].

4.5 Bibliography

[1] The YouTube video is at http://www.youtube.com/watch?v=eCMmmEEyOO0.

[2] P. Gash, The Phys. Teach. **58**, 198 (2020).

[3] H. Aref, S. Hutzier and D. Weaire, Europhysics News **38** (3), 23 (2007).

[4] S. Reif-Acharman, Ingenieria y Competitividad **17**, 111 (2015).

[5] J. Blake and L.N. Smith, Am. J. Phys. **47**, 807 (1979).

[6] J.M. Bowen, Am. J. Phys. **50**, 1145 (1982).

[7] J.T. Cushing, Am. J. Phys. **52**, 925 (1984).

[8] J.T. Cushing, Am. J. Phys. **52**, 933 (1984).

[9] S.Y. Mak, Am. J. Phys. **55**, 994 (1987).

[10] G. Vandegrift, T. Baker, J. DiGrazio, A. Dhohne, A. Flori, R. Loomis, C. Steel and D. Velat, Am. J. Phys. **57**, 949 (1989).

[11] R.A. Young, Am. J. Phys. **61**, 353 (1993).

[12] M.M. Sawicki, The Phys. Teach. **40**, 276 (2002).

[13] J.M. Aguirregabiria, A. Hernandez and M. Rivas, Am. J. Phys. **75**, 583 (2007).

[14] W.G. Unruh, The falling slinky. arXiv:1110-4368v1, 19 October 2011.

[15] R.C. Cross and M.S. Wheatland, Am. J. Phys. **80**, 1051 (2012).

[16] J. Pretz, Eur. J. Phys. **42**, 045008 (2021).

[17] M.G. Calkin, Am. J. Phys. **61**, 261 (1993).

[18] L.A. Pipes and L.R. Harvill, *Applied Mathematics for Engineers and Physicists* (McGraw-Hill, Singapore, 1984) 3rd edn.
[19] M. Mewes, Am. J. Phys. **82**, 254 (2014).

4.6 Exercises

4.1 Show that for a spring falling under gravity from rest in a coordinate system falling with the CoM of the spring, Eq. (4.4) reduces to the usual wave equation and the wave propagates along the length of the spring with a characteristic time $t_{\mathrm{p}} = \sqrt{m/k}$.

4.2 Consider a hanging slinky of mass m and spring constant k shown in Fig. 4.4 where F is the applied force, $a(x)$ is the elongation of a single turn at x and L_0 and L are the lengths of the slinky with $F = 0$ and $F \neq 0$, respectively [11]. $a(x)$ can be treated as the vertical component of a small length of the slinky. Denote $v(x)$ as the speed of wave propagation on the slinky. Show that

$$v(x) = \sqrt{\frac{k}{m}}a(x), \quad a(x) = \left[\left(\frac{F}{k}\right)^2 + \frac{2mg}{k}(L - x)\right]^{1/2}.$$

4.3 Consider the falling slinky with instant collapse of turns. Obtain expressions for the velocity $v(t)$ of top of the slinky for $t \leq t_{\mathrm{c}}$ and also $v(0)$.

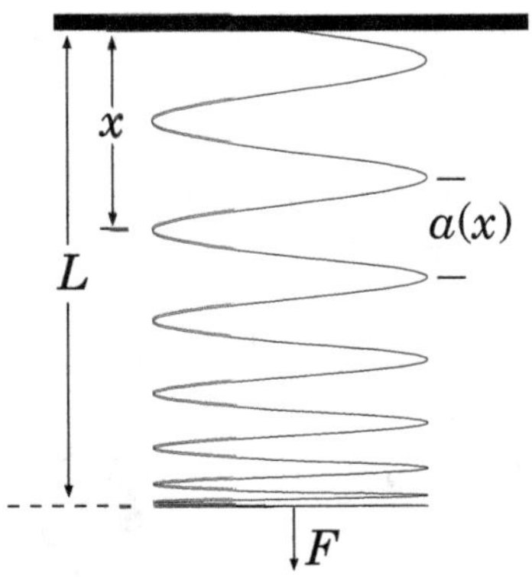

Fig. 4.4: A hanging slinky with its top end fixed while the other end subjected to an external force F.

4.4 One can compute at $t = t_{\mathrm{c}} = \sqrt{\frac{m}{3k}\xi_1^3}$, the values of vertical positions of top of the slinky $(x_{\mathrm{t}}(t_{\mathrm{c}}))$, bottom of the slinky $(x_{\mathrm{b}}(t_{\mathrm{c}}))$ and $(x_{\mathrm{CoM}}(t_{\mathrm{c}}))$ from

$$x(\xi,t) = \begin{cases} l_1\xi + \frac{mg}{k}\xi_{\mathrm{c}}\left(\xi_1 - \frac{1}{2}\xi_{\mathrm{c}}\right), & 0 \le \xi \le \xi_{\mathrm{c}} \\ l_1\xi + \frac{mg}{k}\xi\left(\xi_1 - \frac{1}{2}\xi\right), & \xi_{\mathrm{c}} \le \xi \le \xi_1 \\ l_1\xi + \frac{mg}{2k}\xi_1^2, & \xi_1 \le \xi \le 1. \end{cases} \tag{4.33}$$

and $x_{\mathrm{CoM}}(t) = x_{\mathrm{CoM}}(0) + gt^2/2$. Further, at $t = t_{\mathrm{c}}$, $v(t_{\mathrm{c}}) = -gt_{\mathrm{c}}$ (see the exercise 4.3). Knowing these quantities, how do you determine $(x_{\mathrm{t}}(t))$, $(x_{\mathrm{b}}(t))$, $(x_{\mathrm{CoM}}(t))$ and $v(t)$ for $t_{\mathrm{c}} < t < t'$ where t' is the time at which the slinky falls on the floor? Also, find an expression for t'.

4.5 Consider the equations

$$x_0(\xi) = \begin{cases} l_1\xi + \frac{mg}{k}\xi\left(\xi_1 - \frac{1}{2}\xi\right), & 0 \le \xi \le \xi_1 \\ l_1\xi + \frac{mg}{2k}\xi_1^2, & \xi_1 \le \xi \le 1, \end{cases} \tag{4.34}$$

and $x_{\mathrm{CoM}}(0) = \dfrac{1}{2}l_1 + \dfrac{mg}{2k}\xi_1^2\left(1 - \dfrac{1}{3}\xi_1\right)$. Find the total length of the slinky at its equilibrium position. Hence, find the CoM from the bottom of the slinky. The bottom of the slinky remains stationary till the slinky collapses to the bottom and the CoM falls due to its gravitational force along. Hence, show that the time taken by the CoM to fall to the bottom is equal to $t_{\mathrm{c}} = \sqrt{m\xi_1^3/(3k)}$.

Chapter 5

Walking Slinky

In the previous chapter, the physical principle and the dynamics of the falling slinky are discussed. We have seen why the bottom of the slinky remains stationary till the whole slinky collapses to the bottom. Though it is an interesting demonstration experiment, it cannot be easily demonstrated to the students without a special digital camera able to record high speed motion as the bottom remains stationary only for a fraction of a second. One cannot see the stationary bottom in a falling slinky with our naked eyes. But a slinky walking down a flight of stairs is highly fascinating to see. This walking slinky has made more appeal than the falling slinky. In this chapter, an elementary theory for the walking slinky is discussed. Another interesting toy is the slinky whistlers. By tapping and exciting a slinky, one can hear a *whistler*. A brief note about the slinky whistlers is also given in this chapter.

5.1 Walking Behavior of a Slinky

We start by illustrating the walking behavior of a slinky [1,2]. Place a slinky on an upper step of a stair as shown in Fig. 5.1a. The slinky has an energy stored in it. As its mass is at an elevated position, essentially, the energy is potential energy. A slinky will remain in the rest state of equilibrium (Fig. 5.1a) and will change its position only if the disturbance given to is sufficiently large. Let us lift the top of the slinky and release it in such a way that it will fall on to the stair next below. The slinky rotates about a diameter and uncoils itself on to the lower stair (Fig. 5.1b). The potential energy will be released and transformed as kinetic energy. This energy, as the slinky moves down the stairs, transfers in the form of a wave propagating back and forth along the axis of the spring repeatedly.

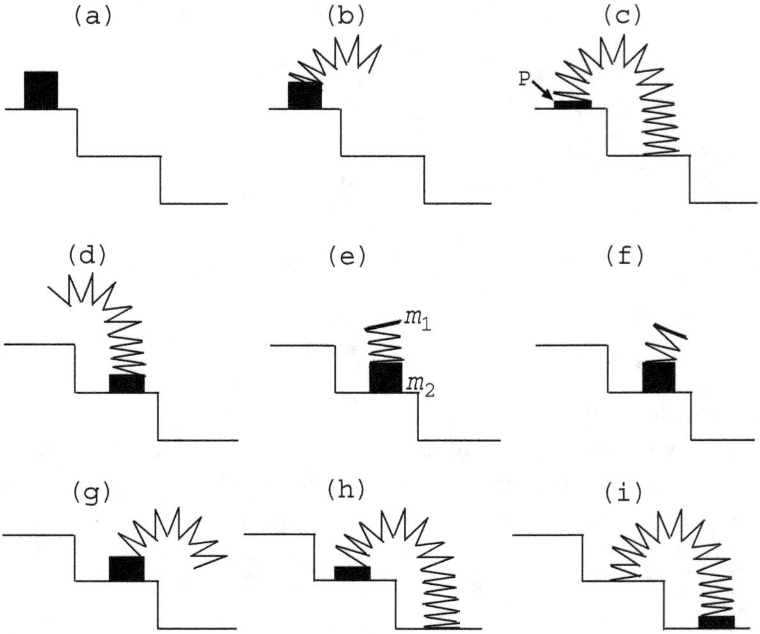

Fig. 5.1: A walking slinky.

The propagation occurs along more than one direction. This is because the axis is not strictly a straight-line but as shown in Fig. 5.1c, it has a U-shape. When the wave moves, first the successive turns of the slinky move upward then along the arc of the spring and towards the next lower stair. In a short time, the wave reach the last turn on the top stair and pulled up with a sufficient speed along the arc (Figs. 5.1d and e). Note that the slinky also has momentum when it moves down the lower stair. As it lands on the lower stair as a result of conservation of momentum, it moves in the opposite direction. This makes the slinky to turn over and travel to the next stair. If the dimensions of the stairs are appropriate, then the upper part of the slinky turns towards the next lower step as shown in Figs. 5.1f and 5.1g and then land on it (Fig. 5.1h and i). The above cycle is repeated.

The walking dynamics of a slinky can be explained applying the concept of energy exchange. As the slinky walks down each step, the energy is converted between potential and kinetic energies continuously. Note that the speed of the slinky walks depends on the various physical properties of the slinky, the height of each step and the velocity of the free end. The energy

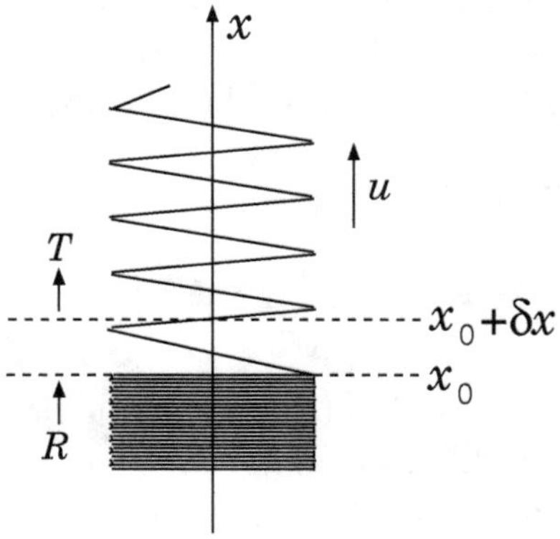

Fig. 5.2: Definition of characteristics of a slinky.

and the momentum of the slinky would decay due to the frictional and damping forces causing it to halt after several stairs.

5.2 Rate of Walking Down the Stairs

It is possible to calculate the rate of walking of the slinky down stairs [1]. The time taken by the slinky for one step down does not depend on the height of the stairs since the rate of uncoiling of the spring depends on its characteristics. Consider a spring as a heavy, elastic fluid flowing along a thin tube. Define x as the distance measured along the axis, u as the component of the velocity parallel to the axis, ρ as the density and T as the component of the reaction parallel to the axis and R is the component of the reaction of the coil as shown in Fig. 5.2. Hooke's law is written as $T = c/\rho$ where c is a constant. As shown in Fig. 5.2 for $x < x_0$, we have $u = 0$, $\rho = \infty$ and $T = 0$. For $x > x_0$ these quantities are finite and nonzero.

As the momentum imparted to $\delta x = u\delta t$ depends on the external forces only,we write $\rho u \cdot u\delta t = (T + R)\delta t$. That is, $\rho u^2 = T + R$. When the spring is uncoiling $R = 0$ and hence $\rho u^2 = T$. Substitution of $T = c/\rho$ gives $\rho^2 u^2 = c$ or $\rho u = \sqrt{c}$. That is, the rate of uncoiling (ρu) is a constant. Consider the Fig. 5.1. At the point P the rate of uncoiling is $\sqrt{c}$. When the coil in the top stair is unwound (Fig. 5.1d) and the end of the coil leaves the top stair, then at the tip suddenly $T = 0$. As a result, a shock wave moves through the spring at the rate $\sqrt{c}$. Suppose t_1 is the time taken by the tail between leaving the top stair and arriving at the lower coil. Then, $m_1 = t_1\sqrt{c}$. The time t_2 taken by the lower coil with mass m_2 to uncoil is $t_2 = m_2/\sqrt{c}$. At time t_2, the slinky will be similar to as shown in Fig. 5.1a but a stair lower. The time $t = t_1 + t_2$ is the time taken by the slinky for one complete cycle. Then,

$$t = \frac{m_1 + m_2}{\sqrt{c}} = \frac{M}{\sqrt{c}} \tag{5.1}$$

and is independent of the height of the stair. The time taken by the slinky to walk over n stairs is nt. The quantity $M/\sqrt{c}$ can be expressed in terms of gravity and the length L of the slinky. For this purpose, integrate the equation $\partial T/\partial x = \rho g$, substitute $T = c/\rho$ and set the integration constant as zero. The result is $\rho = \sqrt{c/(2gx)}$. Then, $M = \int_0^L \rho\,\mathrm{d}x = \sqrt{2cL/g}$. Equation (5.1) gives $t = \sqrt{2L/g}$.

Solved Problem 1:

The loss of energy in an element of a spring in δt is given by

Work done − kinetic energy gained − elastic energy gained.

If the spring is not sticky, then the reaction $R \geq 0$. Show that for a loss less slinky $R = 0$.

Let T be the applied tension. The work done for a displacement δx is $T\delta x$. Since u is the velocity, $\rho\delta x$ is the mass of the element of thickness δx, then we have the kinetic energy (K.E.) and elastic energy (E.E.) gained as

$$\text{K.E. gained} = \frac{1}{2}\rho\delta x u^2\,, \quad \text{E.E. gained} = \frac{1}{2}T\delta x\,. \tag{5.2}$$

Therefore,

$$\text{loss of energy in } \delta\text{t} = T\delta x - \frac{1}{2}\rho u^2\delta x - \frac{1}{2}T\delta x = \frac{1}{2}(T - \rho u^2)\delta x\,. \tag{5.3}$$

As $\rho u^2 = T + R$, we get

$$\text{loss of energy in } \delta\text{t} = -\frac{1}{2}R\delta x = -\frac{1}{2}Ru\delta t\,. \tag{5.4}$$

$u > 0$ and $R \geq 0$ for non-sticky slinky. Hence, $R = 0$ for the loss less slinky.

5.3 Slinky Whistlers

Using a piece of a tape fasten one end of a metal slinky (called *output end*) on a flat door. Hold the other end (called *input end*) in your hand and stand about, say, 2 m from the door. Make sure that the neighboring turns of the slinky are not in contact with one another. However, the turns are close enough to one another. When you tap the input end with a pencil, you will hear a *whistler*. This sound is initially audible with a very high pitch, descends in pitch quickly and in a fraction of a second becomes inaudible. The explanation is that the pencil tap produced a delta function excitation and its high-frequency components propagates more rapidly than the low-frequency components. For more details about slinky whistlers one may refer to the Ref. [4].

5.4 Some Applications of a Slinky

Apart from being a fascinating toy displaying variety of dynamics the slinky has many notable applications in science. Let us mention some of them [2].

1. A set of experiments have been performed with a slinky in the context of analysis of laws of physics of springs by the crew members of the STS-54 Mission of NASA-ST1 Programme. In the new environment, it behaved as a continuously propagating wave.
2. Certain special models of ground heat exchangers for heat pump systems (applied for cooling and heating of buildings) where the pipe is twisted like slinky. This has resulted in advantage in places where the recharge of temperature of the ground is not crucial [5-7].
3. Slinky structures have yielded improved cooling performance parameters in certain MEMS devices [8].
4. Some structural levels of a protein are described using a slinky [9].
5. In the nano-scale the properties of the slinky such as stretching, bending and compressing are found advantages in electronic, photonic and electromechanical devices [10,11].

5.5 Conclusion

Slinky is a very simple toy which does not even require a battery like most of other toys. Still it can be used to teach many physical phenomena like transverse and longitudinal wave propagations. The walking slinky can be easily demonstrated to the students by constructing a flight of stairs. They can measure the time of descent per step for different step height and prove that it does not depend on the height and width of the flight of stairs. As described by Crawford [4] by exciting the slinky in two different ways, two different whistler models can be setup and the nature of the whistler can be studied.

5.6 Bibliography

[1] M.S. Longuet-Higgins, Math. Proc. Camb. Phil. Soc. **50**, 347 (1954).
[2] S. Reif-Acherman, Ingenieria y Competitividad **17**, 111 (2015).
[3] Ai-Ping Hu, Am. J. Phys. **78**, 35 (2010).
[4] F.S. Crawford, Am. J. Phys. **55**, 130 (1987).
[5] Ch.S.A. Chong, G. Gan, A. Verhoef, R. Gonzalez-Garcia and P.L. Vidale, Appl. Energy **104**, 603 (2013).
[6] P. Neuberger, R. Adamovsky and M. Sedova, Energies **7**, 972 (2014).
[7] Z. Xiong, D.E. Fisher and J.D. Spitler, Appl. Energy **141**, 57 (2015).
[8] S. Celen, Opt. Laser Technol. **44**, 2043 (2012).
[9] J.J.W. Baker and L.N. Smith, *A Course in Biology* (Addison-Wesley, Massachusetts, 1979).
[10] F. Xu, W. Lu and Y. Zhu, ACS Nano **5**, 672 (2011).
[11] Y. Sun, W.M. Choi, H. Jiang, Y.Y. Huang and J.A. Rogers, Nature Nanotechnology **1**, 201 (2006).

5.7 Exercises

5.1 Suppose holdup a slinky in air and strike its one end. This produces a metallic tone, the pitch of which sharply lowers. Why?

5.2 On what type of slope (steep/gentler) of a stair a slinky would move faster/slowly? State the reason.

5.3 Suppose a smaller slinky and a larger slinky are allowed to walk down the stairs. Which will win? Why?

5.4 Consider the model of the motion of the walking slinky shown in Fig. 5.3 [3]. The model has three point masses, each with mass m. They are connected by the massless rigid linksof length l as shown in Fig. 5.3. At the junction of the two massless rigid links a linear torsion spring with stiffness coefficient k_1 is placed. Write the potential and kinetic energies of the system.

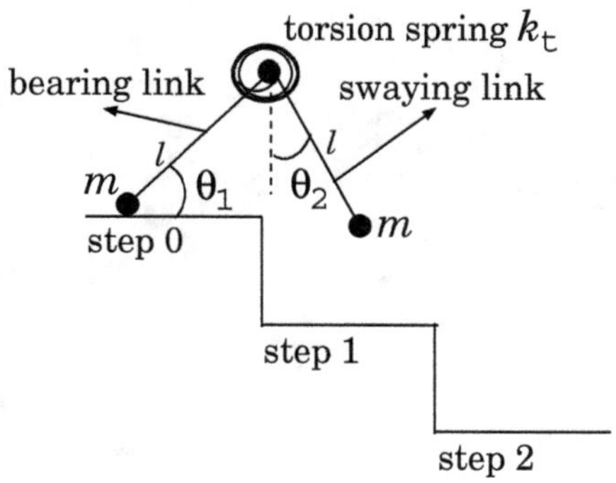

Fig. 5.3: A model of a walking slinky.

5.5 Experimentally study the role of the angle of inclination of an inclined plane on the walking of the slinky.

Chapter 6

Woodpecker

One of the traditional toys that is easy to play with and shows unexpected motions is the woodpecker. This toy is shown in Fig. 6.1. It consists of a pole on which a sleeve (ring) is attached. The sleeve is joined to the body of the woodpecker by means of a small spring. To activate its motion hold its head between your thumb and index finger, curve it back so that the spring bend and then release it. The woodpecker begins to oscillate and its beak hits the pole while it is moving downward uniformly along the pole. This delightful action resembles the behaviour of a real woodpecker which digs for worms from the bark.

When the woodpecker is at the top of a pole it possesses gravitational energy. As it descend downwards the potential energy is converted into

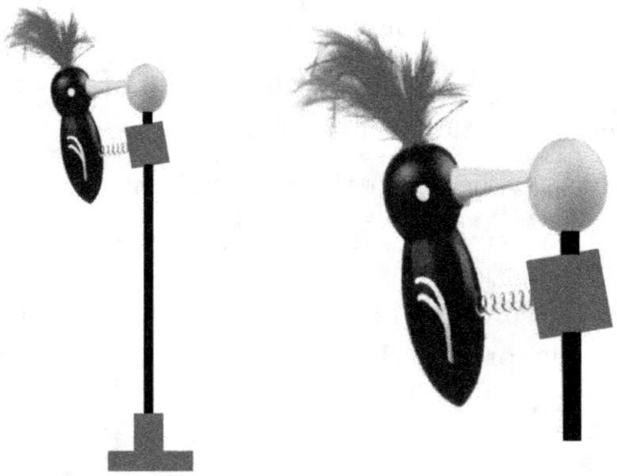

Fig. 6.1: A woodpecker toy and a magnification of the part of the toy clearly showing the wooden sleeve and a small spring establishing the connection between the sleeve and the body of the woodpecker.

kinetic energy. That is, gravitational energy serves as an energy source. This energy makes the woodpecker go downward. When it does so, it undergoes an oscillation up and down. The oscillation in turn interacts through the spring and the sleeve. As it goes downward the sleeve turns and the frictional contact jamming of the sleeve at the pole is switched-on and off. These make the bird to gain energy. Shortly, a stable cyclic process is attained with a balance of the gained kinetic energy per cycle by the downward motion and the frictional energy dissipation [1].

Through a careful watching of the motion of the woodpecker one can clearly observe the following features.

1. The woodpecker oscillates above and below the sleeve under the restoring torque from the spring. The sleeve gets locked on the shaft, because of the friction, during a short time when the bird is above or below the sleeve. The sleeve is unlocked between these points and moves downwards over a short duration.
2. Its oscillation is indeed periodic with a definite constant period.
3. The sleeve descends a short distance twice during each cycle.
4. Its steady state oscillation is independent of the initial conditions.
5. The oscillations are self-excited, that is, the toy does not require any external energy or external force continuously to exhibit its motion.
6. Regular impact and friction between the pole and the sleeve regulate the motion of the bird.
7. The motion appears as a coupling between oscillation and sliding.
8. Each cycle of the motion of the bird consists of a pure oscillation, an oscillation with sliding and a pure sliding.
9. The bird is not revolving over the pole while it is moving downward.
10. In every cycle the sleeve hits the pole twice.

Theories have been proposed to understand the behaviour of the woodpecker toy [1-6]. In the present chapter we present the theoretical model of the toy proposed in the Ref. [6].

6.1 A Theoretical Model

Figure 6.2 depicts a simple model of the woodpecker toy. Assume that the torque $\boldsymbol{\tau}$ exerted on the woodpecker and sleeve by the spring is given by

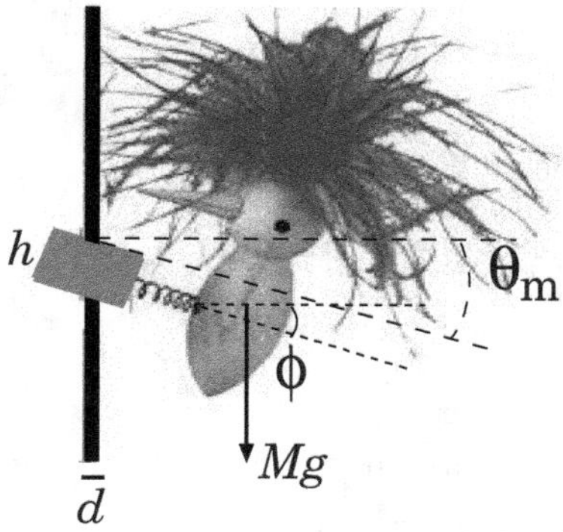

Fig. 6.2: A model of woodpecker toy.

$|\boldsymbol{\tau}| = k\beta$ where β is the deflected angle of the spring and k is the angular stiffness of the spring. Further, $\beta = \phi - \theta_{\mathrm{m}}$ where ϕ and θ are the angles of inclination of the bird and sleeve, respectively, (refer Fig. 6.2), θ_{m} is the maximum value of θ, h is the height (thickness) of the sleeve and l is the length of the spring. Denote the friction coefficient between the sleeve and the pole as μ. M and m are the masses of the woodpecker and the sleeve, respectively.

The sleeve does not bounce back after hitting the pole. Therefore, the underlying collision is inelastic. The vertical displacement y and the angle ϕ form two degrees of freedom of the toy model. The equations of motion are to be setup separately for the three stages of cycle, namely, *oscillation (stage-1), oscillation with sliding (stage-2)* and *pure sliding (stage-3)* as described below [6]. Note that in the first stage an oscillation takes place when the sleeve gets locked on the pole due to friction. In the second stage the friction is unable to holdup the sleeve, the bird descends along the pole and the the oscillation continues around the sleeve. In the third stage the sleeve detaches from the pole and the bird rotates and slides together with the sleeve, for a tiny duration.

6.1.1 *Oscillation*

The sleeve is stationary and the oscillation is due to the torque from gravity and the spring. We write

$$I\ddot{\phi} = Mgl\cos\phi - k(\phi - \theta_{\mathrm{m}}), \quad \dot{y} = 0, \tag{6.1}$$

where I is the moment of inertia of the bird pivoting on the centre of the sleeve.

6.1.2 *Oscillation with Sliding*

The normal force on the sleeve from the equilibrium of torque is given through the equation

$$\frac{Nh}{\cos\theta_{\mathrm{m}}} = k|\phi - \theta_{\mathrm{m}}|. \tag{6.2}$$

The translational equation for the coupled system, *bird+sleeve*, is

$$M(l\ddot{\phi}\cos\phi + \ddot{y}) + m\ddot{y} = (M+m)g - 2f, \tag{6.3}$$

where f is the kinetic friction and can be chosen as $f = \mu N$. $\ddot{y}$ in $m\ddot{y}$ is the acceleration of the sleeve while $l\ddot{\phi}\cos\phi + \ddot{y}$ is the aggregated vertical acceleration of the woodpecker. As the acceleration of the sleeve affects the rotational motion of the woodpecker, we have

$$I\ddot{\phi} = (Mgl - Ml\ddot{y})\cos\phi - k(\phi - \theta_{\mathrm{m}}). \tag{6.4}$$

The term $Ml\ddot{y}\cos\phi$ is the torque of the inertial force.

6.1.3 *Pure Sliding*

We can neglect the deformation of the spring since the woodpecker moves and rotates together with the sleeve. Now, we write

$$M(l\ddot{\phi}\cos\phi + \ddot{y}) + m\ddot{y} = (M+m)g, \tag{6.5a}$$

$$I\ddot{\phi} = (Mgl - Ml\ddot{y})\cos\phi. \tag{6.5b}$$

6.1.4 *Boundary Conditions*

We need to setup appropriate boundary conditions for the motions of the three stages [6]. We assume that there is no difference between static friction and kinetic friction. The amount of friction necessary to hold the sleeve can be computed from the equation

$$Ml\ddot{\phi}\cos\phi = (M+m)g - 2f \tag{6.6}$$

as

$$f = \frac{1}{2}\left[-Ml\ddot{\phi}\cos\phi + (M+m)g\right]. \tag{6.7}$$

The maximum friction is given by (see Eq. (6.2))

$$f_{\max} = \mu N = \frac{\mu k}{h}|\phi - \theta_{\mathrm{m}}|\cos\theta_{\mathrm{m}}. \tag{6.8}$$

As long as f given by Eq. (6.7) is higher than $f_{\rm max}$, the sleeve is not sliding. The start of the sliding motion can be determined by calculating f and $f_{\rm max}$ and checking every instant of time whether $f > f_{\rm max}$ (no sliding) or $f < f_{\rm max}$ (sliding).

Next, when the inclination ϕ of the woodpecker becomes $\theta_{\rm m}$, then it passes either $\phi = \theta_{\rm m}$ from above or $\phi = -\theta_{\rm m}$ from below. Now, the third stage begins. The bird drives the sleeve to the same angular velocity. Both of them rotate with no deflection on the spring. In this driving process the angular momentum of the bird and the sleeve is conserved. So, we write

$$I\dot{\phi} = (I + i)\dot{\phi}' , \tag{6.9}$$

where i is the moment of inertia of the sleeve. After a brief stage-3 motion, the bird enters into stage-2. In this case ϕ leaves the interval $[-\theta_{\rm m}, \theta_{\rm m}]$.

6.1.5 *Numerical Calculation*

First, solve Eqs. (6.1) for stage-1 motion with appropriate initial conditions with a time step size Δt. The sleeve is stationary on the pole. At every instant of time, compute f and $f_{\rm max}$ the sleeve is still on the pole. If at a time t' the value of f is $> f_{\rm max}$ and at $t' + \Delta t$ the value of f is $< f_{\rm max}$, then make a rough estimate of time at which $f = f_{\rm max}$. At this time the sleeve begins to slide. From this time solve the equations of motion for the stage-2 motion (oscillation and sliding). In this stage ϕ value lies out side the interval $[-\theta_{\rm m}, \theta_{\rm m}]$. When ϕ enters into the interval $[-\theta_{\rm m}, \theta_{\rm m}]$ the stage-3 starts. We replace $\dot{\phi}$ by $\dot{\phi}' = I\dot{\phi}/(I+i)$ and starts solving Eqs. (6.5) for pure sliding. When ϕ leaves the interval $[-\theta_{\rm m}, \theta_{\rm m}]$ the system enters into the stage-2 motion. In this way the appropriate stages are identified and the corresponding equations of motion are to be solved.

6.1.6 *Determination of Relevant Parameters*

We need to find the values of the relevant parameters of the toy system in order to perform the numerical computation of the woodpecker motion.

Theoretical calculation of the moment of inertia I of the woodpecker is difficult since its shape is irregular. It can be measured experimentally [6] by hanging it under a spring thereby forming a compound pendulum as shown in Fig. 6.3a. Let I_0 and I are the moments of inertia of the woodpecker with respect to its centre and about a parallel axis separated by a distance r away from one that passing through the CoM, respectively.

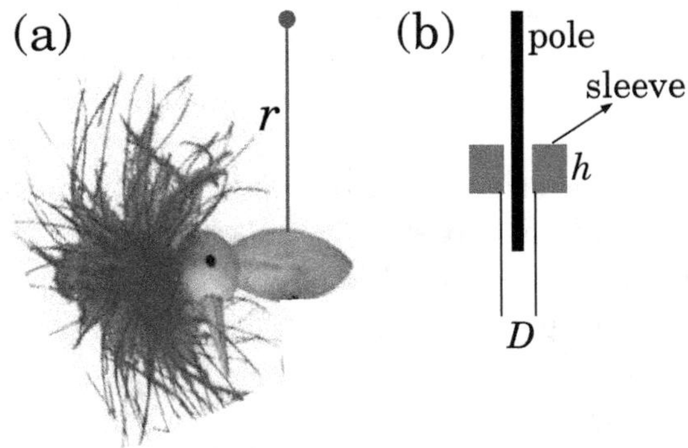

Fig. 6.3: (a) A setup to measure the moment of inertia I of the woodpecker. (b) Definition of the parameters h and D.

Then, $I = I_0 + mr^2$. The period of the compound pendulum under small amplitude oscillation is

$$T = 2\pi \left(\frac{I_0 + mr^2}{mgr} \right)^{1/2} . \tag{6.10}$$

The above equation can be rewritten as

$$\frac{T^2 mgr}{4\pi^2} = I_0 + mr^2 . \tag{6.11}$$

Experimentally for a range of values of r one can measure T and draw a graph of $T^2 mgr/(4\pi^2)$ versus mr^2. The interception of the straight-line on the vertical axis gives the value of I_0.

Solved Problem 1:

In an experiment to determine the moment of inertia of the woodpecker about its centre I_0, Eq. (6.11) was used by measuring the period of oscillation T for different r values. It was found that the intercept of the straight-line curve for $T^2 mgr$ versus $4\pi^2 mr^2$ was found to be $2.974 \times 10^{-5}\,\text{kgm}^2$. Find the value of I_0.

From Eq. (6.11) we find

$$T^2 mgr = 4\pi^2 mr^2 + 4\pi^2 I_0 . \tag{6.12}$$

The intercept is $4\pi^2 I_0$. It is given that $4\pi^2 I_0 = 2.974 \times 10^{-5}\,\text{kgm}^2$. Therefore, $I_0 = 7.53 \times 10^{-5}\,\text{kgm}^2$.

To find the angular stiffness k of the spring, hang a mass one side of the spring of the toy and measure the angular deflection (angle) β. Repeat this for different mass and convert the mass to the torque exerted on one end of the spring. Draw a graph between β (along the x-axis) and the torque (along the y-axis). We expect a linear relation. The slope is the angular stiffness k of the spring.

The friction coefficient μ is $\tan\theta_{\mathrm{c}}$ where θ_{c} is the critical angle at which the sleeve happens to slide. This angle is measured step by step increasing the inclination of the pole and finding the minimum angle at which the sleeve begins to slide. Define the parameters h, d and D as in Figs. 6.2 and 6.3b. The relation between h, d, D and θ_{m} is [6]

$$h\tan\theta_{\mathrm{m}} = D - \frac{d}{\cos\theta_{\mathrm{m}}}\,. \tag{6.13}$$

Thus, θ_{m} can be determined from this relation.

6.2 Conclusion

The woodpecker toy operates in the presence of friction and the occurrence of impacts and jamming. For the motion of the woodpecker the hitting of its beak with the pole is not required. However, this act resembles with the typical woodpecker bird's behaviour. Though the woodpecker is a simple toy and the physical principles involved on which it works is well known, the equation of motion to describe its working is not easily solvable. Its motion is a combination of oscillatory and linear motions. It has three types of motions, namely, oscillation, oscillation with sliding and pure sliding taking place periodically. The equations of motion for all the three types of motion have been discussed. They can be solved only numerically with appropriate boundary conditions. The experimental methods to determine the relevant parameters have been discussed. Students can study the working of the woodpecker and its three motions if the required equipments are available.

6.3 Bibliography

[1] C. Glocker and C. Studer, Multiple System Dynamics **13**, 447 (2005).
[2] R.A. Roy, *The Analytical Dynamics of the Woodpecker Problem* (Ph.D. Thesis, University of Florida, 1974).
[3] R.I. Leine, C. Glocker and D.H. van Campen, Nonlinear dynamics of the woodpecker toy. In *ASME 2001 Design Engineering Technical*

Conference and Computers and Information in Engineering Conference, ASME, Pittsburgh, PA (September 2001).
[4] C. Glocker and C. Studer, The woodpecker toy, Proceedings of the IMES Centre for Mechanics, ETH Zurich (2003).
[5] R.I. Leine and D.H. Van Campen, J. Vib. Control **9**, 25 (2003).
[6] Z. Zhu, W. Gao, S. Wang and H. Zhou, *2012 Problem 6: Woodpecker Toy* in International Young Physicists' Tournament: Problems and Solutions(World Scientific, Singapore, 2014).

6.4 Exercises

6.1 Normally a falling body will be accelerated due to gravity. But the woodpecker motion is steady even though it falls down due to gravity. What is the reason for its steady motion?

6.2 What will be the nature of the segment of the phase-space curve $(\phi - \dot{\phi})$ for pure sliding motion?

6.3 Consider the relation between h, d, D and θ_{m} given by Eq. (6.13). Given $h = 0.5\,\mathrm{cm}$, $d = 0.15\,\mathrm{cm}$ and $D = 0.2\,\mathrm{cm}$ determine θ_{m}.

6.4 Figure 6.4 shows a typical phase portrait sketch of the woodpecker toy depicting ϕ versus $\dot{\phi}$ during a steady motion [3]. The numbers $1 - 8$ specify the sequence of events.The number 1 corresponds to the lower sleeve stick to slip transition, sleeve sliding downward and the body rotating upward. The number 8 corresponds to lower sleeve slip to stick transition, jamming of sleeve and body rotating downward [3]. Enumerate the sequence of the events of the woodpecker toy.

6.5 For a woodpecker toy $D = 0.006\,\mathrm{m}$, $d = 0.005\,\mathrm{m}$ and $h = 0.01\,\mathrm{m}$. Compute the value of θ_{m}.

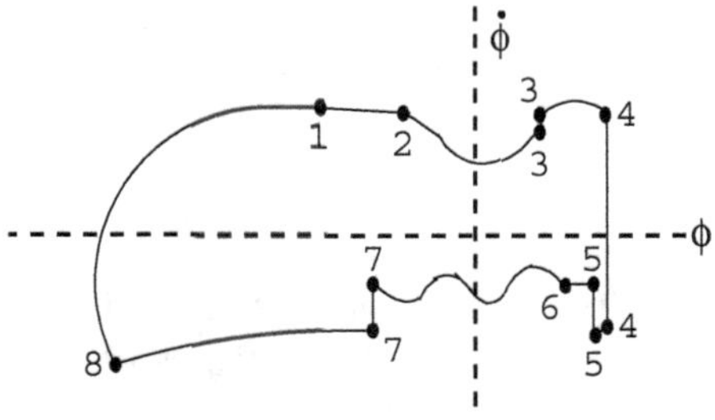

Fig. 6.4: Variation of ϕ and $\dot{\phi}$ during a steady motion of a woodpecker.

Chapter 7

Cartesian Diver

A Cartesian diver is a near-neutrally buoyant that sinks and rises in a water filled closed bottle when changes takes place in volume, pressure, temperature and buoyancy. This simple toy can be used to demonstrate the principle of Archimedes, Pascal's, Boyle's, Charles' and ideal gas laws and relations among volume, pressure, temperature and buoyancy. These concepts have important applications in science, engineering and technology, specifically in scuba diving, fish physiology and many submersibles, high pressure systems such as autoclaves and deep sea drilling.

A typical Cartesian diver toy is shown in Fig. 7.1a. The diver, a thin glass with an attractive design and bottom ends opened, is set to float in a closed water bottle. The diver is initially floating with part of it is above the top surface of the water while the remaining part is below the top surface of the water. When the bottle is gently squeezed the diver sinks and lands on the bottom of the bottle as depicted in Figs. 7.1b-c. When the squeezing is released the diver rises up to the top surface of the water. For the divers one may use test tubes, small syringes fitted with small lead ball, medicine droppers, plastic drinking straws, ketchup bags, small shampoo pockets, paper matches, blown glass bubbles, etc.

The description of a Cartesian diver appeared in the book of Raffaello Magiott in 1648. Itinerant glassworkers used Cartesian divers to entertain audiences in the late 1600s. About 1670 a Cartesian diver appeared in an advertisement for a demonstration of glass work in Poland [1].

René Descartes, a French mathematician and philosopher became famous for his quote *I think therefore I am* (the original in Latin is *Cogito, ergo Sum*). The diver is said to have been named as *Cartesian* because it make us think about weight, pressure, gravity, floating and sinking of matter in fluids, forces acting on floating and sinking objects and explanations

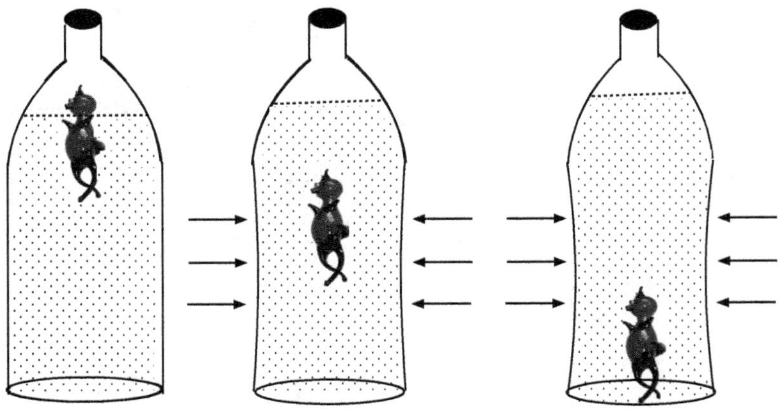

Fig. 7.1: A one littre water bottle version of a Cartesian diver toy. The diver is a decorated thin glass tube with bottom ends are opened and is partially filled with water.

about them. Moreover, we can make the diver to go up and down by thinking. Cartesian diver was referred as a *hydrostatic toy* by Carhart and Chute four hundred years ago [2]. It is also called as water devils or water dancers, bottle imps and Cartesian devils.

In this chapter, we explain the fundamental physics principle involved in the working of a Cartesian diver. Next, we give references for construction of different types of Cartesian divers. Then, the working of a typical Cartesian diver is discussed. An irreversible sink of a diver is different from a normal Cartesian diver. In a normal diver, a floating diver sinks on applying a pressure in the bottle and rises up when the pressure is released. But in an irreversible sink of a diver, the diver does not rise up after the pressure is released. The remaining part of this chapter gives a theory for the irreversible sink of a diver.

7.1 Gas Laws, Buoyant Force and Archimedes' Principle

Gas laws give the relationships between pressure, volume and temperature of a gas. Boyle's law states that the volume V of a gas is inversely proportional to the pressure P when the temperature is constant. That is,

$$V \propto \frac{1}{P} \text{ and } P \propto \frac{1}{V}. \tag{7.1}$$

Therefore, $PV =$ is a constant. When pressure P_1 and volume V_1 changes to P_2 and V_2, respectively, for the same mass of gas, we get

$$P_1 V_1 = P_2 V_2 \,. \tag{7.2}$$

Charles' law states that the volume V is directly proportional to the temperature T, $V \propto T$ at P constant and hence

$$\frac{V}{T} = \text{a constant} \,. \tag{7.3}$$

Gay-Lussac's law states that for a constant volume, pressure is proportional to temperature, $P \propto T$ at V constant. All these three gas laws can be combined to a general gas law $PV = kT$, where k is a constant or

$$\frac{P_1 V_1}{T_1} = \frac{P_2 V_2}{T_2} \,. \tag{7.4}$$

These laws are valid only for ideal gases in which there is negligible intermolecular forces and the collision between the gas particles are elastic.

Pascal's law states that for a fluid in a closed container the pressure applied to the fluid is transmitted equally to each and every point of the fluid and as well as the walls of the container. If P is pressure, F is the force and A is the area, then $F = PA$. The Pascal's law implies that any change in pressure and anywhere in an enclosed fluid is transmitted throughout the fluid equally.

When objects are placed on a fluid, some of them float and others sink. *What is the reason for it?* It is due to a buoyant force acting on the object. As the pressure of a fluid increases with depth of a fluid and as the pressure is the force acting per unit area, the upward force acting on an object at the bottom will be greater than the downward force acting at the top of the object. The difference of these two forces is the buoyant force acting on any object. If the buoyant force is greater than the weight of the object, the object rises to the surface and floats. If the buoyant force is less than the weight of the object, the object sinks.

Archimedes' principle gives the way to find the buoyant force. It states that the buoyant force on an object equals to the weight of the fluid it displaces. If F_{buoy} is the buoyant force and W is the weight of the fluid displaced by the object, then the equation for the Archimedes' principle is $F_{\text{buoy}} = W = mg$, where m is the mass of the fluid displaced and $g = 9.8\,\text{ms}^{-2}$.

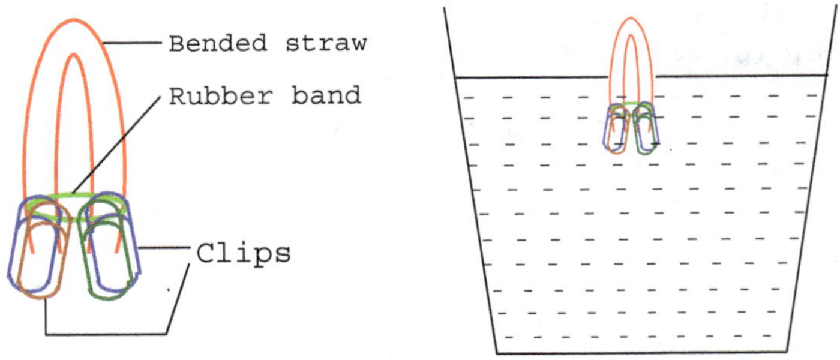

Fig. 7.2: Making a diver with a cool drinks straw and checking its floating in a cup of water.

7.2 Methods of Constructing Cartesian Divers

Different Cartesian diver apparatuses have been described in the literature from a small scale to a large scale and with a wide variety of materials. Construction details and physics behind the observations have been reported in detail. For example, a Cartesian diver with two liquids [3], two droppers [4], extremly sensitive to external pressure applied and small changes in temperature [5] and irreversible sink [6], sea diver and Dyna-diver [7], depth-dependent one [8], diver containing a drop of volatile liquid [9,10], two paper matches as divers [11] and a diver with a pressure head [12] had been studied.

Now, we describe a very simple construction of a Cartesian diver with a straw. Cut a straw and make it into a U-shape with 1 to 1.5 inches long. Hold the two ends together using a rubber band and paper clips as shown in Fig. 7.2. Using additional paper clips make the U-shaped piece of the straw to float vertically with the top being the bend of the straw. You can check its float in the cup of water (see Fig. 7.2). Fill a thin plastic bottle with water. Leave the prepared diver in the bottle in a floating stage and close it. Squeezing the bottle will make the diver to sink to the bottom and releasing the squeeze will make the diver to move to the top.

In a typical Cartesian diver toy the diver is a small rigid glass tube with bottom end opened similar to an eyedropper or a decorated glass tube shown in Fig. 7.1. The diver is partially filled with water and the rest with air. The air is just enough so that it is buoyant enough. It is placed in a thin bottle containing mostly water and remaining filled with air. The bottle is closed with its cap. In a simple case the diver is partly below the top surface of the water and partly above the surface of the water in a vertical position as shown in Fig. 7.1a.

Squeeze the bottle using your hand. The diver would sink to the bottom. *Why?* Now we proceed to explain the physics behind the sinking of the diver [5,13-17]. The inward pressing of the bottle compresses the air (the water is incompressible) leading to increase in the air pressure. According to the Pascal's law a pressure applied at any point on a confined incompressible fluid get transmitted equally throughout the fluid. The pressure thus pushes entire water in the bottle. This forces water into the bottom of the diver. As air can be compressed much more easily than water, the air inside the diver is compressed and the volume of the air inside the diver is decreased. *Why?* Decrease in the volume leads to water taken up into the diver. This in turn increases the density of the diver. As a consequence of the increase in the relative density of the diver, the diver is forced to sink to the bottom of the bottle in accordance with Archimedes' principle. Decreasing the pressure by releasing the squeezing reverse the above process and makes the diver to rise to the top. When the tail part of the glass tube of the diver is made to have a twist, then the flow of the water inside and out of the glass tube makes the diver to spin as it sinks and rises. *What are the other possible simple ways of changing pressure of air above the water surface in the Cartesian diver?* By means of a careful squeeze, it is possible to make the diver to stay in a suspended state nearly in the middle of the water. This will happen when the squeeze makes the buoyant force to balance the weight of the diver resulting in neither floating nor sinking.

7.3 Change in Pressure Necessary for Irreversible Sinking

In a typical Cartesian diver toy when the bottle (say, 1 littre water bottle) is squeezed, the diver which is initially floating sinks to the botton and rises to the top by releasing the squeeze. If the water container is sufficiently long, then irreversible sinking of the diver can be witnessed. That is, the diver will not rise to the top when squeezing is released [6,18]. We now analyse this phenomenon [6].

Consider a glass tube of length, say, about 1.5 m with about 2 cm diameter. The bottom of the tube is closed while the top is opened. The diver is a thin eyedropper of about 2 cm long. A small amount of water is injected into the diver using a syringe. The diver floats in the water glass tube with a very small fraction of its volume is above the surface of the water. When the top of the glass tube is closed by a finger, the diver sinks to a height h

below the surface of the water. If the pressure change is sufficient enough, then the diver may reach a higher h so that removing the finger will not be able to make the diver to rise. That is, the diver has reached an irreversible sinking state.

Denote h_c as the critical height for irreversible sink. Applying simple laws of physics we can obtain an expression for h_c [6]. Let V_T and V_{ext} be the total volume of the diver and the volume above the top of the water surface, respectively, at the atmospheric pressure P_0 above the water surface. m_i is the mass of the water of density ρ inside the diver. m_g is the mass of the glass of the diver and g is the acceleration due to gravity. V is the volume of the air inside the diver at the pressure P.

The weight of the diver is $W = (m_g + m_i)g$. The total buoyancy force is $F_{buoy} = \rho V_T g - \rho V_{ext} g = \rho(V_T - V_{ext})g$. Then, $F_{buoy} = W$ at the equilibrium where the buoyancy on the glass volume and the mass of the air trapped in the diver have been neglected. We write $F_{buoy} = W$ as

$$(m_g + m_i)\, g = \rho\,(V_T - V_{ext})\, g\,. \tag{7.5}$$

At pressure P the volume of the air inside the diver is, say, V and V'_{ext} is the diver volume above the water surface. Then, $m_i = \rho(V_T - V)$ while $m_g = \rho(V - V'_{ext})$. For a change in pressure, from $PV =$ constant,

$$P\mathrm{d}V + V\mathrm{d}P = 0\,. \tag{7.6}$$

That is,

$$\mathrm{d}V = -\frac{\mathrm{d}P}{P}V\,. \tag{7.7}$$

The point is that an increase in pressure causes a decrease in the air volume of the diver resulting in the sinking of the air.

At a constant temperature when $P = P_0$ inside the diver, we have $V = V_0$ and for $P = P_0 + \Delta P$, $V = V_0 - \Delta V$. At P_0, $V_{ext} \neq 0$ and suppose at $P = P_0 + \Delta P$, $V'_{ext} = 0$. We write

$$m_g = \rho\,(V_0 - V_{ext})\,, \quad P = P_0 \tag{7.8a}$$

$$m_g = \rho\,(V_0 - \Delta V - V'_{ext}) = \rho\,(V_0 - \Delta V)\,, \quad P = P_0 + \Delta P\,. \tag{7.8b}$$

Comparision of Eqs. (7.8a) and (7.8b) gives $\Delta V = V_{ext}$.

For smaller ΔV and ΔP

$$\Delta V \approx \frac{\Delta P}{P_0} V_0\,. \tag{7.9}$$

From this equation ΔP_s inside the diver necessary for irreversible sinking of the diver is

$$\Delta V \approx \frac{\Delta P_s}{P_0} V_0\,. \tag{7.10}$$

That is,

$$\Delta P_{\mathrm{s}} \approx \frac{\Delta V}{V_0} P_0 . \tag{7.11}$$

Since $\Delta V = V_{\mathrm{ext}}$, we obtain

$$\Delta P_{\mathrm{s}} \approx (V_{\mathrm{ext}}/V_0)\, P_0 . \tag{7.12}$$

The above equation indicates that for small V_{ext}, a small change in the pressure is enough to drive the diver to sink completely. Moreover, if we neglect the changes in hydrostatic pressure, then the change in the pressure inside the diver ΔP_{s} is approximately equal to the change in the pressure in the top of the glass tube [6].

Solved Problem 1:

Determine the change in the temperature necessary for complete sinking of the diver [6].

Assume that the air inside the diver is at a constant pressure and the temperature is decreased. The initial volume V_0 of air inside the diver at the room temperature T_{R} changes into, say, V_{f} when the temperature is changed into T_{f}. Then, we have

$$V_{\mathrm{f}}/V_0 = T_{\mathrm{f}}/T_{\mathrm{R}} . \tag{7.13}$$

Next, the absolute value of the difference between V_{f} and V_0 is

$$|\Delta V| = |V_{\mathrm{f}} - V_0| = \left| \frac{T_{\mathrm{f}}}{T_{\mathrm{R}}} V_0 - V_0 \right| = \left| \frac{V_0}{T_{\mathrm{R}}} (T_{\mathrm{f}} - T_{\mathrm{R}}) \right| . \tag{7.14}$$

We set $\Delta V = V_{\mathrm{ext}}$ and $\Delta T_{\mathrm{s}} = |T_{\mathrm{f}} - T_{\mathrm{R}}|$ as the decrease in temperature for complete sinking. Then, the above equation gives

$$V_{\mathrm{ext}} = \frac{V_0}{T_{\mathrm{R}}} \Delta T_{\mathrm{s}} . \tag{7.15}$$

Thus,

$$\Delta T_{\mathrm{s}} = T_{\mathrm{R}} \frac{V_{\mathrm{ext}}}{V_0} . \tag{7.16}$$

Therefore, for a complete sinking the decrease in temperature ΔT must be greater than ΔT_{s}.

7.4 Expression for the Critical Height h_c

The pressure change ΔP_s is necessary for the diver to reach the height h_c below the top surface of the water. Once the diver reaches h_c, it will continuously sink to the bottom and the sinking is irreversible. Therefore, $\Delta P_s = \rho g h_c$. Substitution of this in Eq. (7.12) gives

$$h_c \approx \frac{V_{ext}}{\rho g V_0} P_0 \,. \tag{7.17}$$

For $V_{ext}/V_0 \approx 0.1$, $g = 9.8\,\text{ms}^{-2}$, $\rho = 1\,\text{gcm}^{-3}$ and $P_0 = 1.013 \times 10^5\,\text{Pa}$, the value of h_c is of the order of 1 m. *How does one make V_{ext}/V_0 small?*

7.5 Conclusion

Cartesian diver is one of the simplest toys which can be made by the students themselves using commonly available materials. It is a very good demonstration apparatus for explaining Archimedes' buoyancy. The students can be trained to design many types of Cartesian divers such as the irreversible sink of a diver, two-liquid diver and highly sensitive Cartesian diver which can sense even temperature variations. The students can learn how the Archimedes' principle leads to the density-buoyancy relationship. Irreversible sinking can be witnessed in an ordinary test tube through practice [6]. A modified diver related to the so-called Feynman's sprinkler was presented in [19].

7.6 Bibliography

[1] https://blog.cmog.org/2017/04/04/making-cartesian-divers.
[2] A. Privat Deschanel, *Elementary Treatise on Natural Philosophy* (Appleton, New York, 1884) Part I, pp.101-102.
[3] G. Planinsic, M. Kos and R. Jerman, Phys. Edu. **39**, 58 (2004).
[4] H. Fakhruddin, Phys. Teach. **41**, 53 (2003).
[5] R.M. Graham, Phys. Teach. **32**, 182 (1994).
[6] R. De Luca and S. Gance, Phys. Edu. **46**, 528 (2011).
[7] R.C. Turner, Am. J. Phys. **51**, 475 (1983).
[8] S.P. He, S.Y. Mak and E. Zhu, Am. J. Phys. **61**, 938 (1993).
[9] R. Stuart Mackay, Am. J. Phys. **26**, 68 (1958).
[10] R. Stuart Mackay, Am. J. Phys. **26**, 403 (1958).
[11] M. Gardner, Phys. Teach. **28**, 478 (1990).
[12] E.V. Lee, Phys. Teach. **19**, 416 (1981).

[13] http://web.physics.ucsb.edu/ lecturedemonstrations/Composer/Page-/36.37.html.
[14] R.C. Turner, Phys. Teach. **11**, 345 (1973).
[15] R.C. Turner, Am. J. Phys. **51**, 475 (1983).
[16] https://en.wikipedia.org/wiki/Cartesian_diver.
[17] J.C. Siddons, *Experiments in Physics* (Oxford, Blackwell, 1988).
[18] J. Guemez, C. Fiolhais and M. Fiolhais, Am. J. Phys. **70**, 710 (2002).
[19] B.J. Ackerson, The Phys. Teach. **58**, 84 (2020).

7.7 Exercises

7.1 Consider a Cartesian diver toy experiment with two incompressible liquids of densities ρ_1 and ρ_2. The diver with total volume V is set at rest with its volume V_1 in the upper liquid with density ρ_1 while $V_2 = V - V_1$ in the lower liquid with density $\rho_2 > \rho_1$.
(a) Write the expression for the total buoyancy force.
(b) Write $V_1 = \eta V$ and $V_2 = (1 - \eta)V$ where η denotes the fraction of the total volume of the diver that is in the upper liquid. Show that $\eta = \dfrac{\rho_2 - (m/V)}{\rho_2 - \rho_1}$ [3].
(c) Determine the pressure needed to sink the diver completely into the lower liquid.

7.2 A thin eyedropper as a diver is placed in a test tube containing water. The volume of the diver above the free surface of the water is nonnegligible so that closing the top of the tube cannnot lead to sinking of the diver. Describe a way of observing sinking and rising of the diver [6].

7.3 Construct a Cartesian diver apparatus with two divers in a glass bottle as shown in Fig. 7.3 [4]. Blow air into the bottle and suck air out through the tube by an appropriate way. Describe your observation.

7.4 Consider a Cartesian diver with three stages shown in Fig. 7.4 [18]. The original length of the air column in the diver (a glass tube) before placing in the larger vessel is l_0 and the length of the diver is L. V is the volume of the test tube. $l < l_0$ is the length of the air column of the diver after placed in the vessel. P_0 is the atmospheric pressure. x is the height of the air column of the diver above the top surface of the water. ξ is the length of the air column in the diver below the top surface of the water. ρ and ρ_{glass} are the densities of the air and glass, respectively.

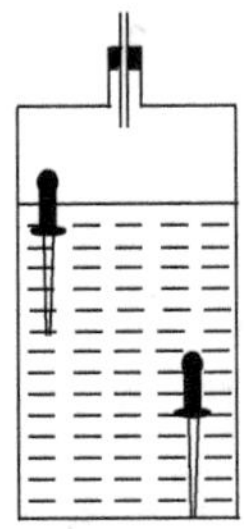

Fig. 7.3: A Cartesian diver apparatus with divers.

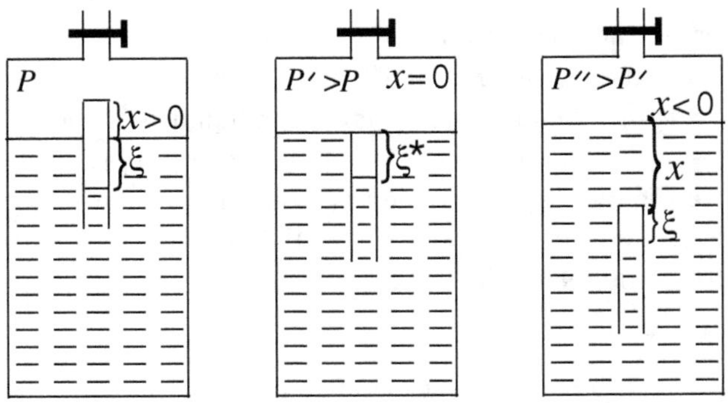

Fig. 7.4: Three stages of a diver apparatus with a small inverted test tube as a diver.

A is the internal cross-sectional area of the diver. d_{int} and d_{ext} are the internal and external diameters of the diver, respectively.
(a) Write the total buoyancy force F_{buoy} for $x > 0$ and $x \leq 0$.
(b) Obtain an expression for ξ^*.
(c) Find the relation between x and ξ using Boyle's and Pascal's laws.

7.5 For $V_{\text{ext}}/V_0 \approx 0.1$, $g = 9.8\,\text{ms}^{-2}$, $\rho = 1\,\text{gcm}^{-3}$ and $P_0 = 10^5\,\text{Nm}^{-2}$, show that the value of h_c given by the Eq. (7.17) is of the order of 1 m.

Part III: Spinning Toys

Chapter 1

Top

There are three noticeable different forms of spinning objects attracted to physicists. They are:

1. A sharply pointed top (spinning top).
2. An elliptical or egg shapped top (PhiTOP).
3. A tippe top.

They display qualitatively quite different behaviours. The term top is used in this book for any rigid object with the rotational symmetry axis passing through its centre of mass (CoM).

Among the variety of simpler childhood toys, the spinning top is a fascinating toy with which most of us in our childhood days played. It is manufactured in a wide variety of shapes, colours and often with wood or plastic. In any case its function is the same. Some of the tops are shown in Fig. 1.1 [1,2]. Tops can be set into spinning by a variety of ways depending on their size and shapes. A very small size top can be spun on a table or floor by giving a flick of our fingers holding firmly a post on it. However, this does not produce fast spinning. Some tops have a support on their top

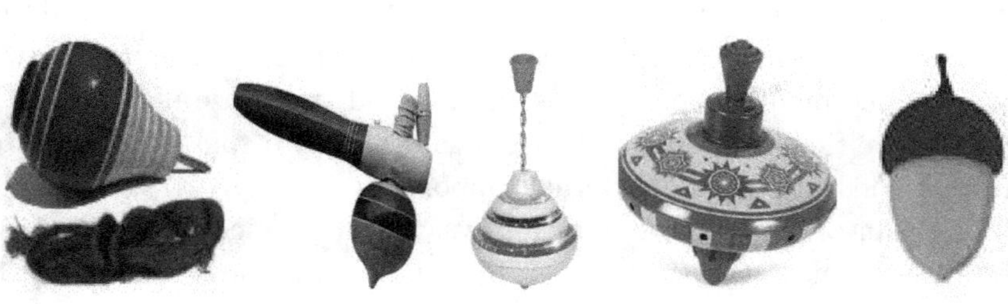

Fig. 1.1: Some of the tops.

part over which a string can be wound and pulled out to spin. Certain tops have hemisphere as head and conical body, called *crown* of the top, with a sharp bottom end called *tip*. A top made of wood often has a nail tip in its bottom. In such a toy, a string is wound around the body from bottom to head. By holding the free end of the string the toy is launched from about waist level to the floor by snapping our wrist as we release it. In any case each type spins in a similar fashion. Tops that are set in motion using a string can spin fastly.

A top will spin with its sharp bottom in contact with the table or floor and about its symmetry axis (also called *spin axis*) passing through the sharp bottom and the CoM. A top, properly set in motion, will spin at a steady angular speed. There is a sliding friction between the tip of the top and the floor. The sliding friction and the friction with air slow down the speed of the top gradually and the top finally falls on the floor.

Spinning tops were invented in various parts of the world almost simultaneously. Turban seashell and the simple acorn (the right-most top in Fig. 1.1) are naturally found top. Tops were found at Ur (Muqayyer, Iraq) dating from 3500 BC, at Troy (Turkey) dating from 3000 BC, in Eqypt dating from 2000 BC, in China, Thebes and Greec dating from 1250 BC [3,4]. In those periods tops were made from clay, terra cotta and wood. In Europe, in 18th century tops were made from iron for playing on the ice of frozen ponds and lakes. In early days seeds, fruits and nuts were developed into tops for prophecy, gampling and gifts to honour gods. A treatment of a spinning top appeared in Jellett's *Treatise on the Theory of Friction* (Macmillan, London, 1872) [5].

Modern technologies and materials are now utilized to manufacture different kinds of tops in a variety of shapes and to minimize friction in order to make it to spin for a very long time. The Fearless Toys (Israel) and the Breaking Toys Ltd (USA) jointly made a battery powered spinning top *LIMBO* which made a Guinnes record time of spinning for 27 hours 9 minutes and 24 seconds in Tel Aviv, Israel on 19 June 2018 [6]. The record title was awarded to Nimrod Back of Fearless Toys. The long time spinning was due to of the minimal energy consumption and small size of the top. In 2016 Lacopo Simonelli made a non-mechanical top to spin for 51 minutes and 3 seconds without any external support [7].

The plan of this chapter is as follows. First, we mention some of the interesting games with tops played by children. Then, we describe the rotation of the top and the occurrence of precession and nutation motions of it. We point out the influence of the height of CoM and weight of the top on its motion. Next, we present a theoretical analysis of the spinning top using Euler angles. We determine separately the conditions for the top to spin with uniform precession and at upright position. Then, we describe the identification of different kinds of motion of the top through numerical simulation. Finally, we point out the stability of the spinning top.

1.1 Playing with the Spinning Top

One can enjoy playing the spinning top games either solo or with a few others. In the hit the target game, for example, a circle is drawn on the ground and a top is released inside it. Points are awarded based on the closeness of the point of the hit of the top to the centre of the circle. Before halting its spin the top has to come out of the circle.

In the catch it game, a player throws a top in air and leans his hand so that the top would land on his palm and continue its spinning for some time [8]. To decide the order of the players to throw the top, before the start all the players can launch their tops on the ground and the order can be fixed based on the duration of the time the tops spin.

Another interesting and challenging game is with tops made of wood and having a nail tip. A string is wrapped around the crown of the top. In the first part of the game called *toss*, all the players simultaneously launch the tops on the (sand) ground. They pick up their respective tops using the string before they come to halt. The players who failed to pick up their tops (or the player(s) who picked it at the last) place their tops inside a small circle drawn on the ground. The players who have completed the toss successfully try to spin their tops, one by one, over the tops kept in the circle and knock them out of the circle. The thrown top should come out of the circle before halt and is to be picked up (as done in the toss) successfully to continue, otherwise the player has to place his top also in the cirlce. If a player is able to knock out a top then he can hold his top in his hand and hit on the knocked out top with his top's nail tip prefixed number of times.

Apart from the above, in YouTube one can see a variety of mesmerizing tricks with a spinning top.

1.2 Rotation of a Spinning Top

When a top is set in motion by spinning it, the force applied to it transforms its potential energy into kinetic energy. As the top spins with its sharp tip

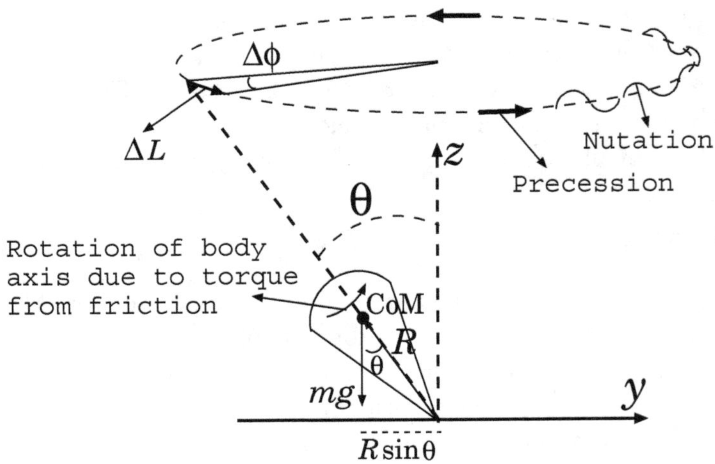

Fig. 1.2: Schematic representation of rotation, precession and nutation of the top. For details see the text.

in contact with the floor, it rotates about its symmetry axis with the spin angular momentum $\mathbf{L} = I\boldsymbol{\omega}$ where $\boldsymbol{\omega}$ is the spin angular velocity and I is the moment of inertia. This angular momentum $\mathbf{L}$ is due to the spinning of the top about its symmetry axis and is along this axis. The top would spin indefinitely due to the conservation of angular momentum if there is no external force acting on it.

What are the possible external forces? In real situations gravity, frictions with the floor and air are the external forces acting on the top. The frictions with the floor and air essentiallly slow down the angular speed of the top and hence reduce the angular momentum and eventually it starts wobbling. In this process the axis of the top gets tilted to the side thereby permits the downward pull of the gravity to exert a torque $\tau = mgR\sin\theta$ on it (refer Fig. 1.2) due to its weight. This torque acts at the CoM of the top. The point is that the gravity in addition to the downward pull (force) also provides a torque that changes the top's angular momentum. That is, this torque produces an additional spin and causes the top to swing (precess) outward. The orientation of the symmetry axis of rotation or spin changes and is called *precession* [9] (for details see Chapter 1 in part I). The precession takes place about the vertical axis (see Fig. 1.2). The head of the top revolves about a vertical direction. The axis of rotation of the top itself rotating about another axis. Note that the contact point of the top on the floor traces out a circle below the CoM whereas the head of it circles about the CoM. The directions of the spin and the precession are the same as shown Fig. 1.2. The rate of precession increases with decrease in spin rate. As a result the top wobbles more and more and finally comes to halt its motion. The up and down wobbling motion of the top is called *nutation* [9] and is indicated in Fig. 1.2. In this process the top nutates the angle θ between two values $\theta_{\rm u}$ and $\theta_{\rm l}$ the upper and lower limits, respectively, of θ. If initially the top is not spun but left out in the upright position then due to the gravitational torque it will fall to the floor.

From Fig. 1.2 an expression for precession angular velocity $\omega_{\rm p}$ can be written. First, the gravitational torque is $\tau = mgR\sin\theta = \Delta L/\Delta t$. We can also say $\boldsymbol{\tau} = \mathbf{R} \times \mathbf{F}$ where $\mathbf{R}$ is the distance from the point of rotation and $\mathbf{F}$ is the applied force. Then with ϕ being the angle of precession

$$\omega_{\rm p} = \frac{\Delta\phi}{\Delta t} = \frac{1}{L\sin\theta}\frac{\Delta L}{\Delta t} = \frac{\tau}{L\sin\theta} = \frac{mgR}{L} = \frac{mgR}{I\omega_{\rm s}}, \tag{1.1}$$

where $\omega_{\rm s}$ is the spin angular velocity and I is the moment of inertia about the point of contact.

One can notice that a top spinning, say, in the clockwise direction often before halting suddenly spins counter-clockwise. *What is actually happening?* This type of phenomenon can be observed in the toy rattleback (see Chapter 6 in part III). The top has unstable rotation in other axis besides the one about which it spins. As the top loses momentum in the key direction and it falls over, the spin in one of the other directions can be transferred into a stable one.

1.3 Effects of Height of the CoM and Weight of the Top

Does the height of the CoM have any effect on the top? When the top is spinning the gravitational force exerts a torque on it and tries to pull it down. This torque is $\tau = mgR\sin\theta = mgl$ where l is the horizontal distance of the CoM from the point of contact with the floor (refer Fig. 1.2). If the height of the CoM (R) is high then the distance l is also high and hence the torque is large. As a result, an increase in the height of the CoM means the top will quickly falls down. In the case of low CoM the gravitational torque requires more time to buildup to make the top fall down.

To find the effect of weight of a top consider two tops of identical shape but one is heavier than the other. Suppose both are set to spin with the same initial speed. We need to spend more energy to set the heavier top to spin with the same speed. This means the heavier top will have more energy than the lighter top. These two tops have the same surface area and hence experience the same air resistance. However, the heavier top has more resistance than the lighter one at the contact point because of its greater weight. But the deceleration is impacted greatly by the momentum rather than the resistance. Consequently, due to greater weight and greater momentum the heavier top will spin longer time. However, if the weight of the top is too heavy then the surface on which the top is spinning got overloaded and the top will dig into the surface and grind to a halt. Therefore, the too heavy top cannot rotate on a fine point on the surface and its sides also get into contact with the surface which increases the rotational friction of the top. If we spend equal amount energy to set the heavier and the lighter tops then the lighter top will spin relatively longer time.

Remember that the CoM of a top is not at its centre. This is one reason that some shapes cannot be used as top. For example, take a pencil and spin it fast placing on a table or a floor. The pencil never climb up like a top.

1.4 Theoretical Analysis

Consider a heavy symmetrical top with mass m. Its motion can be described by the Euler angles [10-14]. As discussed in Chapter 1 of part I, the angular velocity and the angular orientation of a rotating body can be described in terms of the three Euler angles.

Consider the three fixed axes system (X, Y, Z). We get the axes (x', y', z') by a rotation of an angle ϕ about Z-axis first (refer Fig. 1.3), that is, we go from (X, Y, Z) to $(x', y', z'(= Z))$. Then we rotate x' by an angle θ to get the axes $(x''(= x'), y'', z'')$. The axis of rotation of this transformation is termed as the *line of nodes* N. Finally, a rotation of ψ about z'' will lead to the body-fixed axes $(x, y, z(= z'))$. These rotations are illustrated in Fig. 1.3. The associated transformation is discussed in Chapter 1 in part I.

The top is, say, spinning with angular velocity $\boldsymbol{\omega}$. The symmetry axis of the top is passing through its CoM and is the z-axis. The distance between the bottom of the top and its CoM is R and θ is the angle made by the symmetry axis with the vertical as shown in Fig. 1.3. The bottom sharp point is assumed to be fixed on the surface and is chosen as the origin.

As pointed out in Sec. 1.2 the spinning top displays three types of rotations and are described in terms of appropriate Euler angles as specified in the following.

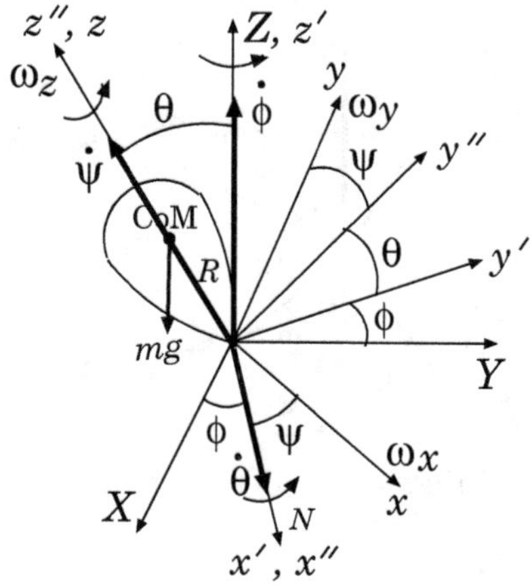

Fig. 1.3: Representation of a spinning top by the Euler angles.

1. **Intrinsic rotation**: It is the spinning or rotation of the top around $z(=z'')$-axis and is represented by ψ. ψ is called the *spin angle*. The rate of the rotation is $\dot{\psi}$.
2. **Precession**: It is the motion about the vertical axis $z'(=Z)$ and is represented by ϕ. That is, it is the motion in ϕ. $\dot{\phi}$ describes the rate of precession of the top.
3. **Nutation**: It is a rotation around the $N(=x')$-axis, represented by θ and is the motion in θ. The nutation rate is $\dot{\theta}$.

Solved Problem 1:

Find the components of the angular velocity $\boldsymbol{\omega}$ in the (x, y, z) coordinate system.

We can write $\boldsymbol{\omega} = \dot{\boldsymbol{\theta}} + \dot{\boldsymbol{\phi}} + \dot{\boldsymbol{\psi}}$. To find the components of $\boldsymbol{\omega}$ we consider Fig. 1.4. $\dot{\boldsymbol{\theta}}$ has both x and y components and has no z-component. The x-component of $\dot{\boldsymbol{\theta}}$ is $\dot{\theta}\cos\psi$ while the y-component is $-\dot{\theta}\sin\psi$. That is,

$$\dot{\boldsymbol{\theta}} = \dot{\theta}\cos\psi\widehat{x} - \dot{\theta}\sin\psi\widehat{y}. \tag{1.2a}$$

$\dot{\boldsymbol{\phi}}$ points in the direction of z'-axis. We note that the projection of z' to z is $\cos\theta$ and the projection onto the corresponding line on the ψ plane of rotation is $\sin\theta$. Then

$$\dot{\boldsymbol{\phi}} = \dot{\phi}\sin\theta\sin\psi\widehat{x} + \dot{\phi}\sin\theta\cos\psi\widehat{y} + \dot{\phi}\cos\theta\widehat{z}. \tag{1.2b}$$

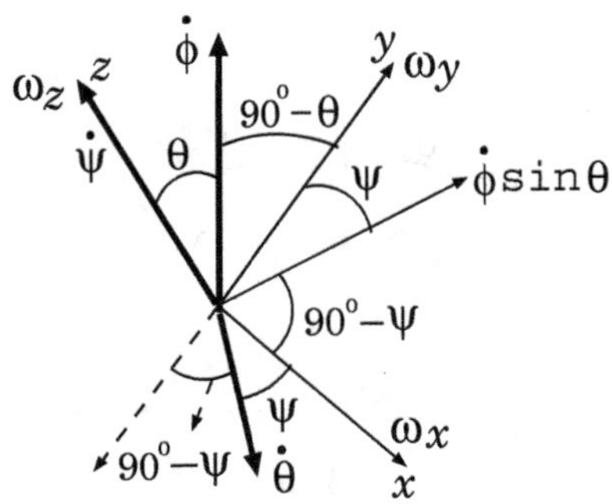

Fig. 1.4: Redrawing of Fig. 1.3 to determine the components of angular velocity $\boldsymbol{\omega}$.

As $\dot{\psi}$ is in the direction of z-axis it has only one component and $\dot{\boldsymbol{\psi}} = \dot{\psi}\widehat{z}$. From the above we write

$$\begin{aligned}\boldsymbol{\omega} &= \dot{\theta}\cos\psi\widehat{x} - \dot{\theta}\sin\psi\widehat{y} + \dot{\phi}\sin\theta\sin\psi\widehat{x} + \dot{\phi}\sin\theta\cos\psi\widehat{y} \\ &\quad + \dot{\phi}\cos\theta\widehat{z} + \dot{\psi}\widehat{z} \\ &= \omega_x + \omega_y + \omega_z, \end{aligned} \tag{1.3}$$

where

$$\omega_x = \dot{\phi}\sin\theta\sin\psi + \dot{\theta}\cos\psi\,, \tag{1.4a}$$

$$\omega_y = \dot{\phi}\sin\theta\cos\psi - \dot{\theta}\sin\psi\,, \tag{1.4b}$$

$$\omega_z = \dot{\phi}\cos\theta + \dot{\psi}\,. \tag{1.4c}$$

1.4.1 *Lagrangian and Energy*

The potential energy of the top is

$$V = mgR\cos\theta\,. \tag{1.5}$$

There are two contributions to the kinetic energy. One is the rapid spinning of the top about the z-axis and is

$$T_1 = \frac{1}{2}I_3\omega_z^2 = \frac{1}{2}I_3(\dot{\psi} + \dot{\phi}\cos\theta)^2\,, \tag{1.6}$$

where I_3 (or I_ψ) is the moment of inertia for rotational motion around the z-axis and ω_z is given by Eq. (1.4c). The second contribution is due to the rotational motion around the contact point at the origin and is

$$T_2 = \frac{1}{2}I_1\left(\omega_x^2 + \omega_y^2\right) = \frac{1}{2}I_1\left(\dot{\theta}^2 + \dot{\phi}^2\sin^2\theta\right)\,, \tag{1.7}$$

where $I_1(= I_2)$ is the moment of inertia of the top about the x-axis and ω_x and ω_y are given by Eqs. (1.4). Then the kinetic energy is

$$T = T_1 + T_2 = \frac{1}{2} I_1 \left(\dot{\theta}^2 + \dot{\phi}^2 \sin^2 \theta \right) + \frac{1}{2} I_3 (\dot{\psi} + \dot{\phi} \cos \theta)^2 . \tag{1.8}$$

The total energy $E = T + V$ is

$$E = \frac{1}{2} I_1 \left(\dot{\theta}^2 + \dot{\phi}^2 \sin^2 \theta \right) + \frac{1}{2} I_3 (\dot{\psi} + \dot{\phi} \cos \theta)^2 + mgR \cos \theta . \tag{1.9}$$

The Lagrangian is

$$\mathcal{L} = \frac{1}{2} I_1 \left(\dot{\theta}^2 + \dot{\phi}^2 \sin^2 \theta \right) + \frac{1}{2} I_3 (\dot{\psi} + \dot{\phi} \cos \theta)^2 - mgR \cos \theta . \tag{1.10}$$

What are the conserved quantities of the system? E and $\mathcal{L}$ are independent of ψ, ϕ and t. This implies that the *total energy E and the angular momenta p_ϕ and p_ψ are constants of motion (conserved quantities).*

The momenta are obtained as

$$p_\psi = \frac{\partial \mathcal{L}}{\partial \dot{\psi}} = I_3 (\dot{\psi} + \dot{\phi} \cos \theta) = I_3 \omega_z = \text{a constant} , \tag{1.11a}$$

$$\begin{aligned} p_\phi &= \frac{\partial \mathcal{L}}{\partial \dot{\phi}} = I_1 \dot{\phi} \sin^2 \theta + I_3 (\dot{\psi} + \dot{\phi} \cos \theta) \cos \theta \\ &= I_1 \dot{\phi} \sin^2 \theta + p_\psi \cos \theta = \text{a constant} , \end{aligned} \tag{1.11b}$$

$$p_\theta = \frac{\partial \mathcal{L}}{\partial \dot{\theta}} = I_1 \dot{\theta} . \tag{1.11c}$$

Use of Eqs. (1.11a) and (1.11b) in Eq. (1.9) leads to

$$E = \frac{1}{2} I_1 \dot{\theta}^2 + \frac{1}{2} \frac{p_\psi^2}{I_3} + \frac{1}{2} \frac{(p_\phi - p_\psi \cos \theta)^2}{I_1 \sin^2 \theta} + mgR \cos \theta . \tag{1.12}$$

1.4.2 *Uniform Precession*

This is a slow circling dynamics without falling over and without wobbling up and down. For this neglect the nutation, that is, $\dot{\theta} = 0$ and $\theta = \theta_0$. Further, the precession rate $\dot{\phi}$ is a constant. We find expressions for $\dot{\phi}$, $\dot{\psi}$ and θ_0. From Eq. (1.11b)

$$\dot{\phi} = \frac{p_\phi - p_\psi \cos \theta}{I_1 \sin^2 \theta} . \tag{1.13}$$

From Eq. (1.11a)

$$\dot{\psi} = \frac{p_\psi}{I_3} - \dot{\phi} \cos \theta = \frac{p_\psi}{I_3} - \frac{p_\phi - p_\psi \cos \theta}{I_1 \sin^2 \theta} \cos \theta . \tag{1.14}$$

To find θ_0 consider the equation $\partial E/\partial\theta|_{\theta=\theta_0} = 0$. From Eq. (1.12) the equation $\partial E/\partial\theta|_{\theta=\theta_0} = 0$ gives, after rearranging the terms,

$$\frac{(p_\phi - p_\psi\cos\theta_0)^2\cos\theta_0}{p_\psi\sin^2\theta_0} - (p_\phi - p_\psi\cos\theta_0) + \frac{mgRI_1\sin^2\theta_0}{p_\psi} = 0\,. \quad (1.15)$$

The solution θ_0 of this equation gives the leaning angle.

Equation (1.15) is a quadratic equation for $(p_\phi - p_\psi\cos\theta_0)$. So, we write

$$(p_\phi - p_\psi\cos\theta_0) = \frac{p_\psi\sin^2\theta_0}{2\cos\theta_0}\left(1 \pm \sqrt{1 - \frac{4mgRI_1\cos\theta_0}{p_\psi^2}}\right). \quad (1.16)$$

Using of this equation for $(p_\phi - p_\psi\cos\theta_0)$ in Eq. (1.13) gives

$$\dot{\phi}_\pm = \frac{p_\psi}{2I_1\cos\theta_0}\left(1 \pm \sqrt{1 - \frac{4mgRI_1\cos\theta_0}{p_\psi^2}}\right). \quad (1.17)$$

For $\dot{\phi}$ to be real we require $p_\psi^2 > 4mgRI_1\cos\theta_0$. As $p_\psi = I_3\omega_z$ we rewrite this condition as $\omega_z > (2/I_3)\sqrt{mgRI_1\cos\theta_0}$. *What will happen if $p_\psi^2 < 4mgRI_1\cos\theta_0$?*

1.4.3 *Nutation*

The kinetic energy associated with the nutation, that is, $\dot{\theta} \neq 0$ is $I_1\dot{\theta}^2/2$. The energy E given by Eq. (1.12) can be rewritten as

$$E = \frac{1}{2}I_1\dot{\theta}^2 + V_{\text{eff}}(\theta)\,, \quad (1.18a)$$

where

$$V_{\text{eff}}(\theta) = \frac{p_\psi^2}{2I_3} + \frac{(p_\phi - p_\psi\cos\theta)^2}{2I_1\sin^2\theta} + mgR\cos\theta\,. \quad (1.18b)$$

The equation of motion for θ is

$$\dot{\theta} = \eta\,, \quad (1.19a)$$

$$\begin{aligned}\dot{\eta} &= -\frac{1}{I_1}\frac{\partial V_{\text{eff}}}{\partial\theta}\\ &= -\frac{1}{I_1}\left[\frac{(p_\phi - p_\psi\cos\theta)\,p_\psi}{I_1\sin\theta} - \frac{(p_\phi - p_\psi\cos\theta)^2\cos\theta}{I_1\sin^3\theta}\right.\\ &\qquad\left. - mgR\sin\theta\right]. \end{aligned} \quad (1.19b)$$

The relevant equations of motion are given by Eqs. (1.13) (for ϕ), (1.14) (for ψ) and (1.19) (for θ). These equations can be numerically integrated.

The range of θ in nutation can be determined. The turning points in the θ-motion occur when $\dot{\theta} = 0$. Consider the Eq. (1.18a). We rewrite it as

$$\dot{\theta}^2 = \frac{2E}{I_1} - \frac{2V_{\text{eff}}}{I_1} = s_1 - s_2 \cos\theta - \frac{(s_3 - s_4 \cos\theta)^2}{\sin^2\theta}, \tag{1.20a}$$

where

$$s_1 = \frac{2E}{I_1} - \frac{s_4^2 I_1}{I_3}, \quad s_2 = \frac{2mgR}{I_1}, \quad s_3 = \frac{p_\phi}{I_1}, \quad s_4 = \frac{p_\psi}{I_1}. \tag{1.20b}$$

Introducing $u = \cos\theta$ Eq. (1.20a) becomes

$$\dot{u}^2 = f(u) = au^3 + bu^2 + cu + d, \tag{1.21a}$$

where

$$a = s_2, \quad b = -s_1 - s_4^2, \quad c = -s_2 + 2s_3 s_4, \quad d = s_1 - s_3^2. \tag{1.21b}$$

For $\dot{\theta} = 0$ we have $\dot{u} = 0$. Then Eq. (1.20a) is cubic in $\cos\theta$ and the Eq. (1.21a) is cubic in u. The turning angles of θ are the roots of the equation $f(u = \cos\theta) = 0$. A cubic equation can admit either three real roots or one real root and a pair of complex conjugate roots. For the present problem the case of three real roots is relevant. As $\cos\theta \in [-1, 1]$ the physically interesting roots are those with $u \in [-1, 1]$. Therefore, in the nutation dynamics out of the three real roots of $f(u) = 0$ the roots with $|u| > 1$ are physically uninteresting. In Fig. 1.3 we note that smaller value of θ means the top being more vertical. Therefore, we designate the turning angle with upper bound as θ_{u} and that with the lower bound as θ_{l} and here $\theta_{\text{u}} < \theta_{\text{l}}$.

Solved Problem 2:

When both θ and $\dot{\theta}$ are 0 then the top spins in its upright position. In this case the top is said to be a *sleeping top*. Determine the conditions for the stability of this special state.

If $\theta = \dot{\theta} = 0$ then $u = \cos\theta = \cos 0 = 1$. That is, one root of $f(u) = 0$ is $u = 1$. $\theta = 0$ should be either a minimum or a maximum of $f(u = \cos\theta) = 0$ where $f(u)$ is given by Eq. (1.21a). Substitution of $\dot{\theta} = 0$ and $\theta = 0$, that is, $\dot{u} = 0$ and $u = 1$ in Eq. (1.21a) gives $p_\phi = p_\psi$. We require that $u = 1$ should be either a minimum or a maximum of $f(u) = 0$. The condition is $f'(u) = 0$ at $u = 1$. This condition together with $p_\phi = p_\psi$ lead to the

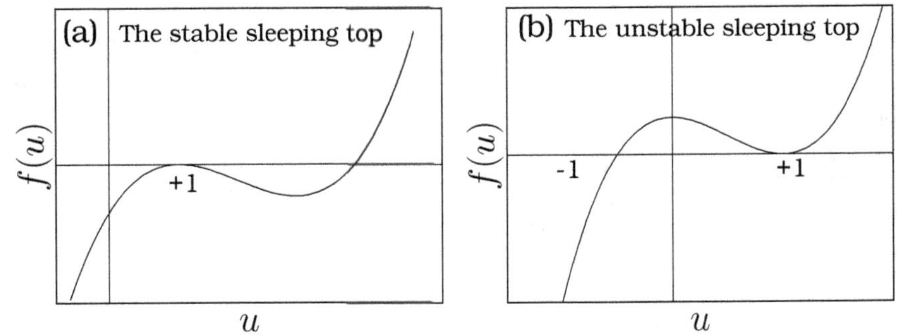

Fig. 1.5: Plot of $f(u)$ for stable and unstable sleeping states of the spinning top.

equation $3a + 2b + c = 0$. This equation gives $s_1 = s_2$. *What are the other roots of* $f(u) = 0$*?* With $\dot{u} = 0$, $s_1 = s_2$ and $s_3 = s_4$ ($p_\phi = p_\psi$) Eq. (1.21a) takes the form

$$(1-u)^2\left(s_2(u+1) - s_4^2\right) = 0\,, \tag{1.22}$$

which indicates that $u = 1$ ($\theta = 0$) is a double root, that is, $u_1 = u_2 = 1$. The third root is

$$u_3 = \frac{s_4^2 - s_2}{s_2}\,. \tag{1.23}$$

Now, there are two cases: $s_4^2 > 2s_2$ and $s_4^2 < 2s_2$.

Case 1: $s_4^2 > 2s_2$

For $s_4^2 > 2s_2$, that is, $p_\psi^2 > 4I_1 mgR$ the root u_3 is > 1 and is physically uninteresting. As $p_\psi = I_3\omega_z$ this condition is stated as

$$\omega_z > \omega_z' = \sqrt{4I_1 mgR/I_3^2}\,. \tag{1.24}$$

In this case the graph of $f(u)$ is like as shown in Fig. 1.5a. If the top is spun with an initial angular velocity $\omega_z > \omega_z'$ then the motion with $\theta = 0$ is stable and is the only possible motion. If we slightly disturb the initial conditions or the motion of the top then $f(u)$ will be slightly perturbed. Since $\dot{u}^2 = f(u)$ and $\dot{u}^2$ is positive, the physically meaningful region is $f(u) > 0$. From the Fig. 1.5a we note that the top will be trapped near $u = 1$. Due to friction the frequency of rotation of the top spinning in its upright state got reduced. When the angular velocity ω_z becomes lower than the critical velocity ω_z' the top begins to wobble and falls down.

Case 2: $s_4^2 < 2s_2$

For $s_4^2 < 2s_2$ ($\omega_z^2 < 4I_1 mgR/I_3^2$) we have $u_3 < 1$. The graph of $f(u)$ is as shown in Fig. 1.5b. Any slight disturbance may give rise a large excursion. The top is thus unstable.

1.5 Motion of the Spinning Top

The spinning motion of a top can be described by plotting ϕ versus θ on a unit sphere. θ will vary between the two turning angles $\theta_{\rm u}$ and $\theta_{\rm l}$. We can solve the Eqs. (1.13), (1.14) and (1.19) numerically, for example, using the fourth-order Runge-Kutta method for a given set of initial conditions and the values of the parameters. p_ϕ and p_ψ are constants of motion. A top can exhibit different types of motion in $\phi - \theta$ plane depending on the sign of $\dot{\phi}$ at the physically meaningful two roots u_1 and u_2. The possibilities are [10-14]:

1. **Cusp**: $\dot{\phi} > 0$ at $\theta = \theta_{\rm l}$ (higher value of turning angle) and $\dot{\phi} = 0$ at $\theta = \theta_{\rm u}$ (lower value of turning angle).
2. **Sinusoidal**: $\dot{\phi} > 0$ at both $\theta = \theta_{\rm u}$ and $\theta_{\rm l}$.
3. **Loop**: $\dot{\phi} > 0$ at $\theta = \theta_{\rm l}$ and $\dot{\phi} < 0$ at $\theta = \theta_{\rm u}$.

For analysis purposes let us consider the top as having a narrow stem of length $R = 0.044\,\text{m}$ and a heavy wheel of radius $0.02\,\text{m}$ and height $0.07\,\text{m}$. The mass of the top is, say, $m = 0.53\,\text{kg}$, $I_1 = 1.7 \times 10^{-4}\,\text{kgm}^2$, $I_3 = 9.8 \times 10^{-5}\,\text{kgm}^2$ and $g = 9.8\,\text{m/s}^2$. Assume that the top is spun with the initial angular velocity $\dot{\psi}(0) = 250\,\text{rads}^{-1}$, $\theta(0) = 30^o \approx 0.52360\,\text{rad}$. Further, we choose $\dot{\theta} = 0$ ($\eta(0) = 0$) and $\psi(0) = \phi(0) = 0$.

Case 1: $\dot{\phi}(0) = 0$.

Equations (1.11a) and (1.11b) give $p_\psi = 2.45 \times 10^{-2}\,\text{J}$ and $p_\phi = 2.12176 \times 10^{-2}\,\text{J}$. The last term in Eq. (1.20a) is $\dot{\phi}^2 \sin^2\theta$ and is zero. As $\dot{\theta} = 0$ Eq. (1.20a) gives $s_1 = s_2 \cos\theta + \dot{\phi}^2 \sin^2\theta = s_2 \cos\theta$. The values of s_1, s_2, s_3 and s_4 are

$$s_1 = 2328.44678, \quad s_2 = 2688.65869, \quad s_3 = 124.80954, \quad s_4 = 144.11764.$$

From Eqs. (1.21b) the values of a, b, c and d are calculated as

$$a = 2688.65869, \; b = -23098.3418, \; c = 33285.8555, \; d = -13248.9746.$$

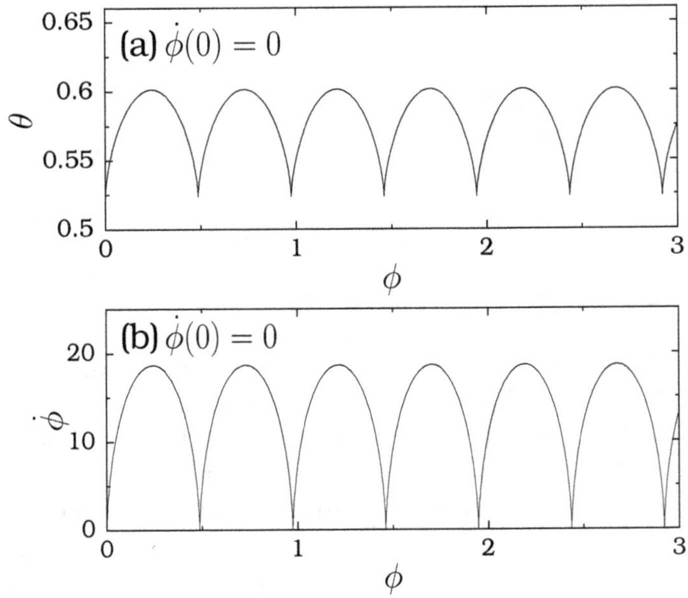

Fig. 1.6: Plot of (a) ϕ versus θ and (b) ϕ versus $\dot{\phi}$ with $\dot{\phi}(0) = 0$.

The roots of a cubic equation can be determined using the formula given in [15]. The roots of the cubic equation $f(u) = 0$, Eq. (1.21a), are

$$u_1 = 0.82459, \quad u_2 = 0.86603, \quad u_3 = 1.69004.$$

Since $u_3 > 0$ this root is physically uninteresting. The turning angles are

$$\theta_{\mathrm{u}} = 0.52360\,\mathrm{rad} = 30^\circ \text{ and } \theta_{\mathrm{l}} = 0.60131\,\mathrm{rad} = 34^\circ 27'.$$

Note that θ_{u} is the initial lean angle ($\theta(0)$). $\theta \in [\theta_{\mathrm{u}}, \theta_{\mathrm{l}}]$ is the range of angle θ of the motion of the nutation.

Figure 1.6 shows the numerically computed ϕ, θ and $\dot{\phi}$ with $\dot{\phi}(0) = 0$. θ varies between θ_{u} and θ_{l}. The top undergoes oscillation (wobbling) between θ_{u} and θ_{l}. In practice, these oscillations die out due to slight frictional loss and the top spins rapidly with precessional movement. In the theoretical treatment we have not taken into account of friction. The cusp in the trajectory at $\theta = \theta_{\mathrm{u}}$ is due to the choice $\dot{\phi}(0) = 0$ at $\theta(0) = \theta_{\mathrm{u}}$.

Case 2: $\dot{\phi}(0) > 0$.

We fix $\dot{\phi}(0) = 3$ and the other initial conditions as in the case 1. Now,

$$p_\phi = 2.15656 \times 10^{-2}\,\mathrm{J}, \quad p_\psi = 2.47546 \times 10^{-2}\,\mathrm{J}.$$

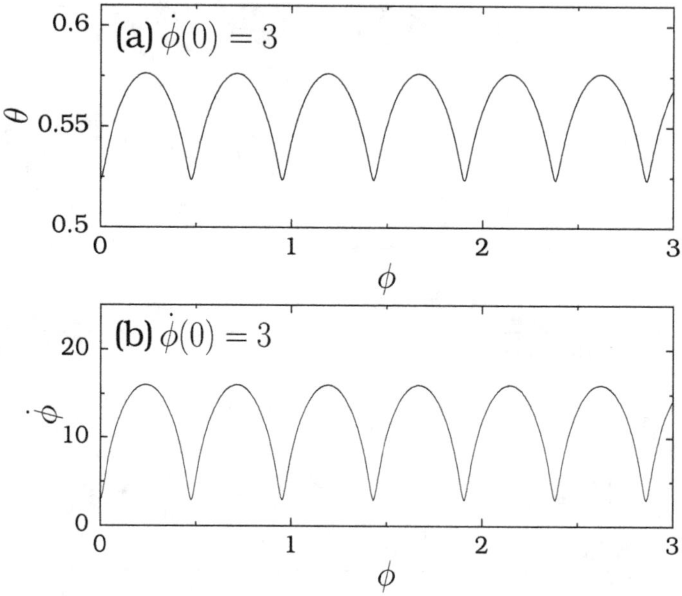

Fig. 1.7: Plot of (a) ϕ versus θ and (b) ϕ versus $\dot{\phi}$ with $\dot{\phi}(0) = 3$.

$$s_1 = 2330.69678, \quad s_2 = 2688.65869, \quad s_3 = 126.85659, \quad s_4 = 145.61536.$$

$$a = 2688.65869, \quad b = -23534.5293, \quad c = 34255.8750,$$
$$d = -13761.8975.$$

$$u_1 = 0.83850, \quad u_2 = 0.86602 \quad u_3 = 7.0487 > 1.$$

The turning angles are

$$\theta_{\mathrm{u}} = 0.52361\,\mathrm{rad} = 30^{\circ} \text{ and } \theta_{\mathrm{l}} = 0.57627\,\mathrm{rad} = 33^{\circ}.$$

θ_{u} is the initial lean angle. Figure 1.7 presents the reults of numerical computation of ϕ, θ and $\dot{\phi}$. The cusp seen in Fig. 1.6 for $\dot{\phi}(0) = 0$ is now disappeared. Further, $\dot{\phi} \neq 0$ at θ_{u}. We clearly see $\dot{\phi} > 0$ at $\theta = \theta_{\mathrm{u}}$ and θ_{l} .

Case 3: $\dot{\phi}(0) < 0$.

The choice $\dot{\phi}(0) = -8$ gives

$$p_\phi = 2.02896 \times 10^{-2}\,\mathrm{J}, \quad p_\psi = 2.38210 \times 10^{-2}\,\mathrm{J}.$$

$$s_1 = 2344.44678, \quad s_2 = 2688.65869, \quad s_3 = 119.35072, \quad s_4 = 140.12373.$$

$$a = 2688.65869, \quad b = -21979.1074, \quad c = 30759.0801,$$

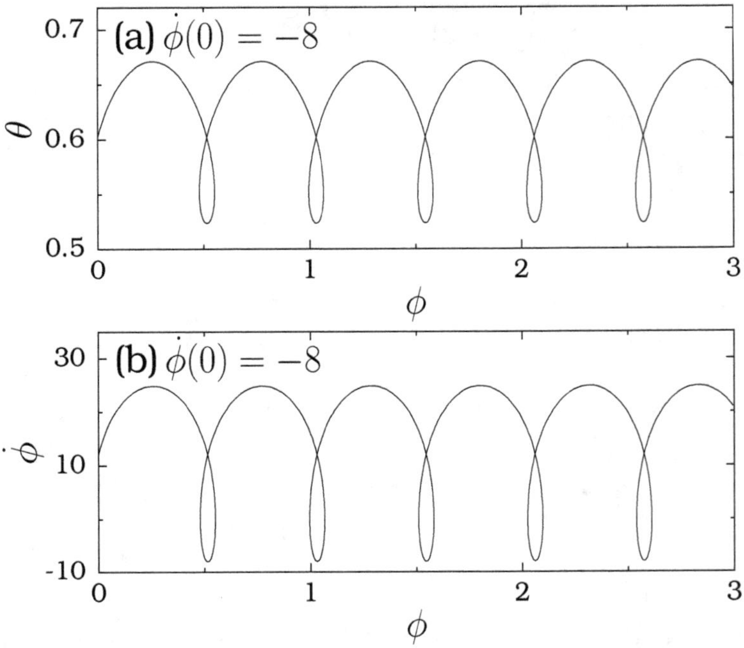

Fig. 1.8: Plot of (a) ϕ versus θ and (b) ϕ versus $\dot{\phi}$ with $\dot{\phi}(0) = -8$.

$d = -11900.1484.$

$u_1 = 0.78319, \;\; u_2 = 0.86603, \;\; u_3 = 6.5255 > 1.$

$\theta_{\rm u} = 0.52360\,{\rm rad} = 30^\circ$ and $\theta_{\rm l} = 0.67101\,{\rm rad} = 38^\circ 27'$.

Figure 1.8 shows the result. The variation of θ for $\dot{\phi}(0) < 0$ is different from $\dot{\phi}(0) = 0$ and $\dot{\phi}(0) > 0$ shown in Figs. 1.6 and 1.7. The trajectory becomes a *loopy* because of $\dot{\phi}$ changes sign periodically.

One can develop a simulation program to visualize the precession and nutation motions in a three-dimensional space [16] and also on the $x-y$ plane as the view from the top of the spinning top. This is left as an exercise to the readers.

1.6 Stability of the Spinning Top

As discussed above, the motion of a top can be described only by an advanced level mathematics. Though it contains all the explanation for the motion of the spinning top, it is better if the stability of a spinning top is explained in simple words. *Why does not a spinning top fall? What makes the spinning top stable? Why does a spinning top start to tilt at a certain speed? Why does a spinning top raise to upright position?*

Suppose a top is kept vertically up without spinning, it falls down immediately. But a spinning top does not fall down for a longer time. The obvious reason is that a spinning top carries an angular momentum. As angular momentum has to be conserved, it keeps its motion stable until it looses its energy due to friction.

A spinning top is governed by the gyroscopic effect. A gyroscope will spin about a constant axis unless acted on by any aother couple. If the top spins faster then it is more resistant to any disturbing couple. When the top is tilted from its verticl axis then the force of gravity and the reaction force at the ground creates a couple. This couple is independent of how fast the top spins. This couple acts in such a direction so as to make the spin axis precess about the vertical direction. The couple acting due to the tilting of the vertical axis leads to precession motion. Hence, the conservation of angular momentum and the gyroscopic effect which tries to keep the spinning axis stable are the physical reasons for the stability of the spinning top.

1.7 Conclusion

A spinning top has three types of rotational motions. It has spin, precession and nutation. The theory involved in getting these types of motion is highly mathematical. Initially spin is given to the top externally. There are in general many types of tops with different shape. They spun in a variety of ways, however, each type has almost similar type of motion.

The top is given an angular momentum about the spinning axis and it will remain constant until some outside torque acts on it. A tilt in the spin axis will lead to a couple between the gravitational force acting through the centre of gravity and the reaction force acting vertically up from the surface at the tip of the top. This couple gives an additional angular momentum which makes the spin axis precess about the vertical axis. If the precession angular momentum is exactly vertical then the top will be ideally balanced. But this type of perfect precession is rarely achieved in practice. There will be disturbance that causes the precession angular momentum to be slightly different from the vertical direction. This leads to the nutation motion of the top. A spinning top sings because it either scratches the surface of the floor on which it is spinning or creates vibration in the air [9].

It may also be observed that the precession will be slower when the spin is faster. As the top begins to slow down it precesses faster and the tilt angle increases. In order to conserve angular momentum the top precesses as it slows down before falling down. All these motions and the physical concepts involved in the spinning top can easily be demonstrated to the students.

Solutions for the case of a top with bottom end being rounded instead of a sharp point can be obtained. In this case the bottom end rolls along the surface on which the top is spinning. The bottom end is not fixed but makes a spiral path on the surface. Through projects students can investigate the effect of smooth, rough and lubricated horizontal surfaces on the spinning of sharp and rounded bottom tops and factors determining the rate of rise or fall of a spinning top.

1.8 Bibliography

[1] https://www.simpledays.in.

[2] https://www.ubuy.co.in/search/?q=spinning+tops.

[3] https://www.artsmendocino.org/wp-content/uploads/sites/www.artsmendocino.org/images/2016/11/Spinning-Top-History-Handout.pdf.

[4] https://www.artofplay.com/blogs/articles/history-of-spinning-tops.

[5] H. Aref, S. Hutzier and D. Weaire, Europhysics News **38** (3), 23 (2007).

[6] https://www.guinnessworldrecords.com/news/2018/10/watch–the–mesmerising-spinning-tops-that-keep-going-for-more-than-27-hours-542573/.

[7] https://recordsetter.com/world-record/spinning-non-mechanical-spinning-top/45510.

[8] https://www.artofplay.com/blogs/articles/5-games-you-can-play-with-your-spinning-tops.

[9] J. Walker, Sci. Am. **244**, 182 (1981).

[10] Ch. Elliott, The spinning top (2009); http://www.phy.pmf.unizg.hr/~npoljak/files/clanci/SpinningTop.pdf.

[11] B. Slade, Classical symmetric top in a gravitational field (2012); https://www.researchgate.netpublication/265624866_Classical_Symmetric_Top_in_a_Gravitational_Field.

[12] H. Goldstein, C. Poole and J. Safko, *Classical Mechanics* (Addison Wesley, New York, 2000) 3rd edition.

[13] N.C. Rana and P.S. Joag, *Classical Mechanics* (Tata McGraw-Hill, New Delhi, 1991).
[14] J. Tatum, The top. https://phys.libretexts.org/Bookshelves/Classical_Mechanics/Book%3A_Classical_Mechanics_(Tatum)/04%3A_Rigid_Body_Rotation/4.10%3A_The_Top.
[15] L.A. Pipes and L.R. Harvill, *Applied Mathematics for Engineers and Physicists* (McGraw-Hill, Singapore, 1984) 3rd edition.
[16] E. Butikov, Eur. J. Phys. **27**, 1071 (2006).

1.9 Exercises

1.1 For the Lagrangian given by Eq. (1.10) using the Lagrange's equation of motion obtain the equation of motion for θ.

1.2 What will happen if a top is set spinning on a slightly inclined surface?

1.3 What will happen when a rounded bottom top spinning at its upright position is slightly toppled by knocking it with a finger?

1.4 For $p_\psi^2 \gg 4mgRI_1 \cos\theta_0$, from Eq. (1.17) obtain
(a) expressions for $\dot{\phi}_{\text{fast}}$ $(= \dot{\phi}_+)$ and rotation speed and
(b) an expression for $\dot{\phi}_{\text{slow}}$ $(\dot{\phi}_-)$ and the corresponding θ_0.

1.5 If the frictional force associated with a rapidly spinning top is $|\mathbf{f}| \approx \mu mg$ where μ is the coefficient of friction then show that an approximate expression for the angular velocity $\dot{\theta}$ of the rising motion of the top is $-\mu\omega_{\text{p}}$.

1.6 Consider a top with a disk of radius $r = 0.03\,\text{m}$ and mass m. The top has a narrow stem with length $R = 3r/4$ and negligible mass. The frictional coefficient is $\mu = 0.1$, the initial inclination of the top is $\theta = 0.64\,\text{rad}$ and the initial spin is $200\,\text{rads}^{-1}$. Assuming that $\dot{\theta} = -\mu\omega_{\text{p}}$ determine the time needed for the top to rise to the vertical and the number of revolutions the top would make during this time.

1.7 For a top with $m = 0.53\,\text{kg}$, $I_1 = 1.7 \times 10^{-4}\,\text{kgm}^2$, $I_3 = 9.8 \times 10^{-5}\,\text{kgm}^2$, $g = 9.8\,\text{m/s}^2$, $R = 0.044\,\text{m}$ and $\theta_0 = 30°$ calculate the minimum spin to be given in order to realize a uniform precession.

1.8 A top with $m = 0.05\,\text{kg}$, $R = 0.04\,\text{m}$, $I_1 = 1.7 \times 10^{-4}\,\text{kgm}^2$, $I_3 = 2 \times 10^{-5}\,\text{kgm}^2$, $\theta(0) = 0$, $\dot{\phi}(0) = 0$ and $\dot{\psi}(0) = 200\,\text{rads}^{-1}$ is spinning uniformly in its upright position. Calculate ω_z and identify the stability of the upright position.

1.9 Develop a program to determine the turning angles by solving the cubic equation $f(u) = 0$ where $f(u)$ is given by Eq. (1.21a).

1.10 Develop a program to solve the equations of motion of ϕ, ψ and θ by the fourth-order Runge–Kutta method (or any other appropriate method) and verify the Figs. 1.6-1.8.

Chapter 2

Tippe Top

As seen in chapter 1 in part III, a spinning top is a remarkable toy for children which when made to spin fast enough, rises up to a vertical state, stays there for some time, stops spinning and finally falls over. On the other hand, if a hard boiled egg is spun fast enough we witness rising of its one end and spinning vertically on another end. A tippe top, a special kind of top toy shown in Fig. 2.1a, displays even more spectacular and unexpected behaviour. When spun fast enough, with its bottom of the rounded part in contact with the floor or the table, at one stage it completely make it upside down (as shown in Fig. 2.1b) and spins. The upside down motion is displayed by several spherical objects in which the centre of mass (CoM) is shifted from the centre of curvature (CoC) (centre of the object). When spun slowly the tippe top is stable and no flipping takes place. The spinning toys moving sideways and upward defy gravity.

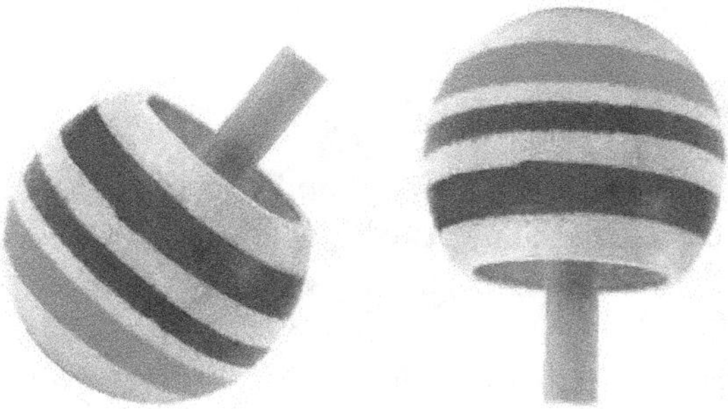

(a) (b)

Fig. 2.1: (a) A tippe top with the rounded bottom part. (b) A tippe top with upside down position.

Commercially available tippe tops come in a variety of forms. The most common model tippe top is a short cylindrical stem mounted to the flat surface of a truncated sphere or cylinder shown in Fig. 2.1. The CoM is shifted slightly below the CoC (centre of the sphere). This toy is also referred as topsy-turvy top, the inverting top and toupie magique.

The tippe top mesmerizes to those who witness first time the unexpected peculiar display of inverting itself. Niels Bohr, Wolfgang Pauli and Winston Churchill were fascinated by the tippe top. In 1980s Sir William Thomson and Hugh Blackburn experimentally observed properties similar to that of the tippe top [1]. In 1891 Helene Sperl patented the top as Wendekreisel in Germany but the patent ran out the year due to nonpayment of the patent fee. The tippe top was reinvented in 1950 by the Danish engineer Werner Ostberg [2]. During his visit to South Africa, he witnessed local people playing with a rounded small fruit. When spun it by a stalk it behaved like a tippe top. The engineer then mass-produced the toy and it became popular in many countries. In 1954 tippe tops were provided freely for those bought sugar rice krinkles [2]. Due to its rapid growth of popularity many theoretical works on it were published in the 1950s [3-9]. In 1977 Cohen reported that the influence of sliding friction is the key to the understanding of the behaviour of the tippe top [10]. David Featonby provided simple levels of explanation of the toy [11]. Investigations of features of inversion dynamics of the tippe top have been reported in [12-22].

Why does the spinning tippe top invert? The theory of the tippe top is complicated. Here, first we bring out the motion and four peculiar features of the tippe top. Next, we present the role of friction on the motion of the tippe top accounting for the occurrence of the unusual behaviour of it. Then, we briefly summarize a theoretical model for its flipping behaviour.

2.1 Motion of a Tippe Top

Let us describe the motion of a tippe top in detail [11].

Hold the top with the stem between your thumb, index finger and middle finger. Spin the toy fast enough on its blunt end on a flat surface in a clockwise direction (viewed from above the top). At the beginning it rotates around a vertical axis passing through the stem in the *clockwise direction*. It starts to lean over away from the vertical and more over bodily rotates in the clockwise direction. The leaning increases while the rotational speed of the top about its own axis decreases. The above process is depicted in Figs.2.2a-c. When it is in horizontal as shown in Fig. 2.2d, it will no longer

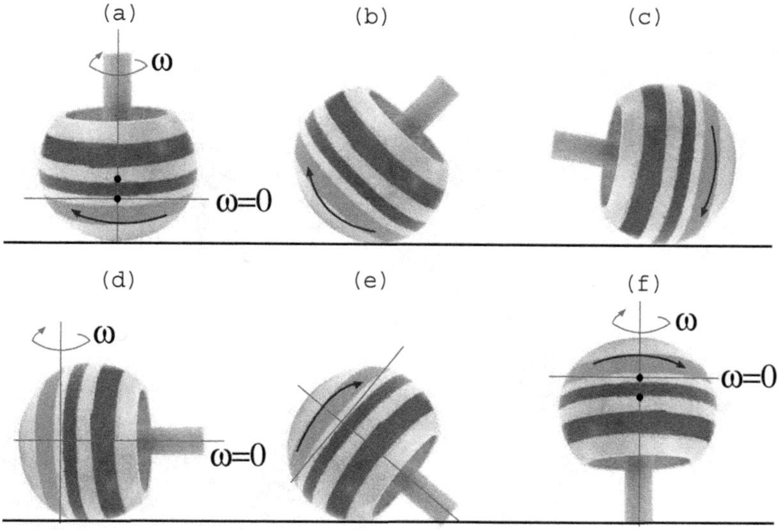

Fig. 2.2: (a)-(e) A spinning tippe top before flipping over to rotate on its stem. (f) The tippe top rotating on its stem. In (a) and (f) the solid circle on the $\omega = 0$ axis represents the CoM while the other solid circle denotes the CoC.

rotate about the axis passing through the stem. Once it leans downwards further, the rotation of the tippe top continues again about the axis going through the stem but in the *counter-clockwise direction*. That is, it has *reversed its direction of spin*, the first feature of it. At one stage the stem touches the surface as shown in Fig. 2.2e and the top quickly stands up on its short stem (Fig. 2.2f). It is now *upside down* from its starting spin position and spins on its stem like a typical sleeping top. This is the second feature of the tippe top. Further, the toy is spinning in the counter-clockwise direction (from the toy's point of view). The third feature of the tippe top is that in its inverted state the *CoM is above its CoC*. The toy continue its spinning on its stem until dissipation causes it to fall over.

Slow run of video record of spinning of a tippe top brought an another feature, fourth of it [22]. There is a short interval between the time at which the stem first touches the surface and turns into the upright position with the stem on the surface. During this time interval the top loses its contact with the surface and becomes *air borne*. The stem then lands on the surface, bounces many times and again becomes air borne. Such jumping

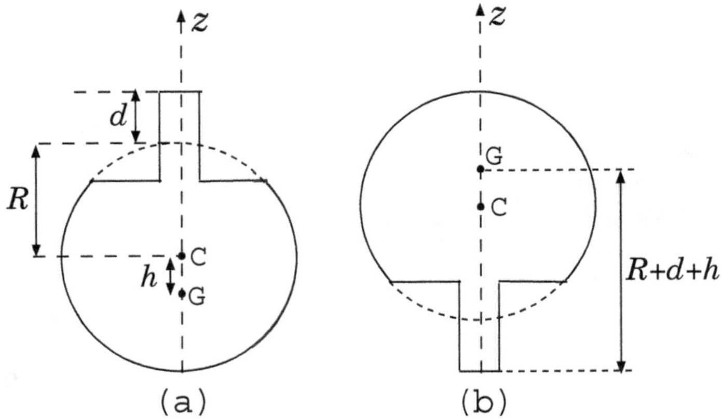

Fig. 2.3: (a) A tippe top at rest with the stem in the upward. (b) A tippe top after turning over with its stem on the surface.

and bouncing happens for a short duration and then remains in contact with the surface. The air borne phase is not caused by the irregularities on the surface since this peculiar behaviour is well observed on smoother surfaces also [22]. Jumping behaviour can be observed in spinning eggs also but the jumping height of a tippe top is relatively much higher than that of spinning eggs. *What about in an ordinary top?*

From the above, the prime features of a tippe top are:

1. The reversal of direction of spin.
2. The upright spinning on its stem.
3. Rising the CoM above the CoC.
4. The occurrence of the air borne phase.

2.2 Role of Friction

Del Campo has shown that friction is important in the case of turning over the tippe top [7]. The simple proof is given below [7].

Consider Fig. 2.3a where the tippe top is at rest with its spherical surface in contact with the surface. When the toy is given a twist it begins in spinning motion from the state depicted in Fig. 2.3a. The final state of the toy is as in Fig. 2.3b. We calculate the energies of the initial and final states and compare them. In both states the moment of inertia I of the toy is about the vertical z-axis. The initial spin given to the toy is say ω_1 while the spin of the toy when it arrives the position in Fig. 2.3b is say ω_2.

The weight of the toy is W. The CoM is at a height $R - h$ from the surface. Set the gravitational potential energy as 0 when the CoM is at $R - h$. Then,

$$\text{initial angular momentum about } z\text{−axis} = I\omega_1 , \tag{2.1a}$$

$$\text{initial kinetic energy of rotation} = \frac{1}{2} I\omega_1^2 , \tag{2.1b}$$

$$\text{initial gravitational potential energy} = 0 , \tag{2.1c}$$

$$\text{initial total energy } E_{\text{i}} = \frac{1}{2} I\omega_1^2 . \tag{2.1d}$$

Assume that there is no loss of energy due to friction in the state of Fig. 2.3b. In this state the CoM has shifted above from $R - h$ to $R + h + d$. Hence, the distance risen by the CoM is $R + h + d - (R - h) = 2h + d$. The resultant gravitational potential energy is $W(2h + d)$ and

$$\text{final angular momentum about } z\text{−axis} = I\omega_2 , \tag{2.2a}$$

$$\text{final kinetic energy of rotation} = \frac{1}{2} I\omega_2^2 , \tag{2.2b}$$

$$\text{final gravitational potential energy} = W(2h + d) , \tag{2.2c}$$

$$\text{final total energy } E_{\text{f}} = \frac{1}{2} I\omega_2^2 + W(2h + d) . \tag{2.2d}$$

Because there is no loss of energy and no supply of energy externally we set $E_{\text{i}} = E_{\text{f}}$ which gives $\omega_2 < \omega_1$. That is, after turning over, the tippe top continues its spinning but with a reduced angular speed. *What does it imply?* $\omega_2 < \omega_1$ means $I\omega_2 < I\omega_1$. Considering that the total angular momentum is almost completely in the vertical z-direction the prediction is that in flipping from the state in Fig. 2.3a to that in Fig. 2.3b the toy has lost certain angular momentum. *What is the source for this loss?* An external torque, a twisting force causing rotation, could be responsible for this. *Which is the associated force?* We consider the external forces such as gravity, normal force from contact with the surface and friction. The first two of them act in the vertical z-direction and they could not give rise the torque required to change the angular momentum in the vertical z-direction. Therefore, we can reason that the responsible torque could be exerted only by friction. Hence, a frictional force acting at the point of contact of the toy on the surface is crucial in the explanation of the peculiar inversion behaviour of the toy.

Now, *what is the source of friction?* In the tippe top the CoM and the CoC of the spherical body of it are not the same. The axis of rotation passes through the CoM. As a result, the toy will slide over the surface setting the source of friction.

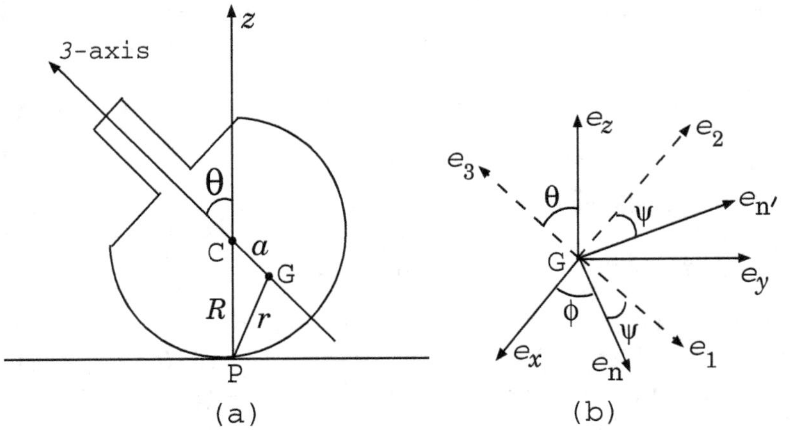

Fig. 2.4: (a) A tippe top and (b) the coordinate axes. C is the CoC, G is the CoM, R is the radius of the spherical bottom, a is the distance from C to G, **r** is the position vector of the contact point from G and 3-axis is the symmetry axis.

2.3 A Simple Model for Turn Over of a Tippe Top

In this section, we develop a simplified theoretical model analysis for accouning the turn over of a tippe top [10].

2.3.1 *Forces and the Coordinate Axes*

What are the forces acting on the tippe top? Neglecting the air drag, as mentioned in the previous section, the forces we can think of are:

1. The gravitational force mg acting through the point G in the downward direction.
2. The normal supportive force $\mathbf{F}_s$ acting upward at the contact point P.
3. The friction force $\mathbf{F}_f$ acting horizontally at the contact point P.

The normal force is actually a constraint force and its magnitude is not known, however, is determined by the motion of the tippe top [10].

Figure 2.4 depicts the coordinate axes of a tippe top. θ in Fig. 2.4b is the angle between the axes $\mathbf{e}_3$ and $\mathbf{e_z}$ [10]. The angle $\theta = 0$ when the top is at rest with stem vertically upward. As the top spins, initially $\theta = 0$ and it increases with time. If it spin fast enough, then at an instant of time, the stem and the rounded bottom part both are in touch with the surface. Till this stage the angle θ increases monotonically from 0. When the top turns over with the stem vertically downward, then the angle θ also increases further. Let us proceed to identify the physical mechanism for the monotonic increase of θ leading to the peculiar behaviour of the tippe top.

The unit vectors $\mathbf{e}_x$, $\mathbf{e}_y$ and $\mathbf{e}_z$ are fixed in the laboratory frame. Introduce the new coordinate system $(\mathbf{e}_{\rm n}, \mathbf{e}_{\rm n'}, \mathbf{e}_3)$ with

$$\mathbf{e}_{\rm n} = \frac{\mathbf{e}_z \times \mathbf{e}_3}{|\mathbf{e}_z \times \mathbf{e}_3|}, \tag{2.3a}$$

$$\mathbf{e}_{\rm n'} = \mathbf{e}_3 \times \mathbf{e}_{\rm n}\,. \tag{2.3b}$$

$\mathbf{e}_{\rm n}$ and $\mathbf{e}_{\rm n'}$ are fixed in the plane where $\mathbf{e}_1$ and $\mathbf{e}_2$ rotate.

Solved Problem 1:

Express $\mathbf{e}_{\rm n}$, $\mathbf{e}_3$ and $\mathbf{e}_{\rm n'}$ in terms of Euler angles.

We have the Euler matrices from Chapter 1 of part I. For ϕ rotation the transformation matrix from Eq. (1.19) in Chapter 1 of Part I is given by

$$T_\phi = \begin{bmatrix} \cos\phi & \sin\phi & 0 \\ -\sin\phi & \cos\phi & 0 \\ 0 & 0 & 1 \end{bmatrix}. \tag{2.4}$$

For the θ rotation the transformation matrix is (refer Eq. (1.21) in Chapter 1 of Part I)

$$T_\theta = \begin{bmatrix} 1 & 0 & 0 \\ 0 & \cos\theta & \sin\theta \\ 0 & -\sin\theta & \cos\theta \end{bmatrix}. \tag{2.5}$$

The transformation matrix from $\begin{bmatrix} \mathbf{e}_n \\ \mathbf{e}_{n'} \\ \mathbf{e}_3 \end{bmatrix}$ to $\begin{bmatrix} \mathbf{e}_x \\ \mathbf{e}_y \\ \mathbf{e}_z \end{bmatrix}$ is given by

$$\begin{bmatrix} \mathbf{e}_n \\ \mathbf{e}_{n'} \\ \mathbf{e}_3 \end{bmatrix} = \begin{bmatrix} T_\theta \end{bmatrix} \begin{bmatrix} T_\phi \end{bmatrix} \begin{bmatrix} \mathbf{e}_x \\ \mathbf{e}_y \\ \mathbf{e}_z \end{bmatrix}. \tag{2.6}$$

The result is

$$\begin{aligned} \begin{bmatrix} \mathbf{e}_n \\ \mathbf{e}_{n'} \\ \mathbf{e}_3 \end{bmatrix} &= \begin{bmatrix} 1 & 0 & 0 \\ 0 & \cos\theta & \sin\theta \\ 0 & -\sin\theta & \cos\theta \end{bmatrix} \begin{bmatrix} \cos\phi & \sin\phi & 0 \\ -\sin\phi & \cos\phi & 0 \\ 0 & 0 & 1 \end{bmatrix} \begin{bmatrix} \mathbf{e}_x \\ \mathbf{e}_y \\ \mathbf{e}_z \end{bmatrix} \\ &= \begin{bmatrix} \cos\phi & \sin\phi & 0 \\ -\cos\theta\sin\phi & \cos\theta\cos\phi & \sin\theta \\ \sin\theta\sin\phi & -\sin\theta\cos\phi & \cos\theta \end{bmatrix} \cdot \begin{bmatrix} \mathbf{e}_x \\ \mathbf{e}_y \\ \mathbf{e}_z \end{bmatrix}. \end{aligned} \tag{2.7}$$

That is,

$$\mathbf{e}_{\mathrm{n}} = \cos\phi\,\mathbf{e}_x + \sin\phi\,\mathbf{e}_y\,, \tag{2.8a}$$

$$\mathbf{e}_3 = \sin\theta\sin\phi\,\mathbf{e}_x - \sin\theta\cos\phi\,\mathbf{e}_y + \cos\theta\,\mathbf{e}_z\,, \tag{2.8b}$$

$$\mathbf{e}_{\mathrm{n}'} = \mathbf{e}_3 \times \mathbf{e}_{\mathrm{n}} = -\cos\theta\sin\phi\,\mathbf{e}_x + \cos\theta\cos\phi\,\mathbf{e}_y + \sin\theta\,\mathbf{e}_z\,. \tag{2.8c}$$

Multiplying Eq. (2.8b) by $\cos\theta$, Eq. (2.8c) by $\sin\theta$ and adding these two give

$$\mathbf{e}_z = \sin\theta\,\mathbf{e}_{\mathrm{n}'} + \cos\theta\,\mathbf{e}_3\,. \tag{2.8d}$$

The angular velocity ($\boldsymbol{\alpha}$) of the coordinate system $(\mathbf{e}_{\mathrm{n}}, \mathbf{e}_{\mathrm{n}'}, \mathbf{e}_3)$ is expressed as

$$\begin{aligned}\boldsymbol{\alpha} &= \dot{\theta}\,\mathbf{e}_{\mathrm{n}} + \dot{\phi}\,\mathbf{e}_z \\ &= \dot{\theta}\,\mathbf{e}_{\mathrm{n}} + \dot{\phi}\sin\theta\,\mathbf{e}_{\mathrm{n}'} + \dot{\phi}\cos\theta\,\mathbf{e}_3 \\ &= \alpha_{\mathrm{n}}\,\mathbf{e}_{\mathrm{n}} + \alpha_{\mathrm{n}'}\,\mathbf{e}_{\mathrm{n}'} + \alpha_3\,\mathbf{e}_3\,,\end{aligned} \tag{2.9a}$$

where we have used

$$\alpha_{\mathrm{n}} = \dot{\theta}\,, \quad \alpha_{\mathrm{n}'} = \dot{\phi}\sin\theta\,, \quad \alpha_3 = \dot{\phi}\cos\theta\,. \tag{2.9b}$$

2.3.2 *Frictional Torque*

Because the tippe top spun rapidly, $\dot{\phi} \gg \dot{\theta}$. It precesses rapidly as it falls over slowly. This implies that the angular velocity $\boldsymbol{\alpha}$ points very close to the positive z-axis. Further, $\dot{\phi} \gg |\dot{\psi}|$. This means that L_z is the dominant component of $\mathbf{L}$. The friction force $\mathbf{F}_{\mathrm{f}}$ acting at the contact point P opposes the sliding motion of the tippe top. If we neglect the translational motion of the top, then $\mathbf{F}_{\mathrm{f}}$ points along the $-\mathbf{e}_{\mathrm{n}}$-axis. This force gives rise a torque $\mathbf{N}_{\mathrm{f}}$ about the point G and is given by

$$\mathbf{N}_{\mathrm{f}} = \mathbf{r} \times \mathbf{F}_{\mathrm{f}}\,. \tag{2.10}$$

From Fig. 2.4 we note that $R\mathbf{e}_z = a\mathbf{e}_3 - \mathbf{r}$. Since $\mathbf{F}_{\mathrm{f}}$ points in the $-\mathbf{e}_{\mathrm{n}}$ direction we write

$$\mathbf{F}_{\mathrm{f}} = -|\mathbf{F}_{\mathrm{f}}|\mathbf{e}_{\mathrm{n}}. \tag{2.11}$$

Equation (2.10) then becomes

$$\begin{aligned}\mathbf{N}_{\mathrm{f}} &= (a\,\mathbf{e}_3 - R\mathbf{e}_z) \times (-|\mathbf{F}_{\mathrm{f}}|\mathbf{e}_{\mathrm{n}}) \\ &= |\mathbf{F}_{\mathrm{f}}|(-a\mathbf{e}_{\mathrm{n}'} + R\mathbf{e}_z \times \mathbf{e}_{\mathrm{n}}) \\ &= |\mathbf{F}_{\mathrm{f}}|R\left[\left(\cos\theta - \frac{a}{R}\right)\mathbf{e}_{\mathrm{n}'} - \sin\theta\,\mathbf{e}_3\right].\end{aligned} \tag{2.12}$$

This torque has components about the axis $\mathbf{e}_{\mathrm{n}'}$ and $\mathbf{e}_3$. *Does there exist any other torque with components about* $\mathbf{e}_{\mathrm{n}'}$ *and* $\mathbf{e}_3$*?* Since the CoM, G, is the origin of the coordinate system $(\mathbf{e}_{\mathrm{n}}, \mathbf{e}_{\mathrm{n}'}, \mathbf{e}_3)$ gravity is not exerting a torque. On the other hand, the torque due to the normal force acts about the $\mathbf{e}_{\mathrm{n}}$-axis. Only the frictional torque $\mathbf{N}_{\mathrm{f}}$ has the components $\mathbf{e}_{\mathrm{n}'}$ and $\mathbf{e}_3$.

2.3.3 *Significance of Frictional Torque*

In the beginning of the motion of the tippe top $\theta = 0$. From the component $\mathbf{e}_{\mathrm{n}'}$ in Eq. (2.12) we see that if $\theta < \cos^{-1}(a/R)$ (with $a/R < 1$), then the torque about $\mathbf{e}_{\mathrm{n}'}$ is > 0 (thus $\alpha_{\mathrm{n}'}$ increases) and that about $\mathbf{e}_3$ is < 0 (thus $\mathbf{e}_3$ decreases). From Eq. (2.9b), $\alpha_3/\alpha_{\mathrm{n}'} = \cot\theta$. Because α_3 is decreasing while $\alpha_{\mathrm{n}'}$ is increasing the ratio $\alpha_3/\alpha_{\mathrm{n}'} = \cot\theta$ decreases monotonically with time. This implies that θ increases with time. When $\theta = \cos^{-1}(a/R)$ we find $\mathbf{N}_{\mathrm{f}} = -R|\mathbf{F}_{\mathrm{f}}|\sin\theta\mathbf{e}_3$. The torque about $\mathbf{e}_3$ is $-\sin\theta$ again negative. For $\theta = \pi/2$

$$\mathbf{N}_{\mathrm{f}} = |\mathbf{F}_{\mathrm{f}}|R\left(-\frac{a}{R}\mathbf{e}_{\mathrm{n}'} - \mathbf{e}_3\right) \tag{2.13}$$

and the torques about $\mathbf{e}_{\mathrm{n}'}$ and $\mathbf{e}_3$ both are negative. That is, for $\theta \leq \cos^{-1}(a/R) \leq \pi/2$ the torques about $\mathbf{e}_{\mathrm{n}'}$ and $\mathbf{e}_3$ both are negative. *What about the magnitudes of these two torques?* As $\sin\theta$ increases and $\cos\theta$ decreases with θ for $0 \leq \theta \leq \pi/2$ we note that the magnitude of the torque about $\mathbf{e}_3$ is larger than that about $\mathbf{e}_{\mathrm{n}'}$. α_3 should decrease more rapidly about $\alpha_{\mathrm{n}'}$ leading to $\alpha_3/\alpha_{\mathrm{n}'} = \cot\theta$ to decrease continuously. Therefore, θ *increases continuously.*

Next, consider the interval $\pi/2 < \theta \leq \pi$. For $\theta = \pi$

$$\mathbf{N}_{\mathrm{f}} = |\mathbf{F}_{\mathrm{f}}|R\left[\left(-1 - \frac{a}{R}\right)\mathbf{e}_{\mathrm{n}'} - 0\,\mathbf{e}_3\right]. \tag{2.14}$$

For $\pi/2 < \theta < \pi$ both the torques are negative, so α_3 and $\alpha_{\mathrm{n}'}$ further decrease with time. As $\alpha_3 = \cos\theta\dot{\phi}$ and $\alpha_{\mathrm{n}'} = \sin\theta\dot{\phi}$, the requirement for the angular velocities to decrease further is to increase θ. The increase in θ finally leads to the stem to touch the surface. Once the stem touches the surface the tippe top is like an ordinary rising top. The friction torque at the stem tends to align with the angular momentum with the vertical z-axis, thereby *pushing upright the top.*

2.4 Conclusion

Everybody will be fascinated the first time they see the tippe top turning upside down and reversing its direction of spin. In the case of a top with sharp bottom, during spinning, the bottom end traces out a circular path of small radius and CoM precesses about a vertical axis and this axis passes through the sharp bottom end. In the case of an elliptical top (PhiTOP) the bottom end can roll approximately circular path with relatively large radius and the top and its CoM precess around a same vertical axis which is located outside the top. A tippe top begins spinning with its stem upward, precesses fastly about a vertical axis, the entire toy slowly rotates about a horizontal axis and ends up spinning upright on the stem.

The tippe top is an wonderful toy to demonstrate to the students the role played by frictional torque in turning over the top. They can see that this top will turn only on a frictional surface and will not turn over in a frictionless surface. As the motion of the tippe top is a combination of gliding, rolling and spinning motion, friction will arise due to each type of motion. We have seen that the CoM rises when the top turns over. This increase of potential energy must come from the decrease of its rotational energy. Its angular velocity and angular momentum decrease when the tippe top turns over. The torque acting on the top due to friction has decreased the angular momentum.

Students can be asked to study the motion of the tippe top on surfaces with different coefficients of friction. They can also be asked to record the time taken by the tippe top to flip over and to relate that to the amount of friction on the surface. They can easily observe that the turn over will be faster if the friction is large. As $\alpha_3 = \dot{\phi}\cos\theta$, the rotation of the $\mathbf{e}_3$ axis (about the stem) keeps on decreasing as θ is increased. At the horizontal position $\theta = \pi/2$ and α_3 becomes zero. As θ increases further, α_3 becomes negative and hence the rotation about the stem changes direction. This change of direction of the spin when the top is nearly horizontal can be demonstrated to the students when the top is spun over a surface covered with either carbon paper or sooted glass. This would trace out a curve on the surface of the top and show how the top changed the direction of rotation when the stem was nearly horizontal.

2.5 Bibliography

[1] J. Perry, *Spinning Tops and Gyroscopic Motion* (Dover, New York, 1957).

[2] https://www.fysikbasen.dk/English.php%3Fpage=Vis&id=79.html.

[3] C.M. Braams, Physica **18**, 503 (1952).

[4] N.M. Hugenholtz, Physica **18**, 515 (1952).

[5] J.L. Synge, Philos. Mag. **43**, 724 (1952).

[6] W.A.Pliskin, Am. J. Phys. **22**, 28 (1954).

[7] A.R. Del Campo, Am. J. Phys. **23**, 544 (1955).

[8] D.G. Parkyn, Physica **24**, 313 (1958).

[9] J.B. Hart, Am. J. Phys. **27**, 189 (1959).
[10] R.J. Cohen, Am. J. Phys. **45**, 12 (1974).
[11] D. Featonby, Phys. Edu. **49**, 11 (2014).
[12] F.F. Johnson, Am. J. Phys. **28**, 406 (1960).
[13] K.W. Ford, The Phys. Teach. **16**, 322 (1978).
[14] A.C. Or, SIAM J. Appl. Math. **54**, 597 (1994).
[15] S. Ebenfeld and F. Scheck, Ann. Phys. **243**, 195 (1995).
[16] C.G. Gray and B.G. Nickel, Am. J. Phys. **68**, 821 (2000).
[17] H. Soodak, Am. J. Phys. **70**, 815 (2002).
[18] N. Bou-Rabee, J.E. Marsden and L.A. Romero, SIAM J. Appl. Dyn. Syst. **3**, 352 (2004).
[19] T. Ueda, K. Sasaki and S. Watanabe, SIAM J. Appl. Dyn. Syst. **4**, 1159 (2005).
[20] M.C. Ciocci and B. Langerock, Regular and Chaotic Dynamics **12**, 602 (2007).
[21] N. Bou-Rabee, J.E. Marsden and L.A. Romero, SIAM Review **50**, 325 (2008).
[22] R. Cross, Am. J. Phys. **81**, 280 (2013).

2.6 Exercises

2.1 A tippe top with a hollow stem and a hollow truncated spherical bottom is depicted in Fig. 2.5. Determine its CoM and represent it on the vertical axis.

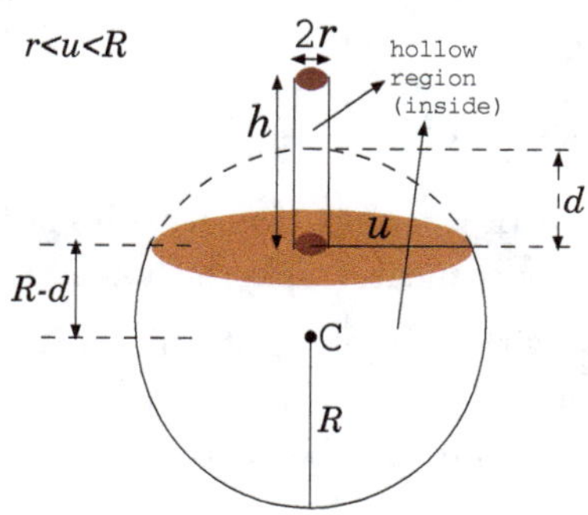

Fig. 2.5: A tippe top with a hollow cylindrical stem and a hollow truncated spherical bottom.

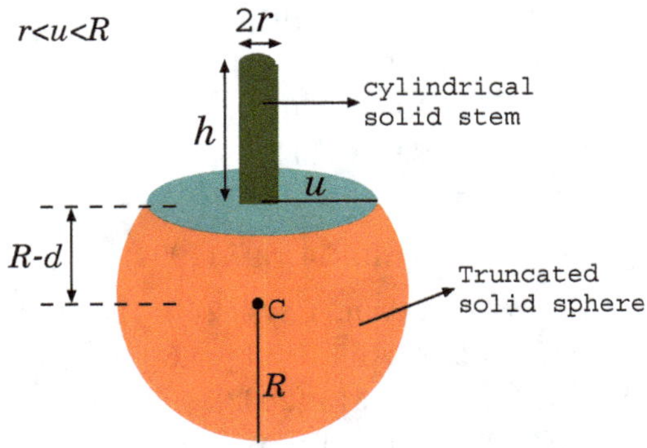

Fig. 2.6: A solid tippe top.

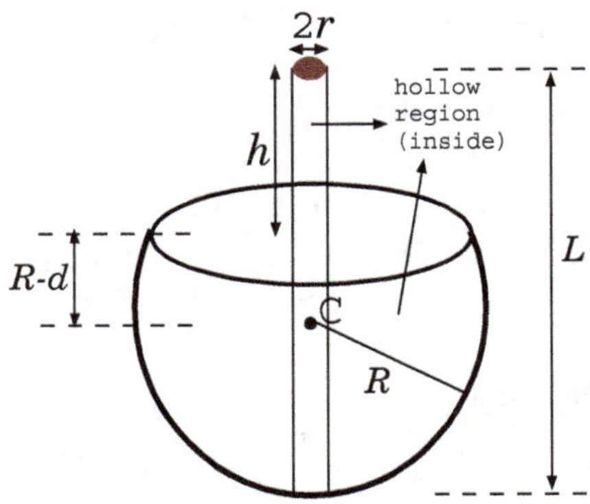

Fig. 2.7: A tippe top with its stem fitted at the inner bottom of the hollow truncated sphere.

2.2 Determine the CoM of the solid tippe top shown in Fig. 2.6.

2.3 Determine the CoM of the tippe top with its stem fitted at the inner bottom of the hollow truncated sphere shown in Fig. 2.7.

2.4 Take a finger ring with a stone on it. Twirl it on the table with the stone down. The ring will turn over and spin with the stone on top. Explain the mechanism of this behaviour.

2.5 Write an expression for the velocity of the contact point P with respect to the CoM of the top in the coordinate system $(\mathbf{e}_{\rm n}, \mathbf{e}_{\rm n'}, \mathbf{e}_3)$. Also, find its projection on the xy plane.

Chapter 3

PhiTOP

Place a hard boiled egg (without removing its shell) on a flat and smooth surface. Keep your thumb of one hand on one end of the long axis of the egg and the index finger on the other end of the egg. Spin it rapidly. You will observe a curious phenomenon. It starts spinning in a horizontal position, rise rapidly spinning on one end at a different spin rate and then gradually settle down to the stationary horizontal position. Usually an object spin about its center of mass (CoM). In the case of an egg or an egg shaped object, its CoM and the point of contact with the surface about which it begins to spin are not aligned. The egg wobbles due to the difference between its CoM and the mid-point. This in turn causes the egg to start rise. A raw egg when it spins will not rise. *Why?*

Experiments with objects with shape similar to an egg have shown that those with length to width ratio about $1.5 - 1.7$ could be able to display the above mentioned curious motion. Objects with the ratio less than 1.4 would display erratic spin while those with the ratio greater than 2 would not stand up and spin steadily [1]. A spheroid or a prolate ellipsoid, elliptical in one cross-section while circular in another with the length to width ratio ≈ 1.618 is referred as *PhiTOP* and it behaves similar to a spinning hard-boiled egg. Figure 3.1 shows a PhiTOP spinning in the upright state on a mirror base [2]. The ratio ≈ 1.618 is called *golden mean* labeled by the Greek letter ϕ and is the positive solution of the equation $1/\phi = \phi - 1$. In a variety of topics in mathematics and physics the golden mean is found to be important. It is also realized in a variety of architectural monuments and art works. Ellipsoids with the length to width ratio lies in the range 1.5–1.7 are called *golden ellipsoids*.

There are certain design considerations of PhiTOPs. The size of a PhiTOP to be such that it would be easy to spin using our fingers. A brass

Fig. 3.1: A PhiTOP with the upright spinning state [2]. (Reproduced from The Phys. Teach. **57**, 74 (2019), with permission of American Association of Physics Teachers. Copyright 2019 K. Brecher and R. Cross, licensed under a Creative Commons Attribution License.)

or aluminum or stainless steel version with about 5 cm length is convenient. PhiTOPs with much larger or smaller sizes are difficult to spin. A typical PhiTOP, started spinning horizontally will stand upright in a few seconds, spin almost twice the speed with which it was spinning horizontally and then due to friction settle down. When a PhiTOP or a hard-boiled egg spins in the upright position, then it is said to be in a *sleeping position*. In this case, its long axis is vertical to the horizontal surface. For a better performance, a PhiTOP is usually set to spin on a convex round shaped mirror platform. When you play with a PhiTOP in a room with a light bulb switched-on, you will see a beautiful variety of painted light circuits [3]. There is an another interesting physical characteristic of it. When a strong magnet is brought near to an upright spinning PhiTOP made of magnetic material it will quickly stop spinning. The PhiTOP can be used for exploration of force, gravity, friction, mass, density and aspects of electromagnetism. It is also used as an aid to relaxation, meditation and stress relieving.

The analysis of spinning objects and rise of their CoM started in the late 19th century. Such analysis are found in the notes of William Thomson written in 1844 and in the book of *Theory of Friction* written in 1872 by J.H. Jellett [4]. Theoretical models have been developed to describe the behaviour of spinning eggs [5-9]. Simple explanation is provided for rising of a PhiTOP [2]. Experiments have been conducted to understand the physics behind the spinning egg as well as to verify the theoretical predictions [2,8,10].

In this chapter, first we describe the geometry and notation used for the analysis of the spinning of a PhiTOP. We setup the equations of motion for

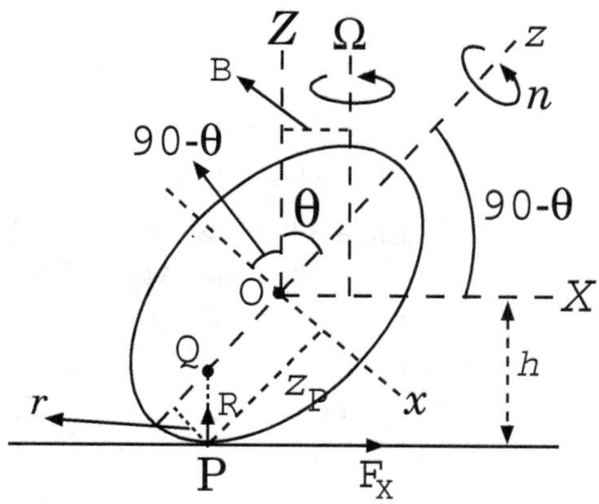

Fig. 3.2: Geometry of a spinning PhiTOP [8]. (Reproduced with permission from Institute of Physics Publishing Ltd.: R. Cross, Eur. J. Phys. **39**, 025002 (2018).)

the components of angular momentum and the angle of rotation. Next, we obtain an expression for the frequency of rotation in the steady precession state. Finally, we explain the rising of a PhiTOP and the spinning of an egg.

3.1 Geometry and Notation of the Spinning Object

Consider a PhiTOP of revolution with a smooth floor or table with P being an instantaneous point of contact as shown in Fig. 3.2. Denote the CoM as O and Oz as the axis of symmetry. Oxyz and OXYZ are the body frame of reference and the rotating frame of reference, respectively. The y and Y axes are identical. The Euler angles of the PhiTOP relative to OZ are θ, ϕ and ψ. The principal moments of inertia of the PhiTOP about the axes through O are $\mathbf{I}_{\mathrm{m}} = (I_1, I_1, I_3)$.

The plane containing OZ and OP processes with the angular velocity Ω about OZ and n is the spin about the axis of symmetry Oz. The position vector of the PhiTOP relative to O is

$$\mathbf{x} = x\mathbf{i} + y\mathbf{j} + z\mathbf{k} = X\mathbf{I} + Y\mathbf{J} + Z\mathbf{K} \tag{3.1}$$

with

$$x = X\cos\theta - Z\sin\theta, \quad y = Y, \quad z = X\sin\theta + Z\cos\theta\,. \tag{3.2}$$

Here $(\mathbf{i}, \mathbf{j}, \mathbf{k})$ and $(\mathbf{I}, \mathbf{J}, \mathbf{K})$ are the unit vectors and

$$\mathbf{i} = \cos\theta\mathbf{I} - \sin\theta\mathbf{K}, \quad \mathbf{j} = \mathbf{J}, \quad \mathbf{k} = \sin\theta\mathbf{I} + \cos\theta\mathbf{K}\,. \tag{3.3}$$

The frame $OXYZ$ rotates with the angular velocity

$$\boldsymbol{\Omega} = \Omega \mathbf{K} \tag{3.4}$$

while the frame $Oxyz$ rotates with the angular velocity

$$\boldsymbol{\Omega}' = \boldsymbol{\Omega} + I_1 \mathbf{j}\,. \tag{3.5}$$

The PhiTOP rotates with the angular velocity $\boldsymbol{\omega}$ in the xyz coordinate system. Referring to Fig. 3.2, we find that $\boldsymbol{\Omega}$ is along the Z-direction. As $z - x$ are in the same plane and θ is the angle between z and $\boldsymbol{\Omega}$, the $\boldsymbol{\Omega}$ component along z-direction is $\Omega \cos\theta$ and the $\boldsymbol{\Omega}$ component along x-direction is $-\Omega \sin\theta$. If $\dot{\psi}$ is the spin given in the z-direction, then $n = \Omega \cos\theta + \dot{\psi} = \Omega \cos\theta + \omega$. So,

$$\boldsymbol{\omega} = -\Omega \sin\theta \mathbf{i} + \dot{\theta}\mathbf{j} + n\mathbf{k}\,. \tag{3.6}$$

Next, we find $\boldsymbol{\omega}$ in the XYZ coordinate system. Substituting (3.3) in (3.6) we get

$$\begin{aligned} \boldsymbol{\omega} &= -\Omega \sin\theta(\cos\theta \mathbf{I} - \sin\theta \mathbf{K}) + \dot{\theta}\mathbf{J} + n(\sin\theta \mathbf{I} + \cos\theta \mathbf{K}) \\ &= (n - \Omega\cos\theta)\sin\theta \mathbf{I} + \dot{\theta}\mathbf{J} + (n\cos\theta + \Omega \sin^2\theta)\mathbf{K} \\ &= \omega \sin\theta \mathbf{I} + \dot{\theta}\mathbf{J} + (\Omega\cos^2\theta + \omega\cos\theta + \Omega\sin^2\theta)\mathbf{K} \\ &= \omega \sin\theta \mathbf{I} + \dot{\theta}\mathbf{J} + n'\mathbf{K}, \quad n' = \Omega + \omega\cos\theta\,. \end{aligned} \tag{3.7}$$

The height of O from the table is, say, $h(\theta)$. The components of $\mathbf{X}_\mathrm{P}$ from O to P are written as

$$\mathbf{X}_\mathrm{P} = (-X_\mathrm{P}, Y_\mathrm{P}, Z_\mathrm{P}) = \left(\frac{\mathrm{d}h}{\mathrm{d}\theta}, 0, -h\right)\,. \tag{3.8}$$

As the centre of origin O is at rest, its velocity $\mathbf{U} = (U, V, W) = (0, 0, 0)$. So, we get

$$\mathbf{U}_\mathrm{P} = \mathbf{U} + \boldsymbol{\omega} \times \mathbf{X}_\mathrm{P} = (U_\mathrm{P}, V_\mathrm{P}, W_\mathrm{P})\,. \tag{3.9}$$

As $\dot{\theta} \approx 0$, substituting Eqs. (3.7) and (3.8) in (3.9) we obtain

$$\begin{aligned} \mathbf{V}_\mathrm{P} &= \boldsymbol{\omega} \times \mathbf{X}_\mathrm{P} \\ &= \left[(n - \Omega\cos\theta)\sin\theta \mathbf{I} + \left(n\cos\theta + \Omega\sin^2\theta\right)\mathbf{K}\right] \times \left(\frac{\mathrm{d}h}{\mathrm{d}\theta}\mathbf{I} - h\mathbf{K}\right) \\ &= h\left[(n - \Omega\cos\theta)\sin\theta + \left(n\cos\theta + \Omega\sin^2\theta\right)\frac{\mathrm{d}h}{\mathrm{d}\theta}\right]\mathbf{J}\,. \end{aligned} \tag{3.10}$$

Then, the components of $\mathbf{V}_\mathrm{P}$ are

$$U_\mathrm{P} = 0, \tag{3.11a}$$

$$V_\mathrm{P} = h\,(n - \Omega\cos\theta)\sin\theta + \left(n\cos\theta + \Omega\sin^2\theta\right)\frac{\mathrm{d}h}{\mathrm{d}\theta}, \tag{3.11b}$$

$$W_\mathrm{P} = 0. \tag{3.11c}$$

3.2 Dynamical Equations

The forces acting on the PhiTOP are:

1. Its weight $-mg\mathbf{K}$ at O where m and g are the mass of the PhiTOP and the acceleration due to gravity, respectively.
2. The normal reaction $\mathbf{R} = R\mathbf{K} = (0, 0, R)$ at P where R is closely equal to the weight mg.
3. The frictional force $\mathbf{F} = F_X\mathbf{I} + F_Y\mathbf{J} = (F_X, F_Y, 0)$ at P.

The torque about O due to $\mathbf{R}$ is mgR.

Next, we find the angular momentum $\mathbf{L}$ of the PhiTOP. In the xyz coordinate system

$$\begin{aligned}\mathbf{L} &= \mathbf{I}_{\mathrm{m}} \cdot \boldsymbol{\omega} \\ &= (I_1\mathbf{i} + I_1\mathbf{j} + I_3\mathbf{k}) \cdot (-\Omega \sin\theta\mathbf{i} + \dot{\theta}\mathbf{j} + n\mathbf{k}) \\ &= -I_1\Omega\sin\theta\mathbf{i} + I_1\dot{\theta}\mathbf{j} + I_3 n\mathbf{k}\,. \end{aligned} \tag{3.12}$$

To find $\mathbf{L}$ in the XYZ system we substitute (3.3) in (3.12). We get

$$\begin{aligned}\mathbf{L} &= -I_1\Omega\sin\theta\,(\cos\theta\mathbf{I} - \sin\theta\mathbf{K}) + I_1\dot{\theta}\mathbf{J} + I_3 n(\sin\theta\mathbf{I} + \cos\theta\mathbf{K}) \\ &= \sin\theta(I_3 n - I_1\Omega\cos\theta)\mathbf{I} + I_1\dot{\theta}\mathbf{J} + (I_1\Omega\sin^2\theta + I_3 n\cos\theta)\mathbf{K} \\ &= L_X\mathbf{I} + L_Y\mathbf{J} + L_Z\mathbf{K}, \end{aligned} \tag{3.13a}$$

where

$$L_X = \sin\theta(I_3 n - I_1\Omega\cos\theta)\,, \tag{3.13b}$$

$$L_Y = I_1\dot{\theta}\,, \tag{3.13c}$$

$$L_Z = (I_1\Omega\sin^2\theta + I_3 n\cos\theta)\,. \tag{3.13d}$$

The Euler angular momentum equation in the XYZ frame is

$$\frac{\partial\mathbf{L}}{\partial t} + \boldsymbol{\Omega} \times \mathbf{L} = \mathbf{X}_{\mathrm{P}} \times (\mathbf{R} + \mathbf{F})\,. \tag{3.14}$$

In the above $\mathbf{X}_{\mathrm{P}} \times (\mathbf{R} + \mathbf{F})$ is the torque acting on the PhiTOP relative to O. Substitution for $\mathbf{L}$, $\boldsymbol{\Omega}$, $\mathbf{X}_{\mathrm{P}}$, $\mathbf{R}$ and $\mathbf{F}$ in the above equation gives

$$\begin{aligned}&\frac{\partial}{\partial t}(L_X\mathbf{I} + L_Y\mathbf{J} + L_Z\mathbf{K}) \\ &\quad = -\Omega\mathbf{K} \times (L_X\mathbf{I} + L_Y\mathbf{J} + L_Z\mathbf{K}) \\ &\qquad\quad + (-X_P\mathbf{I} - h\mathbf{K}) \times (F_X\mathbf{I} + F_Y\mathbf{J} + R\mathbf{K}) \\ &\quad = (\Omega L_Y + hF_Y)\,\mathbf{I} + (-\Omega L_X + RX_P - hF_X)\,\mathbf{J} - X_P F_Y\mathbf{K}\,. \end{aligned} \tag{3.15}$$

In the components form, Eq. (3.14) takes the form

$$\frac{\partial}{\partial t}L_X = \Omega L_Y + hF_Y\,, \tag{3.16a}$$

$$\frac{\partial}{\partial t}L_Y = -\Omega L_X + RX_{\mathrm{P}} - hF_X\,, \tag{3.16b}$$

$$\frac{\partial}{\partial t}L_Z = -X_{\mathrm{P}}F_Y\,. \tag{3.16c}$$

From Eqs. (3.16), we observe that there are three torques act on the PhiTOP. ΩL_X and ΩL_Y represent the change in the angular momentum components as a result of precession about the vertical axis.

The total energy of the PhiTOP is given by

$$E = \frac{1}{2}\mathbf{U}\cdot\mathbf{U} + \frac{1}{2}\boldsymbol{\omega}\cdot\mathbf{L} + h(\theta)\,. \tag{3.17}$$

3.3 Equation of Motion for θ

The equation of motion for the evolution of θ is easy to obtain from Eqs. (3.16) [6]. First, consider Eq. (3.16b). Use of Eqs. (3.13) in Eq. (3.16b) gives

$$I_1\ddot{\theta} + \Omega\,(I_3 n - I_1\Omega\cos\theta)\sin\theta = RX_{\mathrm{P}} - hF_X\,. \tag{3.18}$$

It is reasonable to assume that $|\ddot{\theta}| \ll \Omega^2$ and the terms in the right-side of the above equation are negligible. Then, Eq. (3.18) gives

$$\Omega(I_3 n - I_1\Omega\cos\theta)\sin\theta = 0. \tag{3.19}$$

The choice $\sin\theta \neq 0$ leads to

$$I_3 n = I_1\Omega\cos\theta\,. \tag{3.20}$$

This condition is termed as *gyroscopic balance* [5]. In this case from Eq. (3.13b) we observe that $L_X = 0$. Then, $\mathbf{L}$ in the XYZ system is (refer Eq. (3.13a))

$$\begin{aligned}\mathbf{L} &= I_1\dot{\theta}\mathbf{J} + \left(I_1\Omega\sin^2\theta + I_3 n\cos\theta\right)\mathbf{K}\\ &= I_1\dot{\theta}\mathbf{J} + \left(I_1\Omega\sin^2\theta + I_1\Omega\cos^2\theta\right)\mathbf{K}\\ &= I_1\dot{\theta}\mathbf{J} + I_1\Omega\mathbf{K}\,.\end{aligned} \tag{3.21}$$

From Eq. (3.21), $L_X = 0$, $L_Y = I_1\dot{\theta}$, $L_Z = I_1\Omega$. Then, with $I_3 n = I_1\Omega\cos\theta$ the Eqs. (3.16a) and (3.16c) give

$$I_1\Omega\dot{\theta} = -hF_Y = Z_{\mathrm{P}}F_Y\,, \quad I_1\dot{\Omega} = -X_{\mathrm{P}}F_Y\,. \tag{3.22}$$

Eliminating F_Y in the first equation in (3.22) and substituting $X_\mathrm{P} = -\mathrm{d}h/\mathrm{d}\theta$, $Z_\mathrm{P} = -h$ lead to the equation

$$\Omega\dot{\theta} = \frac{h\dot{\Omega}}{-\mathrm{d}h/\mathrm{d}\theta}\,. \tag{3.23}$$

That is, $\dot{\Omega}/\Omega = -\dot{h}/h$. Integration of this equation gives $\ln\Omega + \ln h = \ln a$. That is,

$$\Omega h = \text{a constant}\,. \tag{3.24}$$

Further, with $I_3 n = I_1\Omega\cos\theta$

$$\mathbf{L}\cdot\mathbf{X}_\mathrm{P} = \left(I_1\dot{\theta}\mathbf{J} + I_1\Omega\mathbf{K}\right)\cdot\left(\frac{\mathrm{d}h}{\mathrm{d}\theta}\mathbf{I} - h\mathbf{K}\right) = -I_1\Omega h\,. \tag{3.25}$$

From Eqs. (3.24) and (3.25) we write

$$-\mathbf{L}\cdot\mathbf{X}_\mathrm{P} = I_1\Omega h = J = \text{a constant}\,. \tag{3.26}$$

Next, from the first equation of (3.22)

$$I_1\Omega h\dot{\theta} = -h^2 F_Y \quad \text{or} \quad J\dot{\theta} = -h^2 F_Y\,. \tag{3.27}$$

Assume that F_Y is the Coulomb friction force given by

$$F_Y = -\mu m g\frac{V_\mathrm{P}}{|V_\mathrm{P}|}\,. \tag{3.28}$$

Then, from Eqs. (3.11b) and (3.20) we obtain

$$V_\mathrm{P} = \frac{J}{I_1 h}\left[\left(\sin^2\theta + \frac{I_1}{I_3}\cos^2\theta\right)\frac{\mathrm{d}h}{\mathrm{d}\theta} + \left(\frac{I_1}{I_3} - 1\right)h\sin\theta\cos\theta\right]. \tag{3.29}$$

We can rewrite Eq. (3.27) as

$$\dot{\theta} = \frac{\tau}{|\bar{V}_\mathrm{P}|}\bar{V}_\mathrm{P}\,, \quad \tau = \frac{\mu m g h^2}{I_1}\,, \quad \bar{V}_\mathrm{P} = \frac{I_1}{J}V_\mathrm{P}\,. \tag{3.30}$$

3.4 Steady Precession

Consider that the object is a spheroid with elliptical cross-section. In Fig. 3.2 denote the length PQ $= A$ and QO $= D$. Then, $r = A\sin\theta$, the horizontal displacement of Q from O is $X_\mathrm{P} = R = D\sin\theta$, $h = A + D\cos\theta$, $z_\mathrm{P} = D + A\cos\theta$. For a prolate spheroid with major radius a and minor radius b we have $h = (a^2\cos^2\theta + b^2\sin^2\theta)^{1/2}$ and $X_\mathrm{P} = -\mathrm{d}h/\mathrm{d}\theta = (a^2 - b^2)\sin\theta\cos\theta/h$ [5].

The PhiTOP set in motion, first spins fastly at an angular velocity Ω about the vertical axis. This motion is due to the torque exerted by the thumb of one hand and the index finger of the another hand. When a spinning top rises, the torque about O due to the force R causes the angular momentum to change in the same direction as the torque [10]. This leads to a precession about the vertical axis. Though friction is present, Ω does not decay to zero but decreases to a constant nonzero value when the PhiTOP stops rising. θ remains constant in time. This is the *steady state precession.*

In the steady state precession $\dot{\theta} = 0$ and from Eq. (3.13c) $L_Y = 0$. Equation (3.16b) then becomes

$$-\Omega L_X + RX_{\mathrm{P}} - hF_X = 0\,. \tag{3.31}$$

If the CoM O rotates about the vertical precession axis with radius B, then $F_X = mB\Omega^2$ (centrepetal force). Substituting this value of F_X, $n = \omega + \Omega\cos\theta$, $R = mg$ and L_X from Eq. (3.13a) in Eq. (3.31), we arrive the equation [8]

$$\Omega^2\left[(I_1 - I_3)\sin\theta\cos\theta - hmB\right] - \Omega I_3\omega\sin\theta + mgX_{\mathrm{P}} = 0\,. \tag{3.32}$$

It is easy to observe that out of the two values of Ω one value is much less than ω and this represents a relatively slow precession of the PhiTOP.

What does happen if the object spins rapidly? In this case it tends to precess about an axis which is through O and $B = 0$, $\omega = R\Omega/r = D\Omega/A$. Then, Eq. (3.32) with $X_{\mathrm{P}} = D\sin\theta$ becomes [8]

$$\Omega^2 = \frac{mgD}{(I_3D/A) - (I_1 - I_3)\cos\theta}\,. \tag{3.33}$$

The above treatment is not applicable when $\theta \to 0$. This is because if $\theta \to 0$, then $L_X \to 0$, $L_Y \to 0$ and the Eqs. (3.16a) and (3.16b) are no longer relevant.

3.5 Rising of a PhiTOP

Rod Cross performed experiments to understand the physics of a spinning egg [8,10] and augmented theoretical approximation. The sliding friction force, particularly, F_Y, is found to play a key role in the rise of the PhiTOP or a spinning egg. Equation (3.16a) is helpful to explore the role of F_Y ($= F_y$). When the PhiTOP is rising, in the absence of any nutation, the numerical solution of Eqs. (3.16) indicated that the value of $\partial L_X/\partial t$ is several times smaller than the term $\Omega L_Y = I_1\Omega\dot{\theta}$. Dropping this term in Eq. (3.16a) gives $hF_Y = -\Omega L_Y$. Because F_Y, h and Ω are positive, $L_Y = I_1\dot{\theta}$ is negative. That is, the rate of change of θ is negative, indicating that θ decreases with time, that is, the object rises. Physically, a torque about O is exerted by a horizontal friction force at the bottom end. Because of this the axis of rotation tilt in a vertical direction leading to the rise of the top [10]. Specifically, a spinning PhiTOP or an egg initially slowly rises and then rises more fastly as it approaches the rolling condition. For a raw egg the angular velocity given to the egg shell diffuses into the interior fluid. In this process the initial kinetic energy given to the egg mostly dissipates. The remaining energy will be insufficient to rise the egg [5].

Solved Problem 1:

Consider a PhiTOP rising from a rotation about the minor axis to a rotation about the major axis. Denote ω_a and ω_b as the angular frequencies of rotation about the major axis and the minor axis, respectively. Applying the conservation of energy obtain the relation between the angular frequencies and find the condition on the frequencies for the rising of the PhiTOP.

The energy of the PhiTOP while it is rotating about the minor axis is $E_b = mgb + \frac{1}{2} I_b \omega_b^2$. When it is rotating about its major axis $E_a = mga + \frac{1}{2} I_a \omega_a^2$. Refering to the exercise 2 we have $I_a = \frac{2}{5} mb^2$ and $I_b = \frac{2}{5} ma^2$. Then, equating of E_a with E_b gives

$$mga + \frac{1}{2}\frac{2}{5} mb^2 \omega_a^2 = mgb + \frac{1}{2}\frac{2}{5} ma^2 \omega_b^2 \,. \tag{3.34}$$

We obtain from this equation

$$\omega_a^2 = \frac{g(b-a) + \frac{2}{10} a^2 \omega_b^2}{\frac{2}{10} b^2} = \frac{5g(b-a)}{b^2} + \frac{a^2}{b^2} \omega_b^2 \,. \tag{3.35}$$

In the right-side of the above equation the first term is linear in a, while the second term is quadratic in a. Therefore, $\omega_a^2 > \omega_b^2$. That is,

$$\frac{5g(b-a)}{b^2} + \frac{a^2}{b^2} \omega_b^2 > \omega_b^2 \,. \tag{3.36}$$

From this equation, we find

$$\omega_b^2 > \frac{5g}{a+b} \,. \tag{3.37}$$

3.6 Rising of a Spinning Egg

The shape of an egg appears as a spheroid, however, it has a thin end and a fat end. More over, there is an air chamber at the fat end. Therefore, it is

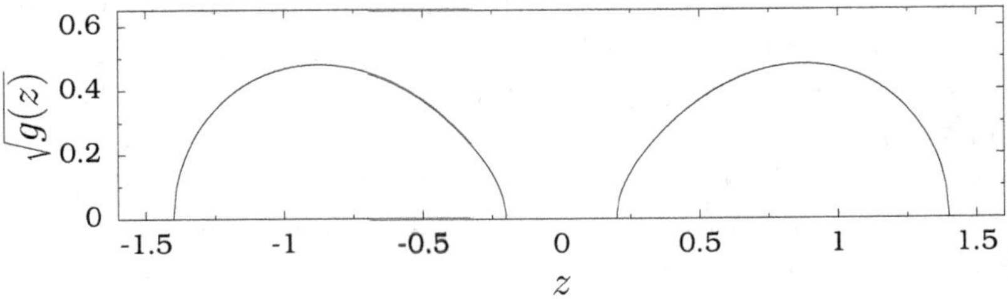

Fig. 3.3: Variation of $\sqrt{g(z)}$ with z in units of a for $\lambda = 0.98$.

natural to ask: *Which end of the spinning egg will rise?* Sasaki theoretically analysed the spinning of egg by considering many oval shape curves with thin and fat ends [6].

Note that the shape of a three-dimensional egg can be reconstructed by means of rotating its two-dimensional cross-section around the axis of symmetry. Suppose the symmetry axis is the z-axis and an axisymmetric body's cross-section is described by $x^2 = g(z)$, with $g(z) > 0$, for $z_{\text{min}} < z < z_{\text{max}}$ and $g(z_{\text{min}}) = g(z_{\text{max}}) = 0$. For the egg shape the Cassini oval with an air chamber is appropriate [6]. In this case

$$g(z) = -\left(z^2 + a^2\right) + a\left(4z^2 + \lambda^4 a^2\right)^{1/2}, \quad \lambda = \frac{b}{a} < 1. \tag{3.38}$$

The roots of $g(z) = 0$ are $z = \pm a\sqrt{1 \pm \lambda^2}$ (see the exercise 3.5 at the end of this chapter). At these four values of z the function $g(z)$ becomes 0. Figure 3.3 show the plot of $\sqrt{g(z)}$ as a function of z in units of a for $\lambda = 0.98$. There are two curves one for $z < 0$ and another for $z > 0$. In both the cases we can clearly notice thin and fat ends. We choose the curve for $z < 0$. In this case $z_{\text{min}} = -a\sqrt{1+\lambda^2}$ and $z_{\text{max}} = -a\sqrt{1-\lambda^2}$. We take $z_{\text{min}} = -\alpha a$ and $\sqrt{1-\lambda^2} < \alpha < \sqrt{1+\lambda^2}$. An empty space exists for $z \in \left[-a\sqrt{1+\lambda^2}, -\alpha a\right]$.

The z-component of the CoM, V, I_1 and I_3 are given by

$$z_{\text{g}} = \frac{\pi}{V}\int_{z_{\text{min}}}^{z_{\text{max}}} z g(z)\,\mathrm{d}z\,, \quad V = \pi\int_{z_{\text{min}}}^{z_{\text{max}}} g(z)\,\mathrm{d}z \tag{3.39}$$

and

$$I_1 = \frac{m\pi}{V}\int_{z_{\text{min}}}^{z_{\text{max}}}\left[\frac{1}{4}(g(z))^2 + g(z)(z - z_{\text{g}})^2\right]\mathrm{d}z\,, \tag{3.40a}$$

$$I_3 = \frac{m\pi}{2V}\int_{z_{\text{min}}}^{z_{\text{max}}} (g(z))^2\mathrm{d}z\,. \tag{3.40b}$$

For $\lambda = 0.98$ and $\alpha = 1.2$ we find $I_1/I_3 = 1.07$ and $z_{\text{g}} = -0.798a$.

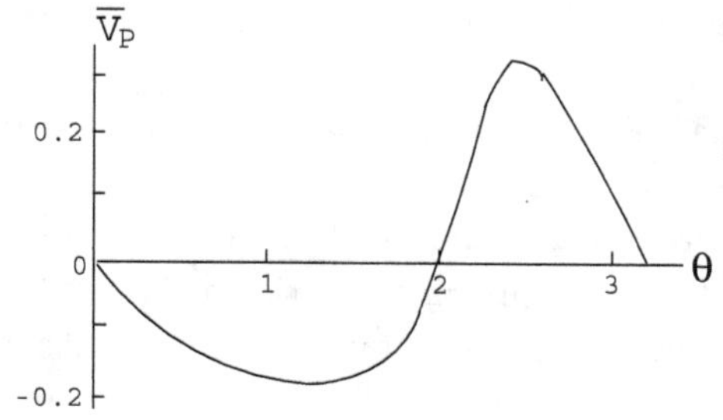

Fig. 3.4: Variation of $\bar{V}_P$ with θ for a Cassini oval with an air chamber [6].

Now, consider the Eqs. (3.30). We infer that the change in the value of θ depends on the sign of $\bar{V}_P$. θ will decrease with time if $\bar{V}_P$ is negative. For positive $\bar{V}_P$, θ will increase with time. Sasaki analysed the variation of $\bar{V}_P$ with θ for the case of a spinning egg with a Cassini oval with an air chamber [6]. Figure 3.4 shows the dependence of $\bar{V}_P$ with θ. At $\theta = 0$ notice that $\bar{V}_P = 0$. As θ increased, $\bar{V}_P$ decreased from 0 and reached a minimum value at a value of θ. Further increase in θ led to increase in $\bar{V}_P$. At θ_c we have $\bar{V}_P = 0$. From θ_c, $\bar{V}_P$ increased from 0, reached a maximum and decreased to 0 at $\theta = \pi$. That is,

$$\bar{V}_P = \begin{cases} < 0, \ \theta < \theta_c \\ 0, \quad \theta = 0, \theta_c, \pi \\ > 0, \theta > \theta_c \end{cases}. \tag{3.41}$$

This implies that if the egg starts spinning with $\theta_{\text{initial}} > \theta_c$, then θ will increase to π. The egg will then spin at the thin end. If $\theta_{\text{initial}} < \theta_c$, then θ will decrease to 0 and the egg will spin at the fat end. θ_c is thus a critical angle. Depending on the initial value of θ, the egg will spin at the fat end or the thin end. When we spin an egg number of times, most of the cases we would observe the egg spinning at the fat end. This is because mostly θ_{initial} is between 0 and θ_c.

3.7 Conclusion

When a hard-boiled egg is spun rapidly on a table with its axis of symmetry horizontal it rises from the horizontal position to vertical position. Similarly, an oblate spheroid also rises vertically for large spin. As the centre of gravity rises in both the cases, this phenomenon looks like a paradoxial one. These strange motions of the boiled egg and the spheroid are explained by taking into account the action of the frictional force between the spinning object and the table.

Many experimental and theoretical results concerning the precession of the spinning eggs and spheroids have shown that their centre of gravity rises due to the torque acting due to the sliding frictional force. But the mathematics involved in explaining the rise of CoM of these spinning objects due to the contact friction is quite involved. A simplified version of this mathematics has been discussed here. It is to be noted that the initial angle θ either greater than or less than a critical angle θ_c determines whether these spinning objects stand upright or not.

3.8 Bibliography

[1] https://www.kickstarter.com/projects/siriusenigmas/the-phitop-a-spinning-top-that-levels-up-executive.
[2] K. Brecher and R. Cross, The Phys. Teach. **57**, 74 (2019).
[3] https://www.vat19.com//item/phitop-top-egg-shaped.
[4] K. Brecher, *Proceedings of Bridges: Music, Art, Architecutre, Culture* (2015) pp.371.
[5] H.K. Moffatt and Y. Shimomura, Nature **416**, 385 (2002).
[6] K. Sasaki, Am. J. Phys. **72**, 775 (2004).
[7] H.K. Moffatt, Y. Shimomura and M. Branicki, Proc. R. Soc. A **460**, 3643 (2004).
[8] R. Cross, Eur. J. Phys. **39**, 025002 (2018).
[9] N.M. Bou-Rabee, J.E. Marsden and L.A. Romero, Z. Angew. Math. Mech. **85**, 618 (2005).
[10] R. Cross, Phys. Edu. **48**, 51 (2013).

3.9 Exercises

3.1 Consider an ellipsoidal PhiTOP in the x-y coordinate system shown in Fig. 3.5 with $\frac{x^2}{a^2}+\frac{y^2}{b^2}=1$. In the new variables $x'=\frac{x}{a}$ and $y'=\frac{y}{b}$ the above equation becomes $x'^2+y'^2=1$. The volume of the disc with thickness $\mathrm{d}x$ is $\mathrm{d}V=\pi y^2\mathrm{d}x$. Determine the volume of the PhiTOP.

3.2 Consider a PhiTOP in the x-y coordinate system shown in Fig. 3.5. Show that the moment of inertia of the PhiTOP around the major and minor axes are $I_a=\frac{2}{5}mb^2$ and $I_b=\frac{2}{5}ma^2$, respectively.

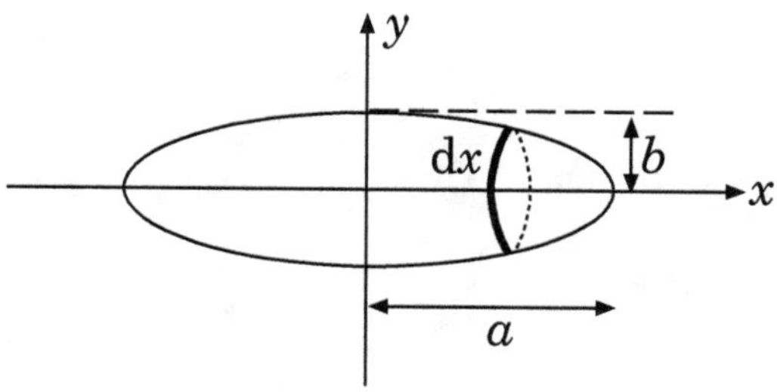

Fig. 3.5: Geometry of a PhiTOP in x-y coordinate system.

3.3 Write the Eqs. (3.16a) and (3.16c) in the xyz coordinate system.

3.4 For the Cassini oval with an air chamber compute $\sqrt{g(z)}$ from Eq. (3.38) for $z \in [-1.4a, -0.199a]$ with $\alpha = 1.2$ and $\lambda = 0.98$ in units of a. Plot $\sqrt{g(z)}$ versus z and mark the air chamber.

3.5 Determine the roots of Eq. (3.38).

Chapter 4

Euler's Disk

Joseph Bendik invented the Euler's disk between 1987 and 1990. He first made an interesting observation of a whirring sound when spun a heavy polished chuck on a desk at Hughes Aircraft (Carlsbad Research Centre). Then, he and his coworker Richard Henry Wyles and his friend Larry Shaw (the inventor of astrojax) worked on a commercial version of the toy (a coin shaped disk) [1]. They filed a patent (US Patent 5863235) for the toy. The motion of spinning of a rigid body was first analysed by Euler and hence the spinning and rolling disk invented by Joseph Bendik was named as *Euler's disk*. The commercially available toy usually has a heavy, thick chrome-plated steel disk and a rigid, concave mirrored base (see Fig. 4.1) [2]. To enhance the visual effect holographic magnetic stickers can be fixed to the disk. The disk can work on a glass-topped table as well. In 2006 the Euler's disk appeared in the film Snow Cake. In 2017 it appeared in the TV show The Big Bang Theory (season 10, episode 16).

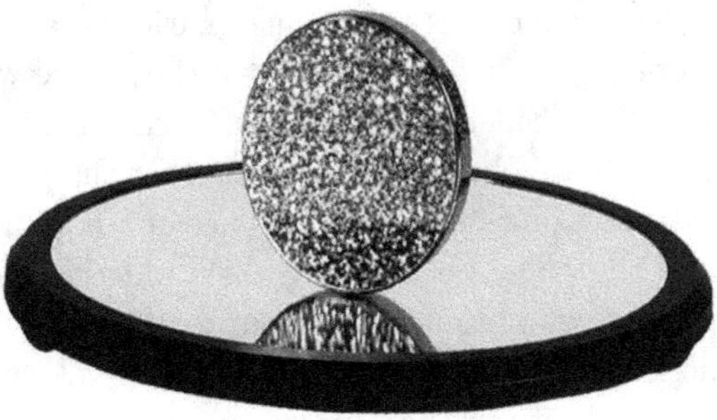

Fig. 4.1: An Euler's disk on a concave mirrored base.

To initiate the motion of the disk, hold it upright on to the mirror base and give a twist to it. A gentle twist is enough to observe its features. It starts spinning and rolling, slowly progresses through different speeds and types of motion. After a long time spinning, it emits a whirring sound of increasing frequency and finally comes to an abrupt halt. It is not only spinning but also rolling (a combination of rotation and translational motion, e.g., a wheel rolling down the road) and hence it is said to be *spolling* on its own. Similar type of motion can be witnessed when a coin is spun on a table or on a smooth floor. But the coin does not spin for a long time. The Euler's disk is larger and heavier than a coin. As the disk spins down the precession rate of its axis of symmetry is accelerated. The mirror base is for the purpose of reducing the friction and a slight concave shape of the mirror keeps the disk from wandering-off the support surface. Note the difference between the spinning egg or PhiTOP (refer Chapter 3 in part III) and the Euler's disk.

There are two prime features with the motion of the Euler's disk. The first is the continuous increasing frequency of the whirring sound as the angle of inclination of the disk declines. Another is the prediction made by Moffatt through a theoretical treatment. In the Euler's disk, though its mechanical energy is dissipating, its speed rapidly increases. *What are the mechanisms of energy dissipation? Which is the dominant energy dissipation?* We can think of energy dissipation due to air friction, sliding friction and rolling friction. Moffatt investigated the above problem by taking into account of viscous damping arising due to the sheared air between the disk and table [3]. The viscous dissipation is found to approach infinity when time t approaches a finite time t_0. In this case, the rate of precession ω_{p} diverges as $\omega_{\mathrm{p}} \propto (t - t_0)^{-1/n}$. Here t_0 is the time at which the disk halts its motion. This is confirmed experimentally [4,5]. Experiments with rings and disks in evacuated chambers ruled out air friction as the prime cause for dissipation energy [6,7]. It has been shown that there was no slip state during the motion of the Euler's disk at least in the early stages [8].

In this chapter, first we obtain an expression for the precession velocity for the Euler's disk and point out the occurrence of a high frequency sound. For small values of the angle between the disk and the horizontal surface on which it is spolling, we obtain expressions for the rate of dissipation of energy due to the rolling friction and the precession velocity ω_{p}. We show the presence of a singularity in the expression for ω_{p}.

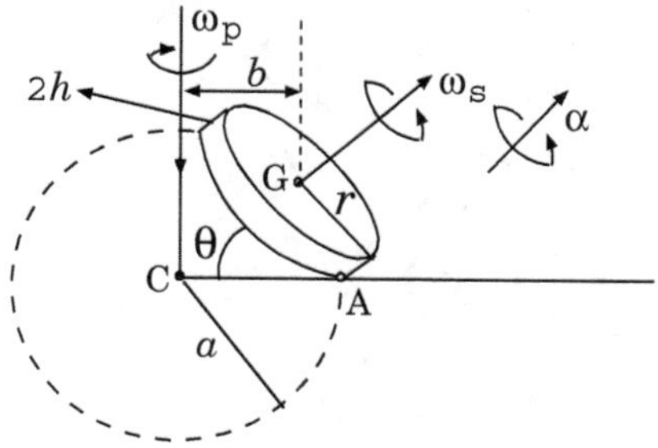

Fig. 4.2: Sketch of the Euler's disk at an instant of time during spolling motion.

4.1 Euler's Disk in Rolling Motion

In this section, we obtain expressions for the rate of spin and the rate of precession of the Euler's disk in the rolling motion without slipping by neglecting all the dissipation effects. We essentially follow the Ref. [9].

4.1.1 *Kinematics*

Consider a uniform solid disk of mass m, radius r and thickness $2h$ with $r \gg 2h$ spolling on a flat horizontal surface. This is schematically represented in Fig. 4.2. At an instant of time θ is the angle between the disk and the horizontal surface with $0 \leq \theta \leq \pi$ and A is the contact point between the disk and the horizontal surface. The contact point A rolls over a circle of radius a with C as the centre at which the axis of precession intersects the horizontal surface. b is the horizontal distance between the centre of mass (CoM) G of the disk and the axis of precession. Denote $\omega_{\rm s}$ and $\omega_{\rm p}$ as the rate of spin and precession, respectively, of the disk in units of radians per second. For simplicity assume that during the rolling motion without slipping the quantities θ, $\omega_{\rm s}$ and $\omega_{\rm p}$ are constants and $b = 0$. With this assumption we can redraw the sketch in Fig. 4.2 as in Fig. 4.3 [9].

Let us introduce X, Y, Z as axes fixed to ground with C as the origin. The unit vectors along X, Y, Z are $\widehat{I}$, $\widehat{J}$ and $\widehat{K}$, respectively. x, y, z are the local axes with G as the centre attached to the disk. x, y, z move with the disk. x-axis is parallel to the X-axis. We denote g as the acceleration due to gravity and a as the distance from the point C to A. $\boldsymbol{\omega}$ and α are the angular velocity and the angular acceleration of the disk, respectively, both with respect to the ground. $\mathbf{F}_{\rm A}$ with components $(F_{{\rm A}X}, F_{{\rm A}Y}, F_{{\rm A}Z})$ is the force at the point A acting on the disk.

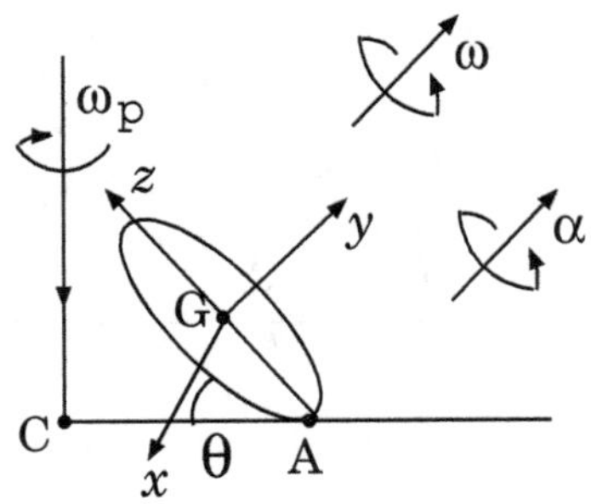

Fig. 4.3: Redrawn schematic of the Euler's disk. For details see the text.

4.1.2 *Expressions for Spin and Precession Rates*

The angular velocity $\boldsymbol{\omega}$ is

$$\boldsymbol{\omega} = \omega_{\rm s} \sin\theta \widehat{J} + (\omega_{\rm s}\cos\theta - \omega_{\rm p})\widehat{K} \tag{4.1}$$

and

$$\begin{aligned}\boldsymbol{\alpha} = \mathbf{J}\frac{\mathrm{d}}{t}(\omega_{\rm s}\sin\theta) + \omega_{\rm s}\sin\theta\frac{\mathrm{d}\mathbf{J}}{\mathrm{d}t} + \mathbf{K}\frac{\mathrm{d}}{\mathrm{d}t}(\omega_{\rm s}\cos\theta - \omega_{\rm p})\\ + (\omega_{\rm s}\cos\theta - \omega_{\rm p})\frac{\mathrm{d}\mathbf{K}}{\mathrm{d}t}\,.\end{aligned} \tag{4.2}$$

In this equation $\mathrm{d}\mathbf{J}/\mathrm{d}t = \omega_{\rm p}$ and all other terms in the right-side are zero. Then,

$$\boldsymbol{\alpha} = \omega_{\rm p}\omega_{\rm s}\sin\theta\widehat{I}\,. \tag{4.3}$$

The velocity $\mathbf{V}_{\rm A}$ of the point A on the disk consists of two terms – $a\omega_{\rm p}$ due to the precession of the disk and $r\omega_{\rm s}$ due to the spin of the disk. These two are acting in opposite directions. Therefore,

$$\mathbf{V}_{\rm A} = a\omega_{\rm p} - r\omega_{\rm s}\,. \tag{4.4}$$

For the case of rolling without slipping $\mathbf{V}_{\rm A} = 0$, that is, $a\omega_{\rm p} = r\omega_{\rm s}$. From Fig. 4.2 we find $a = b + r\cos\theta$. As we assumed that $b = 0$, we have $a = r\cos\theta$. Then, $a\omega_{\rm p} = r\omega_{\rm s}$ gives

$$\omega_{\rm s} = \omega_{\rm p}\cos\theta\,. \tag{4.5}$$

This is the relation between $\omega_{\rm p}$ and $\omega_{\rm s}$.

By the Newton's second law

$$ma_{\rm GX} = F_{\rm AX}\,,\ \ ma_{\rm GY} = F_{\rm AY}\,,\ \ ma_{\rm GZ} = F_{\rm AZ} - mg\,. \tag{4.6}$$

When $b = 0$ the point G is stationary and hence $a_{GX} = a_{GY} = a_{GZ} = 0$. Then, $F_{AX} = 0$, $F_{AY} = 0$ and $F_{AZ} = mg$. Next, we have

$$\omega_x = 0, \quad \omega_y = \omega_s - \omega_p \cos\theta, \quad \omega_z = -\omega_p \sin\theta \tag{4.7}$$

and

$$\alpha_x = \omega_s \omega_p \sin\theta, \quad \alpha_y = 0, \quad \alpha_z = 0\,. \tag{4.8}$$

To setup the Euler equations of motion for the Euler's disk denote $\sum M_{Gx}$, $\sum M_{Gy}$ and $\sum M_{Gz}$ are the sum of the components of the moments about G in the x, y, z directions, respectively. Similarly, I_{Gx}, I_{Gy} and I_{Gz} are the principal moments of inertia of the disk about G about the x, y z directions, respectively. They are given by

$$I_{Gx} = I_{Gz} = \frac{1}{4} m r^2, \quad I_{Gy} = \frac{1}{2} m r^2\,. \tag{4.9}$$

The Euler equations are

$$\sum M_{Gx} = I_{Gx}\alpha_x - (I_{Gy} - I_{Gz})\,\omega_y\omega_z\,, \tag{4.10a}$$

$$\sum M_{Gy} = I_{Gy}\alpha_y - (I_{Gz} - I_{Gx})\,\omega_z\omega_x\,, \tag{4.10b}$$

$$\sum M_{Gz} = I_{Gz}\alpha_z - (I_{Gx} - I_{Gy})\,\omega_x\omega_y\,. \tag{4.10c}$$

From Eqs. (4.7) and (4.8), we find that $\sum M_{Gy} = \sum M_{Gz} = 0$. Therefore, we need to consider only the Eq. (4.10a). $\sum M_{Gx}$ is simply $rF_{AZ}\cos\theta = mgr\cos\theta$. Then, Eq. (4.10a), after making use of the Eqs. (4.7) and (4.9), becomes

$$mgr\cos\theta = \frac{1}{4} m r^2 \omega_p^2 \sin\theta\cos\theta\,. \tag{4.11}$$

That is,

$$\omega_p = \left(\frac{4g}{r\sin\theta}\right)^{1/2}\,. \tag{4.12}$$

For the case of $b \neq 0$, ω_p is calculated as [10]

$$\omega_p^2 = \frac{(g/r)(\cos\theta - (b/r)\sin\theta)}{\left(\frac{1}{4} + \frac{1}{3}(b/r)^2\right)\sin^2\theta\cos\theta - \frac{1}{2}(b/r)\sin^2\theta}\,. \tag{4.13}$$

In the limit $b \to 0$, Eq. (4.13) becomes Eq. (4.12). In arriving the formula we neglected all kinds of dissipation. In a practical situation the system has a friction at the point A and rolling resistance. We need to take into account the loss of gravitational potential energy. Consequently, the CoM G would drop in height. In reality, θ decreases slowly due to damping and hence ω_p slowly increases [8]. When θ becomes so small, the disk comes to an abrupt halt. As $\theta \to 0$, ω_p becomes very large and we have a high-frequency sound.

4.2 Finite-Time Singularity

When the rate of dissipation of energy of a system is sufficiently large enough, then all its motion can cease in a finite time. There are physical systems where a kinetic variable approaches infinity even when the total kinetic energy approaches zero. Such a finite-time singularity occurs in the bouncing ball motion of an inelastic ball on a plane [4], simulations of gases of inelastic particles [11] and motion of a rod sliding in a ring with Coulomb friction [12,13].

A spinning or rolling disk ultimately comes to rest quite abruptly. The final stage of motion is accompanied by a whirring sound of rapidly increasing frequency ω_{p}. As the disk rolls, the point of contact describes a circle with an angular velocity ω_{p}. If there were no energy dissipation ω_{p} will remain constant. But in reality, frictions exist always. So, ω_{p} does not remain constant. There are two main dissipative frictional forces existing for the rolling of the disk. They are the rolling friction present when the disk slips along the surface and air drag from the viscous resistance of air.

Keith Moffatt [3] showed that viscous dissipation in the thin air between the disk and the table results in the abrupt settling process of the disk. He also showed that the motion ended with a finite-time singulaity. So, the motion of the Euler disk appears to cease rather abruptly for small values of the angle θ, corresponding to large angular velocity ω_{p}. The finite-time singularity arises due to the dependence of ω_{p} as a power-law of time due to the frictional forces acting on the system. We follow the analysis of McDonald and McDonald [14] and get the time power-law for ω_{p}.

4.3 Friction at Small Angle and Finite-Time Singularity

We consider the Euler's disk with its CoM G lying along the vertical axis and so $b = 0$. Let the distance from the CoM G to the point C be l. We consider the motion of the Euler disk just before it falls flat on the horizontal plane. At this stage, the angle θ becomes very small corresponding to large precession of the angular velocity ω_{p}. If $\mathbf{F}$ is the force at the point of contact A, then the equation of motion for the centre of the disk is given by

$$m\ddot{l} = \mathbf{F} - mg\mathbf{z}. \tag{4.14}$$

At the last stage of the motion of the disk when $\theta \to 0$ and the rolling motion ceases, the disk seems to float for a moment and then settles on the horizontal surface. So, the upward contact force $\mathbf{F}$ becomes zero and the disk looses contact with the surface.

Solved Problem 1:

For a small angle θ setup an expression for total energy E of the system and determine the power dissipated by the frictional force.

The total energy E of the system is given by the sum of the kinetic energy, potential energy and the rotational energy about the z-axis. We write

$$E = \frac{1}{2}m\dot{l}^2 + \frac{1}{2}I_{\text{G}z}\omega_z^2 + mgl. \tag{4.15}$$

For a small angle θ we approximate l as $r\theta$ and so $\ddot{l} \approx r\ddot{\theta}$. As the acceleration of the CoM is along the $-z$ direction. With $\mathbf{F} = 0$ and $\ddot{l} \approx r\ddot{\theta}$ Eq. (4.14) gives $-mr\ddot{\theta}\mathbf{z} = -mg\mathbf{z}$. That is,

$$E = \frac{1}{2}mr^2\dot{\theta}^2 + \frac{1}{8}mr^2\frac{4g}{r\theta}\theta^2 + mgr\theta = \frac{1}{2}mr^2\dot{\theta}^2 + \frac{3}{2}mgr\theta. \tag{4.16}$$

If P is the power dissipated by frictional forces, then

$$P = \frac{\mathrm{d}E}{\mathrm{d}t} = mr^2\dot{\theta}\ddot{\theta} + \frac{3}{2}mgr\dot{\theta}. \tag{4.17}$$

Using $\ddot{\theta} = g/r$ we obtain

$$P = \frac{5}{2}mgr\dot{\theta}. \tag{4.18}$$

Considering a model for the rolling friction of the disk on the horizontal surface with small bumps on the surface with small spacing Δ and average height $l = \epsilon\Delta$, it is found that the disk dissipates energy $mgl = mg\epsilon\Delta$ when passing over a bump [14]. As the distance traveled by the disk over one period is $2\pi r$, we find the time taken by the disk to pass over one bump is $\Delta/(r\omega_{\text{p}})$. Then, the rate of dissipation of the energy due to the rolling friction in this model is

$$P = -\frac{mg\epsilon\Delta}{\Delta/(r\omega_{\text{p}})} = -\epsilon mrg\omega_{\text{p}}. \tag{4.19}$$

Therefore, for any generalized form of velocity dependent friction, P can be written as

$$P = -\epsilon mrg\omega_{\text{p}}^{\alpha}, \tag{4.20}$$

where drag force varies with angular velocity as $\omega_{\text{p}}^{\alpha-1}$.

From Eqs. (4.18) and (4.20) we get

$$\dot{\theta} = -\frac{2}{5}\epsilon\omega_{\text{p}}^{\alpha}. \tag{4.21}$$

From Eq. (4.12), we get for small θ, $\omega_\mathrm{p} = \sqrt{4g/(r\theta)}$. Substitution of this in Eq. (4.21) and integration of the resultant equation between the limit t_0 to t gives

$$\theta = \left(\frac{\epsilon(\alpha+2)}{5}\right)^{2/(\alpha+2)} \left(\frac{4g}{r}\right)^{\alpha/(\alpha+2)} (t_0 - t)^{2/(\alpha+2)}. \tag{4.22}$$

Use of this expression for θ in $\omega_\mathrm{p} = \sqrt{4g/(r\theta)}$ gives

$$\omega_\mathrm{p} = \left[\frac{20g}{\epsilon(\alpha+2)r(t-t_0)}\right]^{1/(\alpha+2)}. \tag{4.23}$$

We find from Eq. (4.23), if α is greater than -2, then as t approaches t_0, ω_p increases. This is called *finitie-time singularity* at time t_0 [3].

In Moffatt's model, and taking into account of the viscous drag of the air between the disk and the surface, it has been found that $\alpha = 4$ and

$$\theta = \left[\frac{2\pi\eta r}{m}(t_0 - t)\right]^{1/3}, \quad \omega_\mathrm{p} = \sqrt{\frac{4g}{r}}\left(\frac{m}{2\pi\eta r(t-t_0)}\right)^{1/6}, \tag{4.24}$$

where η is the viscosity of air. But it was found [6] that in vacuum and in air the total times of spin of coins are similar. So, the air drag as suggested by Moffatt's model is not the dominant mechanism of energy dissipation. Various other mechanisms have been proposed.

The halting motion of the Euler's disk is not smooth but abrupt. This is partly due to the slipping of the disk. It has been conjuctured that this is due to the vibrations occurring resulting in a loss of contact of the disk with the surface of the base when the angle of inclination becomes small [15]. A theoretical model of a partially deformable Euler's disk has been proposed [16]. Certain examples of using Euler's disk in engineering and applied mathematics were presented in [17].

4.4 Conclusion

Euler's disk is a fascinating simple toy as it is only a heavy disk rolling on a plane or concave smooth surface. Euler's disk uses its angular momentum to hold itself upright. As the disk spins around in a circle, it is held in place by a balance of the gravitational force pulling the disk down and the force applied by the mirror base which holds the disk up. If it were not for friction and vibration the disk would rotate for a very long time. The motion of Euler's disk after spinning it is quite complicated. Frictional forces acting on the system decide the motion of the system at the final stage of its motion.

The most important feature of Euler's disk motion is that it exhibits finite-time singularity which was pointed out by Moffatt. It has been noted that the disk in the late stage of motion precesses faster about the rotation axis as it loses energy. Moffatt assumed the dominant source of energy loss as the viscous drag in the sheared layer of air between the disk and the surface. He found that the precession velocity is proportional to $(t_0 - t)^{-1/n}$ where $n = 6$. But subsequently experiments found that the air drag frictional force is not the primary cause for the power-law of the precession velocity. Studies on a variety of surfaces have found that the power-law of Moffatt is still $(t_0 - t)^{-1/n}$.

Both theoretical and experimental works on the dynamics of the Euler's disk have shown that it behaves drastically on different surfaces. For example, the commercially available Euler's disk on a concave mirror displays its feature motion several times longer duration than on the certain other surfaces. This is because the commercially available Euler's disks have a slopping edge rather than a sharp edge. Note that a sharp edge will lead to a higher pressure on the substrate resulting in an increase in the rolling motion. Therefore, rolling friction due to plastic deformation at the contact point is an important factor in the spinning of the Euler's disk.

4.5 Bibliography

[1] https://en.wikipedia.org/wiki/Euler%27s_Disk.

[2] https://www.weirdbydesign.com/eulers-spinning-disk/?cn-reloaded-=1.

[3] H.K. Moffatt, Nature **404**, 833 (2000).

[4] K. Easwar, F. Rouyer and N. Menon, Phys. Rev. E **66**, 045102 (2002).

[5] H. Caps, S. Dorbolo, S. Ponte, H. Croisier and N. Vandewalle, Phys. Rev. E **69**, 056610 (2004).

[6] G. van der Engh, P. Nelson and J. Roach, Nature **408**, 540 (2000).

[7] H.K. Moffatt, Nature **408**, 540 (2000).

[8] D. Petrie, J. Hunt and C. Gray, Am. J. Phys. **70**, 1025 (2002).

[9] https://www.real-world-physics-problems.com/eulers-disk.html.

[10] L.A. Whithead and F.L. Curzon, Am. J. Phys. **51**, 449 (1983).

[11] S. McNamara and W.R. Yonng, Phys. Rev. E **50**, R28 (1994).

[12] D.E. Stewart, C. R. Acad. Sci. Ser. I Math. **325** 689 (1997).

[13] D.E. Stewart, Philos. Trans. R. Soc. Lond. Ser. A **395**, 2467 (2001).

[14] A.J. McDonald and K.T. McDonald, arxiv:physics/0008227v3 [physics.class-ph] 5 April (2002).

[15] P. Kessler and O.M. O'Reilly, Regular Chaotic Dyn. **7**, 49 (2002).
[16] R. Villanueva, Phys. REv. E **71**, 066609 (2005).
[17] P. Abbott, G. Keady and S. Tyler, ANZIAM J. **51**, C360 (2010).

4.6 Exercises

4.1 Consider the Euler's disk for the case where the CoM lies in the vertical z-axis as shown in Fig. 4.4 and has an inclination angle θ. $(\widehat{1}, \widehat{2}, \widehat{3})$ are the unit vectors of the right-hand coordinate system attached with the Euler disk of radius r. $\widehat{3}$ gives the direction of the line connecting the CoM G and the point of contact A. Find the angular momentum of the system assuming that there is no slipping in the motion of the Euler disk.

4.2 What is the torque associated with the Euler's disk of exercise 4.1 about the CoM?

4.3 Consider Fig. 4.4. The angular velocity about $\widehat{3}$ can be thought of as composed of two parts, that is, $\omega\widehat{3} = \omega_{\mathrm{p}}\mathbf{z} + \omega_{\mathrm{s}}\widehat{1}$ where $\omega_{\mathrm{s}} = \omega_1$ is the angular velocity of the disk along its symmetry axis $\widehat{1}$ and $\omega_{\mathrm{p}} = \Omega$ is the angular velocity about $\mathbf{z}$. Using the angular momentum and torque obtained from the exercises 4.1 and 4.2, respectively, prove the relation $\omega_{\mathrm{p}} = (4g/r\sin\theta)^{1/2}$.

4.4 Setup an expression for the total energy of the Euler's disk spinning without slipping.

4.5 Given that the rate of dissipation energy P is $\propto \omega_{\mathrm{p}}^{\alpha}$, show that the frictional force is proportional to $\omega_{\mathrm{p}}^{\alpha-1}$. If the rolling frictional force is proportional to velocity and air drag force is proportional to square of the velocity, find α values for both the frictional forces.

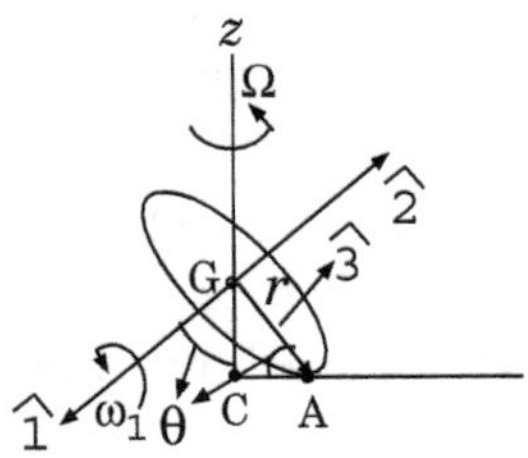

Fig. 4.4: Sketch of the Euler's disk where the centre of gravity lies in the vertical z-axis.

Chapter 5

Fidget Spinner

Fidget spinner is one of the popular toys invented by Catherine Hettinger in 1993 [1,2]. She was suffering from *myasthenia gravis* disorder. For this reason, she had problems in moving her arms and unable to play with her seven year old daughter. This disorder motivated Catherine Hettinger to invent the fidget spinner, a small plastic toy which appeared like a hat and can be spun and balanced on the top of a finger. Later, many versions of it came into the market.

A typical fidget spinner toy (a small pocket sized gadget) (Fig. 5.1) consists of a central ball bearings, a multi-lobed metal or plastic blades (arms) and two removable pads on the top and bottom side of the ball bearings. Without the bearings the toy is very light and cannot spin for a long time. The toy is designed in such a way that the blades move around a central axle. The name *fidget* is because it relieves stress [3]. The toy is often used as a therapeutic aid for children with autism or anxiety or ADD (Attention Deficit Disorder). In the book *Fidget to Focus* [4] the authors Ronald Rotz (a child and adult psychologist) and Sarah Wright (a trained

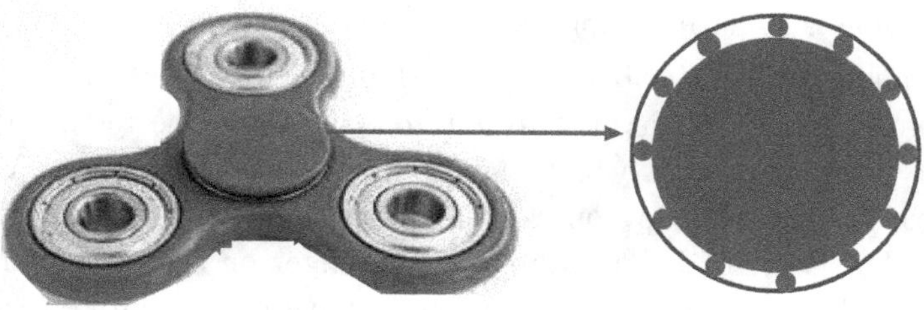

Fig. 5.1: A fidget spinner toy.

Fig. 5.2: A set of fidget spinner toys.

professional coach for people affected by ADD) explained the use of the fidget spinner in preventing the distractions.

Nowadays, fidget spinners are sold in a wide range of models with two or more blades, different types of bearings, materials and colour (see Fig. 5.2). All kinds of fidget spinners have two things in common – they spin and have a ball bearings at the centre. The simple structure of the fidget spinner is a great advantage for easy manufacture with low cost. For details about the structure and design of a fidget spinner one may refer to [5]. On 4 May 2017 fidget spinners became the 17th toy of Amazon's 20 best-sellers in the Toys and Games category [6].

How do we play with a fidget spinner? What do people do with it? Let us mention two of these.

1. Keep the toy on a table. Hold the centre point firmly by a finger. Then, spin it with the other hand. While it is spinning take away the finger. The goal is to have the longest and smoothest spinning.
2. Keep a fidget spinner between your thumb and the forefinger, top and bottom. Slap a blade of it to initiate spinning. The removable pads give grips to prevent your hand from touching the blades. You can place the spinning toy on a table or a smooth floor, shift from one hand to another, balance on top of a finger or nose or forehand or toes and rotate your forehand a few to mention. It can be thrown to another person to catch it. Spinning fidget spinners can be placed one on another.

Interestingly, the fidget spinner toy can be used to explain the moment of inertia, the torque and the angular momentum [7,8]. Experimental measurement moment of inertia of this toy has been presented in [9,10].

In the present chapter, we bring out the physics behind the various features of fidget spinners.

5.1 Physics of the Fidget Spinner

Let us enumerate the physics of fidget spinners [5,11].

1. Forces Causing the Spinner to Spin for a Long Time

To make a fidget spinner to start spinning we need to apply a torque. This is achieved by spinning it by a hand. We need to push it or rotate it with a lot of energy in order to make it to spin for a long time. There are three factors causing the spinner to spin longer times. First is the centripetal force F_{CP} given by

$$F_{\mathrm{CP}} \propto \frac{mv_0^2}{r}, \tag{5.1}$$

where m is the mass of the object, v_0 is the initial speed and r is the radius of the object. The heavier the blade the longer it would spin. The second factor that would reduce the spin is the air friction. The friction is reduced greatly by the shape of the fidget spinner. The third factor is the friction between the rotatory objects. The central part of the fidget spinner has little ball bearings as shown in Fig. 5.1. The inner and outer parts of the fidget spinner are connected by the bearings. These balls rotate between the inner and the outer rings of the central part. The balls are in spherical shape. Since the area of contact with the inner and the outer rings with the balls is very less and negligible the friction is greatly reduced. If there is no ball bearings, then the spinner would not spin for a long time.

2. Source for Balancing a Fidget Spinner on a Finger

A rotating fidget spinner can be balanced on a finger or even on a pen. The source for its balance while spinning is the conservation of angular momentum. We know that a rotating object continue its rotational motion until it is acted upon by an external force. Friction is one external force. This resistive force is minimized by the ball bearings so that the blades spin for a long time.

3. Source for Upright Stay of a Fidget Spinner

The source for the upright spinning of the toy without leaning is due to the law of conservation of angular momentum. We know that the angular velocity of a body will remain constant until an unbalanced torque acts on it. When we neglect the friction forces, then the other force acting on the toy is the gravity. However, the gravity does not exert any unbalanced force because the toy is symmetrical. Therefore, the spinner spins in a direction without leaning.

4. Factors Causing the Fidget Spinners to Slow Down and Stop

The toy is not free from friction forces. The ball bearings of the fidget spinner and the friction force of air are weak. However, they are not zero. There is another source for the loss of energy. When the spinner is spinning, bring it near to your ear. You will hear two kinds of sound. One is the rushing sound due to the contact motion of the ball bearings. The second is the humming sound energy caused by the blades of the spinner. This sound energy leads to minute changes in air pressure. The frequency of this humming sound decreases as the spinner slows down. In these ways the energy of the spinner is slowly lost.

5.2 Spin Time of a Fidget Spinner

How long a spinner can spin? To get an expression for the spin time of a fidget spinner denote $\theta(t)$ as the angular position, $\omega(t)$ is the angular velocity and $\alpha(t)$ is the rate of change of angular velocity of the spinner at time t. Then, we can write

$$\omega = \frac{\Delta\theta}{\Delta t} \tag{5.2a}$$

and

$$\alpha = \frac{\Delta\omega}{\Delta t}\,. \tag{5.2b}$$

From Eq. (5.2b) if T is the spin time, ω_{i} is the initial angular velocity and zero is the final angular velocity, then we have

$$T = \frac{\Delta\omega}{\alpha} = \frac{0-\omega_{\mathrm{i}}}{\alpha} = -\frac{\omega_{\mathrm{i}}}{\alpha}\,. \tag{5.3}$$

To calculate T we need to determine α. It can be determined by measuring the change in angular velocity which is not an easy task. Rhett Allain described an experimental setup using a laser to measure α [10].

Solved Problem 1:

In an experiment with a fidget spinner it was found that the angular frequency decreases as given by $\omega(t) = 40 - 0.5t$. Find the total duration of the spinning of the fidget spinner.

We have $\omega(t) = 40 - 0.5t$. Denote $t = T$ as the time at which the fidget spinner stops spinning. Then, at $t = T$, $\omega(T) = 0$. So, at $t = T$ we have $0 = 40 - 0.5T$. We obtain $T = 80\,\text{sec}$.

5.3 Calculation of the Moment of Inertia

The measurement of the moment of inertia I of a fidget spinner is an interesting experiment. Two methods have been proposed to measure I [9,10]. We present the method described by Allain [10].

For a body consisting of N point particles of masses m_i, $i = 1, 2, \ldots, N$ at perpendicular distances r_i, $i = 1, 2, \ldots, N$ from the axis of rotation the moment of inertia I is given by

$$I = \sum_{i=1}^{N} m_i r_i^2 . \tag{5.4}$$

To determine I using the above equation we need to find the mass and the size of the given fidget spinner. Since the toy does not have simple shape and uniform density, it is not an easy task to determine the mass and its size theoretically. However, I can be estimated with a physical pendulum.

The time period of a small amplitude oscillation of the familiar simple pendulum is $T = 2\pi\sqrt{l/g}$ where l is the length of the string to which a small mass is attached. Suppose, we replace the string by a rigid one, for example, a stick. This is an example of a physical pendulum and is shown in Fig. 5.3.

As the net external torque τ is equal to the moment of inertia I about the axis of rotation (pivot point) times the angular acceleration (α), we write

$$I = I_{\text{support}}\alpha = I_{\text{support}}\frac{\mathrm{d}^2\theta}{\mathrm{d}t^2} . \tag{5.5}$$

Further, τ is given by $\tau = -mgl\sin\theta$. For small θ

$$I_{\text{support}}\frac{\mathrm{d}^2\theta}{\mathrm{d}t^2} = -mgl\sin\theta \tag{5.6}$$

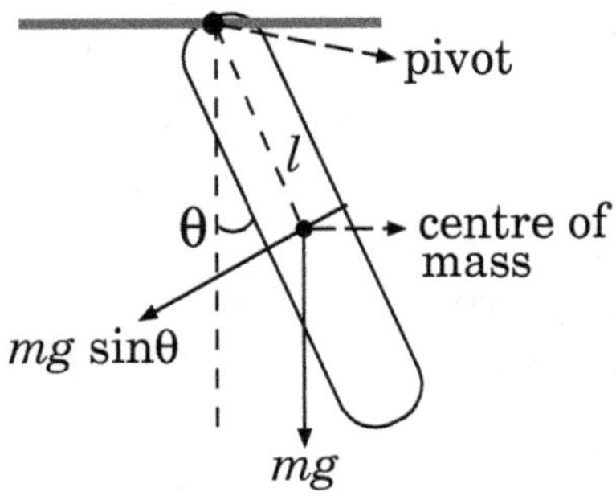

Fig. 5.3: A physical pendulum.

or

$$\ddot{\theta} + \omega^2\theta = 0, \quad \omega^2 = \frac{mgl}{I_{\text{support}}}. \tag{5.7}$$

Its solution for $\theta(0) = \theta_0$ an $\dot{\theta}(0) = 0$ is $\theta(t) = \theta_0 \cos \omega t$. The period of oscillation is

$$T = \frac{2\pi}{\omega} = 2\pi\sqrt{\frac{I_{\text{support}}}{mgl}}. \tag{5.8}$$

The point is that by hanging the fidget spinner with the physical pendulum and measuring the time period of oscillation T, its moment of inertia can be determined. However, in this case we need to find the I of the fidget spinner through another axis. For this we can make use of the parallel-axis theorem discussed in Sec. 1.2. According to this theorem if I_{CoM} is the moment of inertia of an object about an axis passing through its centre of mass (CoM), then I_{parallel} about some other axis parallel to the first at a distance d is given by

$$I_{\text{parallel}} = I_{\text{CoM}} + md^2. \tag{5.9}$$

To estimate the I of the fidget spinner tie it with a light weight stick using a rubber-band and use it as the physical pendulum. Denote l as the distance between the pivot point and the centre of the fidget spinner. Neglecting the masses of the the stick and the rubber-band the time period of the physical pendulum is

$$T = 2\pi\sqrt{\frac{I_{\text{CoM}} + ml^2}{mgl}}, \tag{5.10}$$

where m is the mass of the fidget spinner. This equation gives

$$\frac{T^2mgl}{4\pi^2} = I_{\text{CoM}} + ml^2. \tag{5.11}$$

One can measure T for various values of l, that is, varying the position of the fidget spinner. Draw a graph between $y = T^2mgl/(4\pi^2)$ and $x = ml^2$. We expect the graph as a straight-line. The y-intercept of the straight-line is the moment of inertia of the fidget spinner.

5.4 Conclusion

A fidget spinner can be used to demonstrate to the students important physical concepts such as the moment of inertia, the torque and the angular momentum. Conservation and vector characteristics of angular momentum can be demonstrated using the fidget spinner by doing experiments as discussed in Ref. [8]. All the three Newton's laws of motion can be explained using it. The role of frictional force on opposing the motion can be explained as discussed in ref. [5]. The role of the moment of inertia in determining the duration of spin can be studied experimentally. The moment of inertia can be increased by adding weight at the ends of the rotating blades. It can be experimentally found that increasing the moment of inertia leads to a larger spinning time. Determination of the moment of inertial of the fidget spinner by a low-cost experimental setup is discussed in [12].

5.5 Bibliography

[1] http://time.com/money/4762207/fidget-spinner-inventor-caterinehettinger/.

[2] http://www.fidget-spinners.link/.

[3] https://byjus.com/the-learning-tree/science-behind-fidget-spinners/.

[4] Ronald Rotz and Sarah Wright, *Fidget to Focus* (iUniverse, Lincoln, 2005).

[5] V. Kaushik, Int. J. Res. Appl. Sci. & Eng. Tech. **6**, 4346 (2018).

[6] https://www.nytimes.com/2017/05/06/style/fidget-spinners.html.

[7] Y. He and L. You, Int. J. Instr. Control Systems **8**, 1 (2018).

[8] J. Linden, Phys. Edu. **53**, 023004 (2018).

[9] V.L.B. de Jesus and D.G.G. Sasaki, The Phys. Teach. **56**, 639 (2018).

[10] R. Allain, *Let's explore the physics of rotational motion with a fidget spinner*; https://www.wired.com/2017/05/physics-of-a-fidget-spinner/.

[11] M. Richard, *Fidgetting for physics: Spinner science in six steps*; https://www.teachingchannel.org/blog/2017/05/19/spinners-science-in-six-steps.

[12] E. Pereira, The Phys. Teach. **59**, 577 (2021).

5.6 Exercises

5.1 A fidget spinner toy spinning between your thumb and a finger is tilted. What is externally applied by you? What is its cause?

5.2 Why does a fidget spinner revolve on a circle [5]?

5.3 List a few ways of making the fidget spinner to spin for a longer time.

5.4 What are the main physical quantities of the rotating fidget spinner? Write their relations.

5.5 Calculate average time duration of a spinning of a fidget spinner over, say, 5 trials. Next, tape the coin of same size on each arm of the spinner. Calculate the average time duration of spinning. Explain the observation using Newton's second law of motion.

Chapter 6

Rattleback

A rattleback toy is in the shape of an elongated boat with a smooth curved bottom (Fig. 6.1). It displays the curious property known as the *rattleback effect* if spun on a smooth horizontal surface. When it spins in the wrong direction (say clockwise) about the vertical axis, it will slow down, wobble or rattle in the longitudinal direction and stop spinning, then it will begin to spin in the opposite direction (counter clockwise). The spin reversal occurs only if it starts in clockwise direction. If it starts spinning in the counter clockwise direction, then it spins until it comes to rest. Essentially, when rotation takes place in the clockwise direction, a pitching instability occurs. This gets much energy from the spin and reverses in sign. The reversal in spinning appears to violate the conservation of the angular momentum.

The rattleback effect, also called *chiral behaviour*, can be clearly seen when the motion of the rattleback toy is initiated by tapping a long edge. This induces it to oscillate. The amplitude of oscillation decays as the rattleback acquires rotation in the counter clockwise. This rotation slows down, because of loss of energy due to the presence of friction, and after some time stops. The rattleback effect is due to the misalignment between distribution of mass and the ellipsoidal shape of the bottom of the toy.

Fig. 6.1: A rattleback toy.

A slice of a hard-boiled egg with shell and with asymmetrical masses added can behave like a rattleback [1]. Note that a slice of a hard-boiled egg with shell will not reverse its direction of rotation. Commercially available rattleback toys are made of plastic. A rattleback toy is known by other names such as celt, rattlerock and wobble stone. The name rattleback was coined by A.D. Moore for the object called a *celt*. The name celt came from celtic priests who have used the rattleback in their rites. Celts displaying spin-reversal are found in ancient celtic and Eqyptian sites.

The first satisfactory mathematical treatment of celt (rattleback) was provided by Sir Gilbert Walker in 1896 [2]. An analytical model was proposed by Sir Hermann Bondi in 1986 [3]. A great deal of interest has been focused on the investigation of the dynamics of rattleback since 1980s. For example, the analysis of growth rate of instability [4], setting of equations of the spin motion [5-7], expressions of the averaged torques and formula for spin reversal time [8], making of a rattleback [9], theoretical analysis and numerical simulation on a model of rattleback [10,11], analysis of rotational dynamics [12], simple physical explanations [13-15] and numerical analysis and an intuitive explanation for spin reversal [16] were reported.

In this chapter first we list out the requirements for an object to display spin reversal. We present a simple mathematical model for the behaviour of the rattleback without dissipation and slipping. We setup the equations of motion. Next, we describe the oscillations of symmetric and asymmetric rattlebacks and bring out the source of torque turning the rattleback counter clockwise during longitudinal oscillations. Then, we point out why does an oscillatory rattleback usually acquire counter clockwise spin.

6.1 Spin Reversal Requirements

Let us list out the requirements for an object to show spin reversal [10].

1. The surface in contact with the table should not be spherically symmetric.
2. The distribution of mass should be asymmetrical with respect to the axes of symmetry of the ellipsoid.
3. The two principal curvatures of the lower surface should not be the same.
4. The two horizontal principal moments of inertia should not be the same.
5. The principal axes of inertia should be misaligned to the principal direction of curvature.

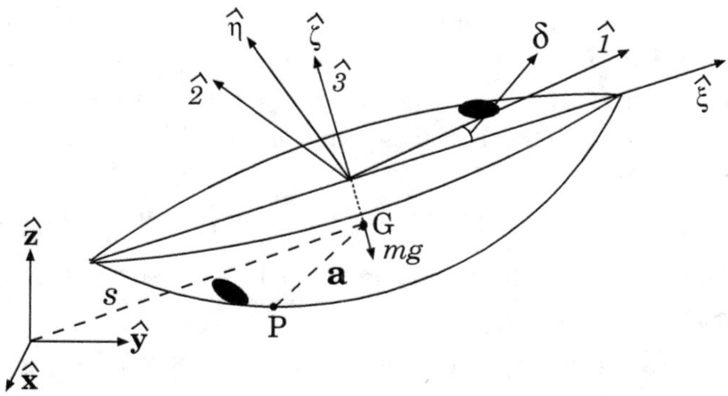

Fig. 6.2: Geometry of the rattleback [11]. For details see the text. (Reproduced with permission from Springer Nature: S.R. Wojciechowski and M. Przybylska, Reg. Chaotic Dyn. **22**, 368 (2017).)

A rattleback displays spinning motion and two kinds of oscillations, namely, the pitching taking place about the short horizontal axis and the rolling that occurs about the long horizontal axis. The above listed characteristics lead to asymmetric coupling between the oscillations, pitching and rolling and the spinning motions. The oscillations produce torque around the spin axis with opposite signs.

6.2 Equation of Motion

We present the treatment in [11]. Consider a rattleback with the shape of an half-ellipsoid and its motion is on a horizontal plane. Assume that it is always in contact with the plane without slipping. The contact point at an instant of time is denoted as P. Figure 6.2 depicts the relevant geometry of the rattleback. We use three frame of references. The reference frames $(\hat{\boldsymbol{\xi}}, \hat{\boldsymbol{\eta}}, \hat{\boldsymbol{\zeta}})$ and $(\hat{\mathbf{1}}, \hat{\mathbf{2}}, \hat{\mathbf{3}})$ are fixed in the body. $(\hat{\boldsymbol{\xi}}, \hat{\boldsymbol{\eta}}, \hat{\boldsymbol{\zeta}})$ are along the main geometrical semi-axes $b_1 > b_2 \gg b_3$ of the ellipsoid. $(\hat{\mathbf{1}}, \hat{\mathbf{2}}, \hat{\mathbf{3}})$ axes point along the directions of the principal moment of inertia. The frame $(\hat{\mathbf{x}}, \hat{\mathbf{y}}, \hat{\mathbf{z}})$ is a fixed inertial frame with the $\hat{\mathbf{x}}$ and $\hat{\mathbf{y}}$ in the plane of the support and $\hat{\mathbf{z}}$ is the vertical axis. I_x, I_y and I_z are the moments of inertia about the $\hat{\mathbf{x}}$, $\hat{\mathbf{y}}$ and $\hat{\mathbf{z}}$ axes, respectively. $\hat{\mathbf{3}}$ coincides with $\hat{\boldsymbol{\zeta}}$. $\hat{\mathbf{1}}$ and $\hat{\mathbf{2}}$ are rotated by an angle $\delta > 0$ counter clockwise with respect to $\hat{\boldsymbol{\xi}}$, $\hat{\boldsymbol{\eta}}$. Therefore,

$$\hat{\mathbf{1}} = \hat{\boldsymbol{\xi}} \cos\delta + \hat{\boldsymbol{\eta}} \sin\delta, \quad \hat{\mathbf{2}} = -\hat{\boldsymbol{\xi}} \sin\delta + \hat{\boldsymbol{\eta}} \cos\delta, \quad \hat{\mathbf{3}} = \hat{\boldsymbol{\zeta}}. \tag{6.1}$$

$\mathbf{s}$ and $\mathbf{a}$ are the position vector of the centre of mass (CoM) G with respect to the frame $(\widehat{\mathbf{x}}, \widehat{\mathbf{y}}, \widehat{\mathbf{z}})$ and the vector from G to the contact point P. $\boldsymbol{\omega}$ is the angular velocity. Choose z-axis as the one close to the rotating axis pointing downward. The other axes x and y are with $I_x > I_y$, that is, the x-axis is the short horizontal axis.

As we neglected the dissipation, the forces acting on the rattleback are:

1. The contact reaction force $\mathbf{F}$ at P exerted by the supporting horizontal surface.
2. The gravitation force $-mg\widehat{\mathbf{z}}$ where $\widehat{\mathbf{z}}$ is the unit vertical vector in the upward direction.

At every instant of time the force $\mathbf{F}$ keeps the rattleback rolling without slipping. The equations of motion of $\mathbf{s}$ and the angular momentum $\mathbf{L} = I\boldsymbol{\omega}$ about CoM are

$$m\ddot{\mathbf{s}} = \mathbf{F} - mg\widehat{\mathbf{z}}\,, \quad \dot{\mathbf{L}} = \mathbf{a} \times \mathbf{F}\,. \tag{6.2}$$

The sliding velocity is

$$\mathbf{v}_{\mathrm{P}} = \dot{\mathbf{s}} + \boldsymbol{\omega} \times \mathbf{a}\,. \tag{6.3}$$

For a pure rolling $\mathbf{v}_{\mathrm{P}} = 0$ which gives $\dot{\mathbf{s}} = -\boldsymbol{\omega} \times \mathbf{a} = \mathbf{a} \times \boldsymbol{\omega}$. Then, Eq. (6.2) takes the form

$$\mathbf{F} = m\ddot{\mathbf{s}} + mg\widehat{\mathbf{z}} = m\frac{\mathrm{d}}{\mathrm{d}t}(\mathbf{a} \times \boldsymbol{\omega}) + mg\widehat{\mathbf{z}}\,, \tag{6.4a}$$

$$\dot{\mathbf{L}} = \mathbf{a} \times \mathbf{F} = m\mathbf{a} \times \left(\frac{\mathrm{d}}{\mathrm{d}t}(\mathbf{a} \times \boldsymbol{\omega}) + g\widehat{\mathbf{z}}\right)\,. \tag{6.4b}$$

The total energy of the system is

$$E = E_{\mathrm{CoM}} + E_{\mathrm{rot}} + E_{\mathrm{pot}}\,, \tag{6.5}$$

where $E_{\mathrm{CoM}} = m\dot{\mathbf{s}}^2/2$ is the energy of the translational motion of CoM, $E_{\mathrm{rot}} = I\omega^2/2$ is the rotational energy and $E_{\mathrm{pot}} = mgs_z$ is the potential energy. E is then given by

$$E = \frac{1}{2}m\dot{\mathbf{s}}^2 + \frac{1}{2}I\omega^2 + mgs_z\,. \tag{6.6}$$

Solved Problem 1:

Show that the energy of the rattleback given by Eq. (6.6) is conserved.

Differentiating Eq. (6.6) with respect to t we find

$$\dot{E} = m\dot{\mathbf{s}} \cdot \ddot{\mathbf{s}} + I\boldsymbol{\omega} \cdot \dot{\boldsymbol{\omega}} + mg\dot{s}_z. \tag{6.7}$$

Substitutions of Eq. (6.2a) for $\ddot{\mathbf{s}}$ and $I\dot{\boldsymbol{\omega}} = \dot{\mathbf{L}} = \mathbf{a} \times \mathbf{F}$ in the above equation give

$$\begin{aligned}\dot{E} &= \dot{\mathbf{s}} \cdot (\mathbf{F} - mg\hat{\mathbf{z}}) + \boldsymbol{\omega} \cdot (\mathbf{a} \times \mathbf{F}) + mg\dot{s}_z \\ &= \dot{\mathbf{s}} \cdot \mathbf{F} + \boldsymbol{\omega} \cdot (\mathbf{a} \times \mathbf{F}). \end{aligned} \tag{6.8}$$

From Eq. (6.3) we write $\dot{\mathbf{s}} = \mathbf{v}_{\mathrm{P}} - \boldsymbol{\omega} \times \mathbf{a}$. Substitution of this expression for $\dot{\mathbf{s}}$ and use of the identity $(\mathbf{B} \times \mathbf{C}) \cdot \mathbf{A} = (\mathbf{C} \times \mathbf{A}) \cdot \mathbf{B} = \mathbf{B} \cdot (\mathbf{C} \times \mathbf{A})$ in Eq. (6.8) give

$$\begin{aligned}\dot{E} &= (\mathbf{v}_{\mathrm{P}} - \boldsymbol{\omega} \times \mathbf{a}) \cdot \mathbf{F} + \boldsymbol{\omega} \cdot (\mathbf{a} \times \mathbf{F}) \\ &= \mathbf{v}_{\mathrm{P}} \cdot \mathbf{F} - (\boldsymbol{\omega} \times \mathbf{a}) \cdot \mathbf{F} + \boldsymbol{\omega} \cdot (\mathbf{a} \times \mathbf{F}) \\ &= \mathbf{v}_{\mathrm{P}} \cdot \mathbf{F} - \boldsymbol{\omega} \cdot (\mathbf{a} \times \mathbf{F}) + \boldsymbol{\omega} \cdot (\mathbf{a} \times \mathbf{F}) \\ &= \mathbf{v}_{\mathrm{P}} \cdot \mathbf{F} \end{aligned} \tag{6.9}$$

For pure rolling $\mathbf{v}_{\mathrm{P}} = 0$. Therefore, $\dot{E} = 0$ and hence the total energy is conserved.

6.3 Rattleback with a Symmetric Distribution of Mass

Before considering the rattleback with an asymmetric distribution of mass, we take up the case with mass distribution being symmetrical about the vertical plane passing through the ellipsoid's long axis [11].

Let the rattleback made to oscillate *longitudinally*. There is no rotational motion and is rocking. We start with the rattleback standing on the left-side as shown in Fig. 6.2. It first moves leftwards. The speed $\dot{\mathbf{s}}$ of the CoM decreases and it stops. This is the first phase of the motion. In the second phase of oscillation $\dot{\mathbf{s}}$ increases in the positive $\hat{\mathbf{x}}$-direction and it becomes maximum when CoM passes through the contact point P. During these phases the force $\mathbf{F}$ points upward within the angle between the $\hat{\mathbf{z}}$-axis and the line determined by $\mathbf{s}$. From Eq. (6.2a) the vertical component of $\mathbf{F}$ is written as

$$F_z = m\ddot{s}_z + mg \tag{6.10}$$

while the horizontal component of $\mathbf{F}$ is

$$F_x = m\ddot{s}_x\,. \tag{6.11}$$

In F_z, the term $m\ddot{s}_z$ is due to positive (or negative) acceleration of CoM. In phase-1 F_x decelerate the CoM and in phase-2 accelerate the CoM in the positive direction of $\hat{\mathbf{x}}$. In these two phases $\mathbf{F}$ points in the right-side. The torque is $\boldsymbol{\tau} = \mathbf{a} \times \mathbf{F}$. In the next two phases-3 and 4 the oscillation of $\mathbf{a}$ and $\mathbf{F}$ are happened to be the mirror images of the phases-1 and 2.

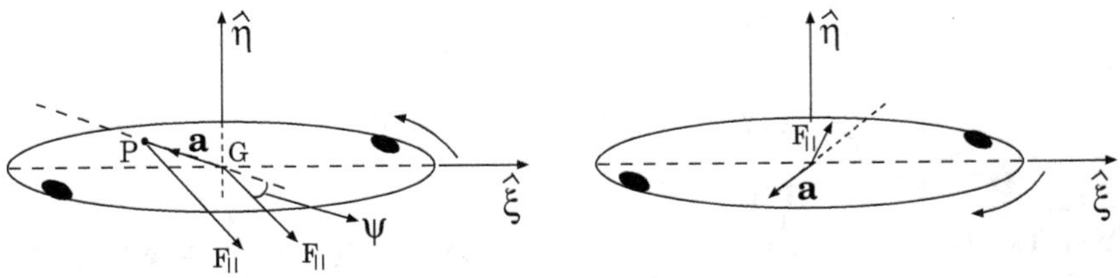

Fig. 6.3: (a) Top view of geometry in the case of stable counter clockwise direction and (b) transversal oscillations generating torque in the clockwise direction [11]. (Reproduced with permission from Springer Nature: S.R. Wojciechowski and M. Przybylska, Reg. Chaotic Dyn. **22**, 368 (2017).)

6.4 Rattleback with an Asymmetric Distribution of Mass

For simplicity, assume that two additional masses are added asymmetrically as in Fig. 6.3a. When the additional weights are present, then during the *longitudinal* oscillation in the left-side in the $\widehat{\boldsymbol{\xi}}$ direction the contact line of the ellipsoid's bottom got deflected on the body surface along a dashed line depicted by **a** in Fig. 6.3a. Suppose the rattleback is accelerated by a torque in the counter clockwise direction. Then, in the $(\widehat{\boldsymbol{\xi}}, \widehat{\boldsymbol{\eta}})$ plane the force $\mathbf{F}_{\|} = F\widehat{\boldsymbol{\xi}} + F\widehat{\boldsymbol{\eta}}$ is directed below the dashed line extending **a**. This leads to a positive vertical component of the torque

$$\tau_z = (\mathbf{a} \times \mathbf{F}_{\|})_z > 0\,. \tag{6.12}$$

This turns the rattleback in the counter clockwise direction since the angle between $\widehat{\boldsymbol{\zeta}}$ and $\widehat{\mathbf{z}}$ is small. It happens similarly when the rattleback inclines to the down-right-side. In this case also the torque has a component contributing to counter clockwise rotation.

If the rattleback begins with *transversal* oscillations along the ellipsoid's short axis, then as shown in Fig. 6.3b the contact trajectory of the body got deflected left. As shown in Fig. 6.3b $\mathbf{F}_{\|} = F\widehat{\boldsymbol{\xi}} + F\widehat{\boldsymbol{\eta}}$ is directed left of the dashed line extending **a**. In this case the vertical component of the torque $\tau_\zeta = (\mathbf{a} \times \mathbf{F}_{\|})_\zeta < 0$ points towards the plane of support. This causes the rattleback to turn in the clockwise direction.

As mentioned above a rattleback can display rotation in the clockwise and counter clockwise directions depending on the initial longitudinal or transversal oscillatory initial conditions. There is a neighbourhood for each initial condition in which $\mathbf{F}_{\|} = F\widehat{\boldsymbol{\xi}} + F\widehat{\boldsymbol{\eta}}$ still occurs below (or left) of the dashed line extending **a** in Figs. 6.3a and 6.3b. There is a direction of initial oscillation where $\mathbf{F}_{\|}$ is pointing along **r** and the vertical torque is ≈ 0. This direction is actually the boundary between the zones where a rattleback turns counter clockwise or clockwise.

A detailed analysis shows that the region of initial conditions corresponding to counter clockwise rotation is very wide. This is a reason for an oscillatory rattleback usually acquire counter clockwise direction of rotation. However, there is a region of initial conditions for which the rattleback first rotates, starts to wobble and undergoes longitudinal rotation. Even though two directions of rotation are allowed, *what is the reason for spin reversal in the case of motion started in the clockwise direction?* This is because when a rattleback is rotated in the clockwise direction it is also slightly oscillating. Because the angle of counter clockwise oscillatory regime is much wider than that of clockwise oscillatory region, the rattleback started spinning in the clockwise direction is more likely to acquire sufficient longitudinal oscillations. As pointed out in [11] this becomes a source for the counter clockwise torque transferring energy from rotation to increase longitudinal oscillations and to reverse rotation of the rattleback.

6.5 Inertial Asymmetry and Rotation of the Principal Moment of Inertia Axes

Rattleback tends to spin smoothly in one sense, when spun in the opposite sense it may even reverse its spin for suitable initial conditions. This chiral behaviour was discussed first by Walker [2]. He found that the structure of the equations changes for an elliptic paraboloid having dissipation free rolling without slip when internal asymmetry exists. He showed that internal asymmetry of the elliptic paraboloid leads to the noncoincidence of the horizontal principal axes of maximum and minimum curvature at the equilibrium contact point and it plays a major role in explaining the chiral behaviour of the reversal of spin in asymmetric rattleback. The analysis was extended to determine the dependence of stability of the vertical spinning solutions on the moments of inertia and curvature of bottom surfaces of rattleback [3].

Following the analysis of Ref. [11] we have seen in Sec. 6.3 in the present chapter that symmetric distribution of mass in rattleback does not produce torque to turn it in the counter clockwise direction. Asymmetry is introduced [11] by considering two extra weights of masses m_{p} each placed asymmetrically to the otherwise symmetrical rattleback as shown in Fig. 6.2.

Due to this asymmetrically added extra masses the axes of the moment of inertia tensor are rotated with respect to the main geometrical axes by an angle δ as shown in Fig. 6.2. Addition of the two masses asymmetrically has made the rattleback acquires chirality which tends to turn in the counter clockwise direction. Section 6.4 explains qualitatively how this change of angles of principal moments of inertia axes leads to a positive vertical components of torque which turns the rattleback in the counter clockwise direction.

In this section, we follow the analysis given in Ref. [11] to find the change of the principal moments of the inertia tensor and their axes and obtain the angle of rotation δ in terms of the principal moments of the inertia tensor.

Let us consider a homogeneous triaxial half-ellipsoid with semi-axis $b_1 > b_2 > b_3$ and mass m. Two identical masses m_p are paced at the ellipsoidal edge of the flat top plane of the rattleback at positions $r_0(x_0, y_0, 0)$ and $-r_0(-x_0, -y_0, 0)$. x_0 and y_0 satisfy the ellipse equation $(x_0^2/b_1^2) + (y_0^2/b_2^2) = 1$. The homogeneous half-ellipsoid has its CoM located at a distance $(3/8)b_3$ from the top flat surface on the z-axis. The principal axes of the inertia tensor without the two masses m_p are $(\hat{\boldsymbol{\xi}}, \hat{\boldsymbol{\eta}}, \hat{\boldsymbol{\zeta}})$ with principal moments $I_1 = (m/5)(b_2^2 + b_3^2)$, $I_2 = (m/5)(b_3^2 + b_1^2)$ and $I_3 = (m/5)(b_1^2 + b_2^2)$. The moment of inertia tensor due to the two masses m_p placed at $r_0 = (x_0, y_0, 0)$ and $-r_0 = (-x_0, -y_0, 0)$ for the same axes is found as (see the exercise 6.1 at the end of this chapter)

$$\mathbb{I}_\mathrm{p} = 2m_\mathrm{p} \begin{bmatrix} y_0^2 & -x_0 y_0 & 0 \\ -x_0 y_0 & x_0^2 & 0 \\ 0 & 0 & x_0^2 + y_0^2 \end{bmatrix}. \tag{6.13}$$

Adding the two masses at the top flat surface shifts the CoM to a new position $d = \dfrac{3}{8}\dfrac{mb_3}{m + 2m_\mathrm{p}}$ (see the exercise 6.2 at the end of this chapter) from the top flat surface. So, we get the moment of inertia for this asymmetrical rattleback as

$$\mathbb{I} = \begin{bmatrix} I_{11} & -I_{12} & 0 \\ -I_{12} & I_{22} & 0 \\ 0 & 0 & I_{33} \end{bmatrix}, \tag{6.14a}$$

where (see the exercise 6.3 at the end of this chapter)

$$I_{11} = \frac{m}{5}\left(b_2^2 + b_3^2\right) - \frac{9}{64}\frac{m^2 b_3^2}{m + 2m_\mathrm{p}} + 2m_\mathrm{p} y_0\,, \tag{6.14b}$$

$$I_{12} = 2m_\mathrm{p} x_0 y_0 \tag{6.14c}$$

and

$$I_{22} = \frac{m}{5}\left(b_3^2 + b_1^2\right) - \frac{9}{64}\frac{m^2 b_3^2}{m + 2m_{\rm p}} + 2m_{\rm p} x_0\,, \tag{6.14d}$$

$$I_{33} = \frac{m}{5}\left(b_1^2 + b_2^2\right) + 2m_{\rm p}\left(x_0^2 + y_0^2\right). \tag{6.14e}$$

For the modified tensor given by Eq. (6.14a) the principal axes have changed from $(\hat{\boldsymbol{\xi}}, \hat{\boldsymbol{\eta}}, \hat{\boldsymbol{\zeta}})$ to $(\hat{\mathbf{1}}, \hat{\mathbf{2}}, \hat{\mathbf{3}})$ with $(\hat{\mathbf{1}}, \hat{\mathbf{2}}, \hat{\mathbf{3}})$ given by Eq. (6.1). Here, δ is the rotation angle shown in Fig. 6.2 and is found to be

$$\delta = \frac{1}{2}\tan^{-1}\left(\frac{2I_{12}}{I_{22} - I_{11}}\right), \tag{6.15}$$

where I_{11}, I_{12} and I_{22} are given by Eqs. (6.14b)-(6.14d). So, asymmetry of rattleback leads to a change of the angles of the principal axes of the inertia tensor from that of the symmetric configuration.

6.6 Conclusion

The rattleback is a rigid mechanical top displaying cértain curious dynamical behaviour. When it is spun in one direction, it will reverse the direction of spin. But if it is spun in other direction, it will not reverse its direction of spin. This strange behaviour may appear to violate the law of conservation of angular momentum. It is found that chirality arising out of the asymmetrical structure of the rattleback is the main cause for this phenomenon of reversal of the spin for a particular direction of the initial spin.

Asymmetry leads to change of the moments of the inertia tensor and their principal axes are rotated by an angle δ from that of symmetrical structure. This leads to a torque for a specific sense of spin and it changes the spin direction of the rattleback. We have given the analysis following the Ref. [11] and explained the basic theory for the working of the rattleback toy. Following the model in [11] the angle of rotation of the principal axes arising out of asymmetry is found by considering two small masses placed asymmetrically on the flat surface of the rattleback. It is to be remembered that this model is only a simplified model which does not take into account of other factors like friction. Coriolis effect, etc. More realistic models lead to nonlinear equations and solving them is quite difficult.

6.7 Bibliography

[1] H.R. Crane, The Phys. Teach. **29**, 278 (1991).
[2] G.T. Walker, Quart. J. Pure Appl. Math. **28**, 175 (1896).

[3] H. Bondi, Proc. R. Soc. Lond. Ser. A **405**, 265 (1986).
[4] W. Case and S. Jalal, Am. J. Phys. **82**, 654 (2014).
[5] M. Paskal, Prikl. Mat. Mekh. **47**, 321 (1983).
[6] A.P. Markeev, Prikl. Mat. Mekh. **47**, 575 (1983).
[7] H.K. Moffatt and T. Tokieda, Proc. Roy. Soc. Edin. Ser. A **138**, 361 (2008).
[8] A. Garcia and M. Hubbard, Proc. Roy. Soc. Lond. Ser. A **418**, 165 (1988).
[9] A.B. Pippard, Eur. J. Phys. **11**, 63 (1990).
[10] Y. Konda and H. Nakanishi, Phys. Rev. E **95**, 062207 (2017).
[11] S.R. Wojciechowski and M. Przybylska, Reg. Chaotic Dyn. **22**, 368 (2017).
[12] L. Franti, Cent. Dur. J. Phys. **11**, 162 (2013).
[13] H. Crane, The Phys. Teach. **29**, 278 (1991).
[14] R. Edge and R. Childers, The Phys. Teach. **37**, 80 (1999).
[15] R. Cross, The Phys. Teach. **51**, 544 (2013).
[16] S. Jones and H.E.M. Hunt, Am. J. Phys. **87**, 699 (2019).

6.8 Exercises

6.1 Obtain the moment of inertia tensor $\mathbb{I}$ for a system with two indentical masses m_p placed at positions $r_0 = (x_0, y_0, 0)$ and $-r_0 = (-x_0, -y_0, 0)$.

6.2 The homogeneous half-ellipsoid of mass m has its CoM at the z-axis at a distance $(3/8)b_3$ from the top flat surface. Show that adding the two masses m_p at $r_0 = (x_0, y_0, 0)$ and $-r_0 = (-x_0, -y_0, 0)$ moves the CoM to the distance $d = \dfrac{3}{8}\dfrac{mb_3}{(m + 2m_\mathrm{p})}$ from the top flat surface.

6.3 Using the moment of inertia given by Eq. (6.13) find the coefficients of the moment of inertia given by Eqs. (6.14).

6.4 The parameters of a symmetrical rattleback are $b_1 = 7.5\,\mathrm{cm}$, $b_2 = 2.0\,\mathrm{cm}$, $b_3 = 1.5\,\mathrm{cm}$ and $m = 18\,\mathrm{g}$. Write the moment of inertia tensor $\mathbb{I}_\mathrm{S}$ and its eigenvalues.

6.5 In an experiment with an asymmetrical rattleback, the values of the parameters used are $b_1 = 7.5\,\mathrm{cm}$, $b_2 = 2\,\mathrm{cm}$, $b_3 = 1.5\,\mathrm{cm}$, $m = 18\,\mathrm{gm}$, $m_\mathrm{p} = 3\,\mathrm{gm}$ and $r_0 = (6, 1.2, 0)$. Determine the inertia matrix and its eigenvalues and the angle δ.

Chapter 7

Hurricane Balls

A *hurricane balls* also called a *double-sphere* is a spinning-top toy made of two steel balls joined together by welding (as shown in Fig. 7.1) or a short steel or a wooden rod. Place this toy on a smooth surface of a floor or a table or on a flat metal pan. Spin it with your fingers. It will spin around at fast rotational rates. *What is amazing about this toy?* Blowing a jet of air on one side of the spinning hurricane balls through a straw or an air hose enormously increases the spin rate, more than several thousands of rotations per minute (rpm) [1,2]. In practice, without much difficulty it is possible to make the double-sphere to rotate with 5000-8000 rpm. When the toy started spinning, initially both the balls are in contact with the flat surface. Then, one ball alone is in contact with the surface, while the other ball detaches it from the surface and makes an angle θ with the surface. This angle is between 0 and $\pi/2$. That is, *it is not spinning with its symmetry axis vertically oriented*. The toy continues its spinning for several minutes with a very little friction. Due to the friction with the surface and air drag, the toy loses its energy slowly and hence its spin rate slowly decreases.

The detailed motion of the double-sphere is not observable with our naked eye due to very high rotational rate. When the spinning rate is

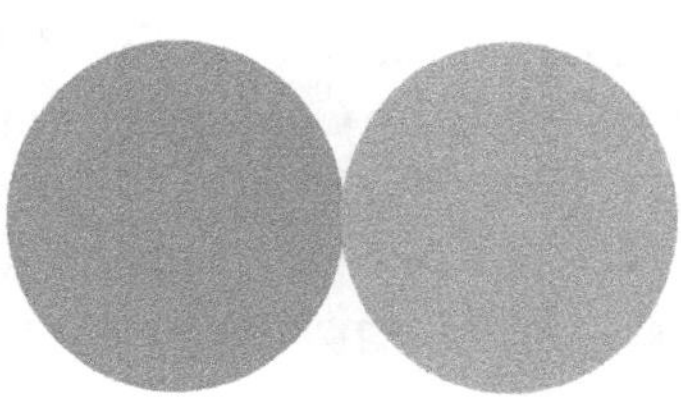

Fig. 7.1: A hurricane balls.

sufficiently high, we can hear the toy hum like a jet engine. For a beautiful and colourful video record and demonstration of the nontrivial rotation of a hurricane balls see [3].

The steady state motion of the double-sphere in which only one ball is in touch with the surface of a floor or a table essentially has only a spinning motion about its centre of mass (CoM) with out slipping. This motion can be dealt using a Lagrangian approach with Euler angles. A qualitative explanation for the above steady state motion can be achieved. Some of the interesting results obtained from the theoretical and experimental analysis with the hurricane balls are [1]:

1. The steady state motion is not torque free.
2. The steady state motion is not sustained if the spin rate is below a critical spin rate.
3. The angle θ between its symmetry axis and the horizontal axis depends on the rate of rotation.
4. The angle θ increases with the spin rate, however, after reaching a maximum value θ remains the same with further increase in the spin rate.

In this chapter we present the theory of the hurricane balls toy and describe its motion [1].

7.1 Qualitative Analysis

First, consider the ideal case of the double-sphere where it is made to spin about its symmetry axis with both balls in contact with the surface of, say, the table [1]. Assume that its CoM is at rest. Let us bring out the forces in action. The prime forces are: (1) gravity, (2) the normal force of the table and (3) kinetic friction. We neglect the air resistance. The normal force balances the gravitational force. As a result the total force points in the direction of friction and is out of the page in Fig. 7.2. Consequently, as the toy begins to roll along the table its CoM gets accelerated.

Denote $\mathbf{N}_{\mathrm{A}}$ and $\mathbf{N}_{\mathrm{B}}$ be the upward normal forces acting at the points A and B in Fig. 7.2. $\mathbf{N}_{\mathrm{A}}$ and $\mathbf{N}_{\mathrm{B}}$ give equal and opposite torques and points into and out of the page, respectively (see the solved problem 1 in the present chapter). Further, there are torques $\mathbf{\Gamma}_{\mathbf{F}_{\mathrm{A}}}$ and $\mathbf{\Gamma}_{\mathbf{F}_{\mathrm{B}}}$ due to the frictional forces $\mathbf{F}_{\mathrm{A}}$ and $\mathbf{F}_{\mathrm{B}}$ acting at the points A and B, respectively. The total torque about the CoM points in a direction that is opposite to

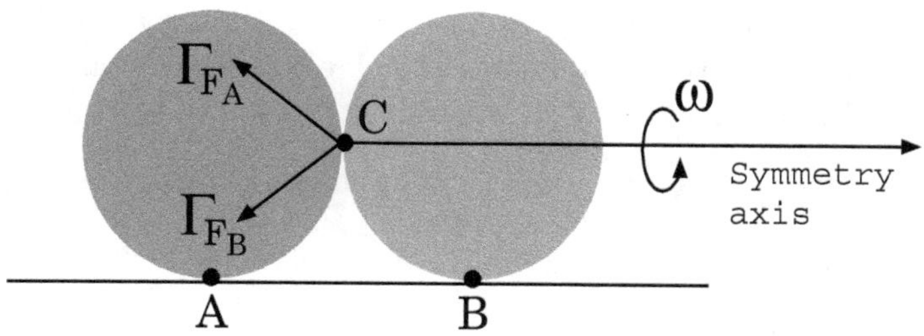

Fig. 7.2: A schematic representation of a double-sphere rotating about its symmetry axis. For details see the text [1]. (Reproduced from Am. J. Phys. **83**, 959 (2015), with permission of American Association of Physics Teachers. Copyright 2015 D.P. Jackson, D. Mertens and B.J. Pearson, licensed under a Creative Commons Attribution License.)

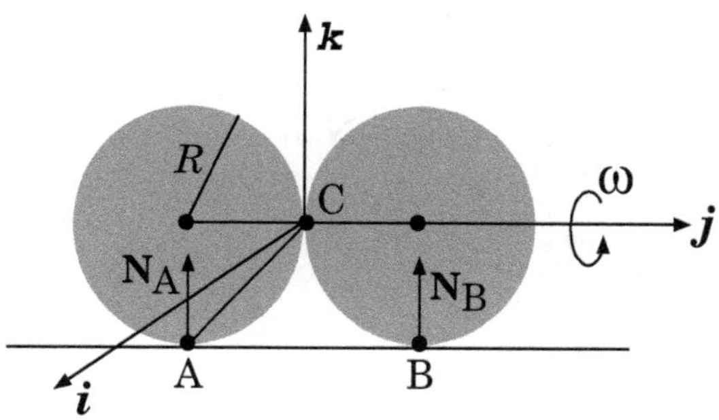

Fig. 7.3: Representation of the upward normal forces $\mathbf{N}_A$ and $\mathbf{N}_B$ acting at the contact points A and B and the coordinate axes of the hurricane balls system.

the angular momentum **L** (see the exercise 7.1 at the end of this chapter). Here $\mathbf{L} = I\boldsymbol{\omega}$ where $\boldsymbol{\omega}$ is the angular velocity and I is the moment of inertia about the symmetry axis. Therefore, the total torque decreases **L** and the spin rate got reduced.

Solved Problem 1:

Considering Fig. 7.3 show that $\mathbf{N}_A$ and $\mathbf{N}_B$ give equal and opposite torques and points into and out of page, respectively.

Consider the coordinate axes as shown in Fig. 7.3. If R is the radius of the sphere, then the position coordinates of A are $(0, -R, -R)$ and that of B are $(0, R, -R)$. We have

$$\mathbf{N}_\mathrm{A} = |\mathbf{N}_\mathrm{A}|\mathbf{k}, \quad \mathbf{N}_\mathrm{B} = |\mathbf{N}_\mathrm{B}|\mathbf{k}, \quad |\mathbf{N}_\mathrm{A}| = |\mathbf{N}_\mathrm{B}|. \tag{7.1}$$

The torque due to $|\mathbf{N}_\mathrm{A}|$ is given by

$$\begin{aligned}\mathbf{\Gamma}_{\mathbf{N}_\mathrm{A}} &= \mathbf{CA} \times \mathbf{N}_\mathrm{A} \\ &= (-R\mathbf{j} - R\mathbf{k}) \times |\mathbf{N}_\mathrm{A}|\mathbf{k} \\ &= -R\mathbf{j} \times |\mathbf{N}_\mathrm{A}|\mathbf{k} \\ &= -R|\mathbf{N}_\mathrm{A}|\mathbf{i}.\end{aligned} \tag{7.2}$$

So, $\mathbf{\Gamma}_{\mathbf{N}_\mathrm{A}}$ points in the $-\mathbf{i}$ direction (into the page).

The torque due to $\mathbf{N}_\mathrm{B}$ is given by

$$\begin{aligned}\mathbf{\Gamma}_{\mathbf{N}_\mathrm{B}} &= \mathbf{CB} \times \mathbf{N}_\mathrm{B} \\ &= (R\mathbf{j} - R\mathbf{k}) \times |\mathbf{N}_\mathrm{B}|\mathbf{k} \\ &= R\mathbf{j} \times |\mathbf{N}_\mathrm{B}|\mathbf{k} \\ &= R|\mathbf{N}_\mathrm{B}|\mathbf{i}.\end{aligned} \tag{7.3}$$

$\mathbf{\Gamma}_{\mathbf{N}_\mathrm{B}}$ acts in the $+\mathbf{i}$ direction (out of the page). From Eqs. (7.1)-(7.3) we find

$$\mathbf{\Gamma}_{\mathbf{N}_\mathrm{A}} = -\mathbf{\Gamma}_{\mathbf{N}_\mathrm{B}}. \tag{7.4}$$

Therefore, the torques are equal and opposite.

Next, assume that the contact of the right balls at B with the table is lost. For simplicity, say the part of the table under the right ball disappeared suddenly. Now, $\mathbf{N}_\mathrm{B}$, $\mathbf{F}_\mathrm{B}$, $\mathbf{\Gamma}_{N_\mathrm{B}}$ and $\mathbf{\Gamma}_{F_\mathrm{B}}$ are 0. However, say, the toy remains oriented as in Fig. 7.2. Then, consider the torques about the CoM. Due to the frictional force $\mathbf{F}_\mathrm{A}$ there is a vertical component of the frictional torque and also due to $\mathbf{N}_\mathrm{A}$ there is a component of the torque pointing into the page. The total torque pointing to the left opposes $\mathbf{L}$. This leads to decrease in the spin rate. However, interestingly, the direction of $\mathbf{L}$ is changed by the vertical component of the torque. Because the left ball cannot move downward there is a net increase in $\mathbf{N}_\mathrm{A}$. There is a rise in the CoM. The right ball is *lifted-off* the table. The toy begins to precess. The spin rate decreases continuously due to dissipation of energy. This continues as long as the left ball rolls without slipping.

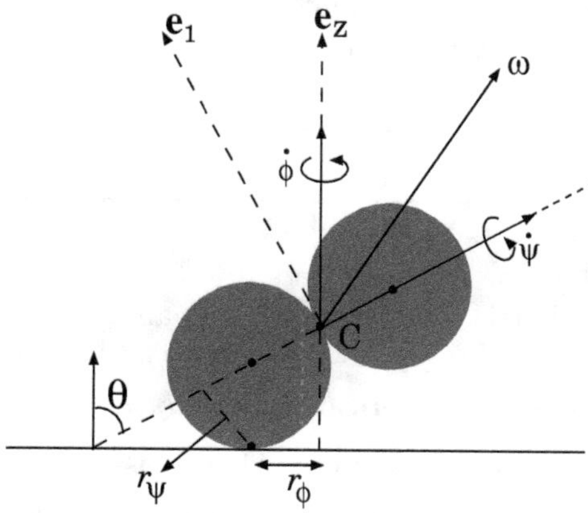

Fig. 7.4: The geometry of a double-sphere in its steady state motion [1].

7.2 Quantitative Analysis

In the steady state motion one of the balls of the double-sphere rises up from the table while the other is in contact with the table. The double-sphere spins about its symmetry axis.

Consider Fig. 7.4 presenting the geometry of the double-sphere in its steady state motion [1]. Each ball has mass m and radius R. $\theta \in [0, \pi/2]$ is the angle between the symmetry axis and the vertical axis. C is the CoM while P is the point of contact with the table. $\{\mathbf{e}_x, \mathbf{e}_y, \mathbf{e}_z\}$ be an inertial laboratory coordinate system. $\{\mathbf{e}_1, \mathbf{e}_2, \mathbf{e}_3\}$ be the rotating frame coordinate system, it remains parallel to the principal axis of the body and is rotating about the z-axis with angular velocity Ω. Here, $\mathbf{e}_z = \sin\theta \mathbf{e}_1 + \dot{\phi}\cos\theta \mathbf{e}_3$.

7.2.1 *Lagrangian of the System*

The kinetic energy T is

$$T = \frac{1}{2}\left(I_1\omega_1^2 + I_2\omega_2^2 + I_3\omega_3^2\right) , \tag{7.5}$$

where ω_i's are the components of the angular velocity vector along the principal axes. From Fig. 7.4 we have

$$\begin{aligned}
\boldsymbol{\omega} &= \dot{\phi}\,\mathbf{e}_z - \dot{\theta}\,\mathbf{e}_2 + \dot{\psi}\mathbf{e}_3 \\
&= \dot{\phi}\sin\theta\,\mathbf{e}_1 - \dot{\theta}\,\mathbf{e}_2 + \left(\dot{\psi} + \dot{\phi}\cos\theta\right)\mathbf{e}_3 \qquad (7.6)\\
&= \omega_1\mathbf{e}_1 + \omega_2\mathbf{e}_2 + \omega_3\mathbf{e}_3 \,. \qquad (7.7)
\end{aligned}$$

Choose the gravitational potential energy be zero at $\theta = \pi/2$. Then, the Lagrangian of the system $\mathcal{L} = T - V$ is

$$\mathcal{L} = \frac{1}{2} I_1 \dot{\phi}^2 \sin^2\theta + \frac{1}{2} I_2 \dot{\theta}^2 + \frac{1}{2} I_3 \left(\dot{\psi} + \dot{\phi}\cos\theta\right)^2 - 2mgR\cos\theta\,. \tag{7.8}$$

Solved Problem 2:

With the moment of inertia of a sphere through an axis passing through its centre as $(2/5)mR^2$ prove that $I_1 = 2\left(\frac{2}{5}mR^2 + mR^2\right)$ and $I_3 = 2\left(\frac{2}{5}mR^2\right)$ where m is the mass of each spherical ball and R is its radius.

Since I_3 is the moment of inertia of the axis $\mathbf{e}_3$ passing through the centre of the two spheres, we get $I_3 = 2\left(\frac{2}{5}mR^2\right)$. For the axis passing through the point of contact C, perpendicular to the symmetry axis, we find the moment of inertia of one sphere using the parallel axis theorem as $\left(\frac{2}{5}mR^2 + mR^2\right)$. Therefore, $I_1 = 2\left(\frac{2}{5}mR^2 + mR^2\right)$.

7.2.2 *Condition for Lift-Off*

Now, we obtain the condition for the right ball to be lifted-off from the table [1]. This will happen when $\theta = \pi/2$ (the double-sphere is horizontal) and $\ddot{\theta} < 0$. From

$$\frac{\mathrm{d}}{\mathrm{d}t}\left(\frac{\partial \mathcal{L}}{\partial \dot{\theta}}\right) - \frac{\partial \mathcal{L}}{\partial \theta} = 0 \tag{7.9}$$

we find

$$\frac{\mathrm{d}}{\mathrm{d}t}\left(I_2\dot{\theta}\right) = I_1\dot{\phi}^2\sin\theta\cos\theta - I_3\left(\dot{\psi} + \dot{\phi}\cos\theta\right)\dot{\phi}\sin\theta + 2mgR\sin\theta\,. \tag{7.10}$$

At $\theta = \pi/2$, Eq. (7.10) becomes

$$I_2\ddot{\theta} = -I_3\dot{\psi}\dot{\phi} + 2mgR\,. \tag{7.11}$$

The condition $\ddot{\theta} < 0$ gives $2mgR - I_3\dot{\psi}\dot{\phi} < 0$ or $2mgR < I_3\dot{\psi}\dot{\phi}$. That is,

$$\dot{\psi}\dot{\phi} > \frac{2mgR}{I_3}, \quad I_3 = 2\left(\frac{2}{5}mR^2\right). \tag{7.12}$$

This is the condition for the right ball to be lifted-off.

7.2.3 Steady State Solution

In the steady state, θ is a constant and hence $\dot{\theta} = 0$. Further, the toy rolls without slipping. From Fig. 7.4 we note that the contact point P moves with the speed $v_\phi = \dot{\phi} r_\phi$ with $r_\phi = R \sin\theta$ about the vertical axis and also with speed $v_\psi = \dot{\psi} r_\psi$ with $r_\psi = R\sin\theta$ about the symmetry axis. When the double-sphere rolls without slipping, then $v_\phi = v_\psi$, that is, $\dot{\phi} = \dot{\psi}$ and let it be Ω.

Now, the $\mathcal{L}$ given by the Eq. (7.8) becomes

$$\mathcal{L} = \frac{1}{2} I_1 \Omega^2 \sin^2\theta + \frac{1}{2} I_3 \Omega^2 (1+\cos\theta)^2 - 2mgR\cos\theta\,. \tag{7.13}$$

Then, as $\dot{\theta} = 0$ the Lagrange Eq. (7.9) becomes $\partial\mathcal{L}/\partial\theta = 0$. For the $\mathcal{L}$ given by Eq. (7.13) $\partial\mathcal{L}/\partial\theta = 0$ is obtained as

$$I_1\Omega^2 \sin\theta\cos\theta - I_3\Omega^2(1+\cos\theta)\sin\theta + 2mgR\sin\theta = 0. \tag{7.14}$$

This equation can be rewritten as

$$\sin\theta\left[\cos\theta - \frac{(2mgR/I_3\Omega^2)-1}{1-(I_1/I_3)}\right] = 0\,. \tag{7.15}$$

The roots of the above equation are the steady state solutions of θ $(=\theta^*)$. Substituting $I_1 = 2(\frac{2}{5}mR^2 + mR^2) = 2(\frac{7}{5}mR^2)$ and $I_3 = 2(\frac{2}{5}mR^2)$ the Eq. (7.15) becomes

$$\sin\theta\left[\cos\theta - \frac{2}{5} + \frac{g}{R\Omega^2}\right] = 0\,. \tag{7.16}$$

This equation is true only if either

$$\sin\theta = 0 \tag{7.17}$$

or

$$\cos\theta = \frac{2}{5} - \frac{g}{R\Omega^2} = \alpha\,. \tag{7.18}$$

Therefore, the steady state solutions (equilibrium points) are

$$\theta_1^* = 0, \quad \theta_2^* = \cos^{-1}\left[\frac{2}{5} - \frac{g}{R\Omega^2}\right], \quad \theta^* \in [0,\pi/2]\,. \tag{7.19}$$

The stability of the the equilibrium points of a two-dimensional system can be analysed as described in Sec. 1.8. A detailed analysis (see the exercise 7.9 at the end of this chapter) shows that $\theta_1^* = 0$ corresponding to one ball being directly above the other is an unstable state. The existence of θ_2^* and its stability depend on the value of Ω. In the next section we discuss the stability of θ_2^* for sufficiently large Ω.

7.3 Role of Ω on the Steady State Solution

First observe that for a sufficiently large Ω, $\theta_2^* = \cos^{-1}(2/5)$ which corresponds to that of a free body motion ($g = 0$). For $0 < \theta_2^* < \pi/2$ we require

$$\Omega^2 > \Omega_c^2 = \frac{5g}{2R}. \tag{7.20}$$

This implies that for a steady state with $\theta_2^* \in [0, \pi/2]$ the angular velocity must be above a certain critical frequency given by Eq. (7.20).

When $\Omega \leq \Omega_c$ the quantity α becomes negative and there is no steady state solution for θ with $0 < \theta_2^* < \pi/2$. That is, both the balls are in contact with the surface of the table and the toy is spinning. When $\Omega = \Omega_c$ we have $\theta_2^* = \cos^{-1} 0 = \pi/2$ and $\alpha = 0$. Note that θ is the angle between the fixed vertical axis and the symmetry axis (refer Fig. 7.4). When both the balls are in contact with the table, then $\theta = \pi/2$ while the angle between the symmetry axis and the horizontal is 0. When the ball is lifted-off, then $\theta < \pi/2$ while the angle $\delta = \pi/2 - \theta > 0$. When Ω increases from Ω_c, α increases from 0 and θ_2^* decreases from $\pi/2 = 90°$. α is not a linear function of Ω. Interestingly, in the limit of very large Ω ($\Omega \to \infty$), $\alpha \to 2/5$ ($= 0.4$) and $\theta_{2,\text{min}}^* = \cos^{-1}(0.4) \approx 66°$. In this limit $\delta_{\text{max}} = 90° - 66° = 24°$. In this case the steady state motion is practically indistinguishable from free body motion ($g = 0$). The possible range of θ_2^* is $[66°, 90°]$. Jackson and his collaborators [1] performed experiments with hurricane balls of different diameter. The theoretically predicted δ_{max} and θ_{min} are in very good agreement with the experimentally calculated values.

7.4 Two Balls Separated by a Distance

In the hurricane balls toy considered so far the two balls are joined together, that is, the spacing between the two balls is zero. Suppose the spacing between the balls is, say, $2Rd$. In other words, the centres of the two balls are separated by the distance $2R(1 + d)$ with $|d| \ll 1$ where d is a small dimensionless parameter [1].

The moment of inertia along the symmetry axis can be approximated as $I_3 \approx 2(\frac{2}{5}mR^2)$ since $|d| \ll 1$. Next,

$$I_1 = 2\left[\frac{2}{5}mR^2 + mR^2(1+d)^2\right]. \tag{7.21}$$

Neglecting the term containing d^2 we get

$$\begin{aligned} I_1 &\approx 2\left(\frac{2}{5}mR^2 + mR^2 + 2mR^2 d\right) \\ &\approx 2\left(\frac{7}{5}mR^2\right)\left(1+\frac{10}{7}d\right). \end{aligned} \tag{7.22}$$

CoM is at a distance $R(1+d)$ from the centre of each ball. In this case $V = 2mgR(1+d)\cos\theta$, $v_\phi = \dot{\phi}R(1+d)\sin\theta$ and $v_\psi = \dot{\psi}R\sin\theta$. The condition $v_\phi = v_\psi$ for rolling without slipping gives $\dot{\phi}(1+d) = \dot{\psi}$. With $\dot{\phi} = \Omega$ the result is $\dot{\psi} = \Omega(1+d)$. Then, Eq. (7.6) becomes

$$\boldsymbol{\omega} = \dot{\phi}\sin\theta\,\mathbf{e}_1 - \dot{\theta}\,\mathbf{e}_2 + \Omega(1+d+\cos\theta)\mathbf{e}_3\,. \tag{7.23}$$

Proceeding further for $d \neq 0$, similar to $d = 0$ case, we arrive at [1] (verify)

$$\theta_2^* = \cos^{-1}\left[\left(\frac{2}{5} - \frac{g}{R\Omega^2}\right)(1-d)\right]. \tag{7.24}$$

The detailed analysis presented for the case of the balls can be extended to other forms of objects, like disks, cylinders etc.

7.5 Conclusion

A theoretical model given in [1] for the motion of the hurricane balls has been presented. Condition for the lift-off of one of the balls has been obtained. The product of $\dot{\phi}\dot{\psi}$ has to be fairly large for one of the balls to be lift-off. The usual method of starting the hurricane balls will give large $\dot{\phi}$ with negligible $\dot{\psi}$. Asymmetries in friction may lead to sufficiently large $\dot{\psi}$ to satisfy the lift-off condition. As suggested in [1] one of the best ways to start the stable motion of the double-sphere is to press down one of the two balls until the double-sphere shoots out from underneath the finger. With suitable equipments the students can perform many experiments to study the motion of the hurricane balls. By using double-sphere glued together and also separated by a small distance with a rod students can study the change in motion of the double-balls.

7.6 Bibliography

[1] D.P. Jackson, D. Mertens and B.J. Pearson, Am. J. Phys. **83**, 959 (2015).

[2] W.L. Andersen and S. Werner, Eur. J. Phys. **36**, 055013 (2015).

[3] https://www.theawesomer.com/hurricane-balls/4421/.

7.7 Exercises

7.1 Find the torques $\mathbf{\Gamma}_{\mathbf{F}_\mathrm{A}}$ and $\mathbf{\Gamma}_{\mathbf{F}_\mathrm{B}}$ due to the frictional forces $\mathbf{F}_\mathrm{A}$ and $\mathbf{F}_\mathrm{B}$, respectively. Prove that the resultant torque acts opposite to $\mathbf{L}$ reducing $\boldsymbol{\omega}$.

7.2 Consider Fig. 7.2. Assume that the sphere on the right-side is without contact at B but still almost horizontal. By finding out the torque acting on the system prove the following:
(a) The spin rate along the symmetry axis reduces.
(b) The direction of $\mathbf{L}$ along the symmetry axis is no more horizontal.
(c) The normal force $\mathbf{N}_\mathrm{A}$ increases.
(d) The centre of mass (CoM) raises.

7.3 Show that the angular momentum $\mathbf{L}$ is not aligned with $\boldsymbol{\omega}$ for the steady state motion of the hurricane balls.

7.4 For a hurricane balls toy with each ball of mass 0.05 kg and radius 0.01 m find the condition on $\dot{\psi}\dot{\phi}$ for lift-off.

7.5 For a hurricane balls toy with each ball of mass 0.05 kg and radius 0.01 m calculate the minimum rotation frequency for steady state solution.

7.6 If $\mathbf{N}$ is the only source for the total torque $\mathbf{\Gamma}$ about the CoM then write the expression for $\mathbf{\Gamma}_{\mathrm{Lab}}$.

7.7 Refering to the Fig. 7.4 write an expression for the normal reaction force $\mathbf{N}$ in the laboratory frame at the point P.

7.8 Plot θ as a function of $\bar{\Omega} = \Omega\sqrt{R/g}$ for $\bar{\Omega} \in \left[\bar{\Omega}_\mathrm{c}, 11\right]$ and determine $\theta^*_{2,\mathrm{min}}$.

7.9 If the gap between the balls is $0.1R$ find $\theta^*_{2,\mathrm{min}}$.

7.10 Determine the stability of the equilibrium points of the hurricane balls system.

Chapter 8

Gee-Haw Whammy-Diddle (The Notched Stick)

An intriguing rhythm toy is the *gee-haw whammy-diddle* also called *notched stick*. It produces a characteristic sound when a stick or a rod is rubbed over the notches back and forth in the notched stick. The rubbing causes a propeller loosely attached to the notched end of the stick to revolve. The direction of rotation of the propeller can be easily controlled. The gee-haw whimmy-diddle is a simple mechanical toy consisting of two wooden (or plastic or metal) sticks and a propeller (spinner) [1-5]. The main stick is approximately of 18-20 cm long, usually of square or rectangular or circular cross-section, notched with a rasp or file for about 6-8 cm at one end and each notch being of V-shape with 3-5 mm wide and about 2 mm gap between the notches. The spacing between the notches need not be uniform. The propeeler is usually a strip of wood roughly of 5 cm long and 1.5-2 cm wide. The propeller is drilled with a round hole twice the diameter of a nail and held loosely to the notched end of the stick by the nail. One can use a pencil attached with an eraser as the main stick, make notches using a paper cutter, a thin circular cardboard as the spinner, a hole puncher as a nail and an ice-cream stick as the driving stick.

Figure 8.1 shows a few notched stick toys. Hold a notched stick stationary in one hand with the notches up. Rub the notches with the driving stick (the second stick). This action will cause the spinner to rotate. The direction of the rotation of the spinner can be reversed by rubbing your finger against the notched stick while rubbing and hence the name gee-haw. The material and cross-sectional features of the toy do not have influence on the basic working principle, however, affect many details of the behaviour of it. Experimental and theoretical studies have been performed to understand the mechanism of working of the notched stick toy [6-9].

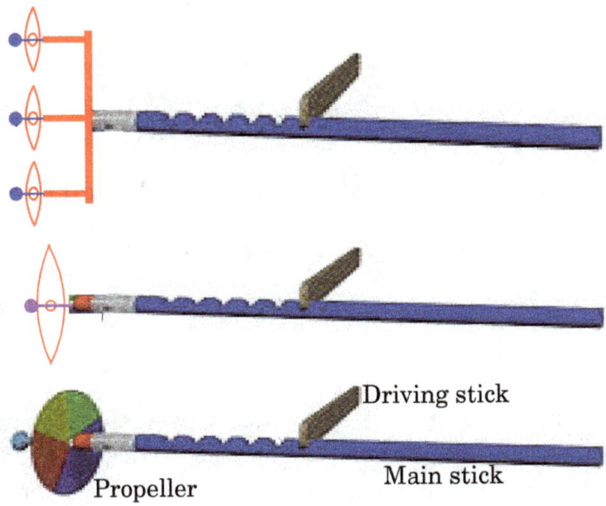

Fig. 8.1: A few notched stick toys.

The world champion competition considers the number of times the spinner is reversed its direction of rotation in twelve seconds. The other names of this toy are girigiri-garigari, bozo-bozo, voodoo and ouija windmill and hoodoo.

How does the motion of the driving stick over the notches produce rotary motion of the spinner? We answer this question in the next section. Then, we present a theoretical model which points out that the working principle of the toy can be understand by the occurrence of parametric and/or non-parametric resonance. We briefly point out an another toy called gravity-spinning magnet where the rotational motion of a propeller is induced by the vibration of the cycle spoke caused by the ring magnet moving down.

8.1 Physical Explanation

The main stick vibrates when it is rubbed by the driving stick. The grip used and the pressure (force) exerted by the performer determine the properties of these vibrations. When the main stick vibrates the propeller collides with the nail resulting in its rotation. The notches serve as a means of producing forced vibrations in the stick. Simple rubbing of the main stick with the driving stick normal to the length of the main stick without notches will not make the propeller to rotate.

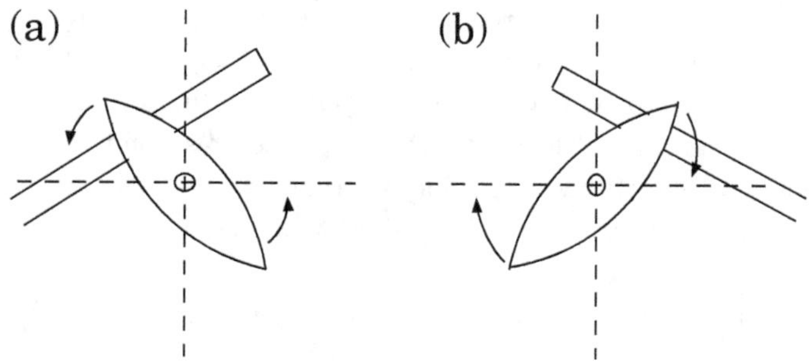

Fig. 8.2: Two different inclinations of the driving stick and the corresponding direction of rotation of the propeller.

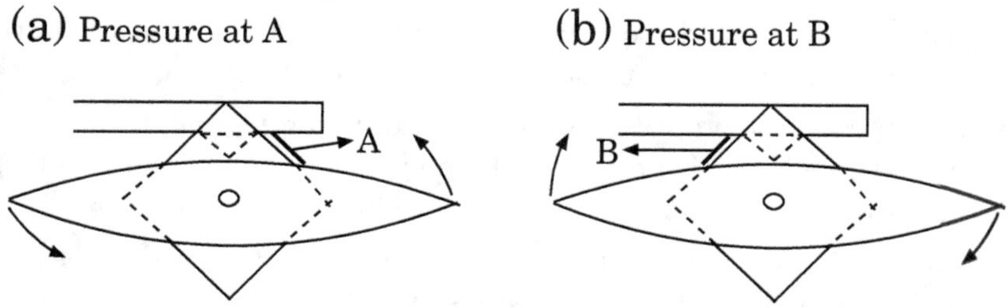

Fig. 8.3: A pressure applied at two different points leading to different direction of rotation.

For simplicity, consider the main stick with square cross-section. The performer can apply two simple tricks for operating the notched stick: incline the driving stick with the main stick or allow the finger (or thumb) to rub the side of the main stick while rubbing it by the driving stick [10]. The direction of rotation of the propeller depends on the way the stick is rubbed. For the first trick Fig. 8.2 shows the two possible inclination of the driving stick to the smaller cross-sectional dimension of the main stick and the corresponding resulting direction of rotation of the propeller [1]. In the second trick the thumb of the hand grasping the driving stick rubs the main stick thereby apply a pressure at A or B as illustrated in Fig. 8.3 [10,11]. The pressure applied at A (B) leads to clockwise (counter clockwise) rotation of the propeller. When the performer changes the point of pressure applied alternately, then the direction of rotation is also reversed alternately.

As pointed out by David Scott [10], the elliptical or circular motion of the nail is the direct cause of the rotation of the propeller. Simple rubbing of the notched stick produces periodic impulses leading to linear vibration. Application of additional force via the trick one or second to the notched stick essentially distorts the force field wherein it moves. Clockwise (counter clockwise) rotation is produced if the distortion increases the forces towards A (B).

We list out some of the features of the notched stick [1-4, 10-12].

- The propeller rotates only if the nail de-touches from it during the movement. That is, the propeller is able to rotate only if it gets separated from the nail during its motion. For this to happen the nail has to undergo acceleration downwards faster than the gravitational acceleration.
- By means of a simple harmonic motion of the stick, the propeller is able to accelerate from rest to high frequencies. This stabilized speed of the propeller strongly depends on the amplitude of the oscillations.
- The speed of the propeller can be changed by varying the speed of rubbing. A rapid spinning can be realized by means of very rapid rubbing. Similarly, the propeller can be made to spin with a constant speed.
- The effects can be pronounced by clamping the notched end of the stick in vise or holding it firmly against a solid table.
- The rotation of the propeller does not depend on the position of the fingers holding the stick.
- The strength and the frequency of rubbing and the geometry of the stick do not have influence on the basic working mechanism of the toy.

8.2 A Theoretical Model

The contact of the hole of the propeller and the moving nail can be either tight or loose. In the case of tight contact, the external boundary of the hole does not slip on the nail due to the static friction forces. The loose contact case is like a ring hanging on a stake. Assume that the nail at rest is placed horizontally. The nail may exhibit the following kinds of motion:

1. A horizontal sinusoidal motion.
2. A vertical sinusoidal motion.
3. A sinusoidal motion in both horizontal and vertical directions.

We consider the model presented in [9].

8.2.1 *Tight Contact*

The associated system has a flat rigid body P rotating freely around a point C (the position of the nail) of it in a fixed plane. Assume that Oxy is an inertial frame of reference. Ox and Oy are the horizontal frame of axes with unit vectos $\mathbf{e}_x$ and $\mathbf{e}_y$, respectively. The law for the movement of C is

$$C - O = \xi(t)\mathbf{e}_x + \eta(t)\mathbf{e}_y\,, \tag{8.1}$$

where $\xi(t)$ and $\eta(t)$ are given functions of time. The centre of mass (CoM) of P is G. If G and C are different, then denote the distance between them as a and the angle between the segment CG and the downward vertical straight-line passing through C as ϕ.

Define the mass of P as m, the moment of inertia of P about the axis passing through G orthogonal to Oxy as I and g as the acceleration due to gravity. Then, the Lagrangian of the system is

$$L = \frac{1}{2}\left(I + ma^2\right)\dot{\phi}^2 + ma\left(\cos\phi\dot{\xi} + \sin\phi\dot{\eta}\right) + mga\cos\phi. \tag{8.2}$$

With the viscous term as $-\beta\dot{\phi}$ where β is the friction constant, the Lagrange equation of motion

$$\frac{\mathrm{d}}{\mathrm{d}t}\left(\frac{\partial L}{\partial\dot{\phi}}\right) - \frac{\partial L}{\partial\phi} = -\beta\dot{\phi} \tag{8.3}$$

becomes

$$\left(I + ma^2\right)\ddot{\phi} + mga\sin\phi + ma\left(\cos\phi\ddot{\xi} + \sin\phi\ddot{\eta}\right) = -\beta\dot{\phi}\,. \tag{8.4}$$

When $a = 0$, that is G coincides with C, then Eq. (8.4) becomes $I\ddot{\phi} + \beta\dot{\phi} = 0$. All the configurations of P are in stable equilibria. The motion of the nail is damped and not significant in setting the propeller to undergo rotation. On the other hand, when the gravity is neglected and $a \neq 0$, Eq. (8.4) takes the form

$$\left(I + ma^2\right)\ddot{\phi} + ma\left(\cos\phi\ddot{\xi} + \sin\phi\ddot{\eta}\right) = -\beta\dot{\phi}\,. \tag{8.5}$$

This is a damped and parametrically driven nonlinear system. This system can exhibit parametric resonance when ξ and η are periodic functions of time. Actually, resonance refers to a realization of a maximum response of a dynamical system. Parametric resonance is a kind of resonance induced by a periodic variation of a parameter of the system [13].

Next, consider the full Eq. (8.4). For $\ddot{\xi} = \ddot{\eta} = 0$, this equation is the unforced but damped pendulum system. The rotational equilibrium points are $\phi_1 = 0$ and $\phi_2 = \pi$. The point $\phi_1 = 0$ is stable while $\phi_2 = \pi$ is unstable as shown below in the solved problem 1.

Solved Problem 1:

Consider Eq. (8.4) with $\ddot{\xi} = \ddot{\eta} = 0$. Determine its equilibrium points and their stability for $a \neq 0$.

Equation (8.4) with $\ddot{\xi} = \ddot{\eta} = 0$ is

$$\ddot{\phi} + \frac{\beta}{(I + ma^2)}\dot{\phi} + \frac{mga}{(I + ma^2)} \sin\phi = 0. \tag{8.6}$$

Define $d = \beta/(I + ma^2)$ and $\alpha = mga/(I + ma^2)$. Then, Eq. (8.6) becomes

$$\ddot{\phi} + d\dot{\phi} + \alpha \sin\phi = 0. \tag{8.7}$$

We rewrite this equation as a system of first-order equation as

$$\dot{\phi} = \psi = P(\phi, \psi), \quad \dot{\psi} = -d\psi - \alpha \sin\phi = Q(\phi, \psi). \tag{8.8}$$

Its phase space coordinates are ϕ and ψ. For a system $\dot{\mathbf{X}} = F(\mathbf{X})$, any admissible solution of $F(\mathbf{X}) = \mathbf{0}$ gives an equilibrium point. This is because if, say, $\mathbf{X}_1$ is a solution of $\dot{\mathbf{X}} = F(\mathbf{X})$ at a given time and $\dot{\mathbf{X}}_1 = 0$, then it continues to be so for all times and hence is an equilibrium point. The equilibrium points of (8.8) obtained from $P(\phi, \psi) = 0$ and $Q(\phi, \psi) = 0$ are

$$(\phi_1, \psi_1) = (0, 0) \text{ and } (\phi_2, \psi_2) = (\pi, 0). \tag{8.9}$$

An equilibrium point is said to *stable* if the neighbouring trajectories approach it asymptotically (as $t \to \infty$). For the two-dimensional system (8.8), an equilibrium point is stable if the real parts of both the eigenvalues of the so-called Jacobian matrix of (8.8) given by

$$J = \begin{pmatrix} \partial P/\partial\phi & \partial P/\partial\psi \\ \partial Q/\partial\phi & \partial Q/\partial\psi \end{pmatrix} \tag{8.10}$$

are negative (refer Sec. 1.8). The eigenvalues are the roots of the equation

$$\left| \begin{matrix} \partial P/\partial\phi - \lambda & \partial P/\partial\psi \\ \partial Q/\partial\phi & \partial Q/\partial\psi - \lambda \end{matrix} \right|_{(\phi^*, \psi^*)} = 0, \tag{8.11}$$

where (ϕ^*, ψ^*) denote the equilibrium point. For the system (8.8), the above equation gives $\lambda(d + \lambda) + \alpha \cos\phi^* = 0$. For $\phi^* = \phi_1 = 0$ the eigenvalues are

$$\lambda_\pm = \frac{1}{2}\left[-d \pm \sqrt{d^2 - 4\alpha}\right]. \tag{8.12}$$

As d and α are positive the real part of λ_+ and λ_- are negative and hence $(\phi_1, \psi_1) = (0, 0)$ is stable. For $(\phi_2, \psi_2) = (\pi, 0)$ the eigenvalues are

$$\lambda_\pm = \frac{1}{2}\left[-d \pm \sqrt{d^2 + 4\alpha}\right]. \tag{8.13}$$

We have $\lambda_+ > 0$ and $\lambda_- < 0$. Since one of the eigenvalues is positive the equilibrium point $(\phi_2, \psi_2) = (\pi, 0)$ is unstable.

Determination of equilibrium points and their stability for $a = 0$ is left as an exercise to the readers.

Onset of rotational motion can be realized in the following three parametrically driven cases:

1. $\ddot{\eta} = 0$ and $\ddot{\xi} \neq 0$. In this case the nail undergoes horizontal motion. For small ϕ we have

$$\left(I + ma^2\right)\ddot{\phi} + \beta\dot{\phi} + mga\phi = -ma\ddot{\xi} \tag{8.14}$$

which is the damped linear harmonic oscillator with the external driving force $-ma\ddot{\xi}$. The equilibrium points $\phi_1 = 0$ and $\phi_2 = \pi$ of the unexcited system disappear.

2. $\ddot{\eta} \neq 0$ and $\ddot{\xi} = 0$. This means that the nail moves vertically. The equation of motion is

$$\left(I + ma^2\right)\ddot{\phi} + \beta\dot{\phi} + ma(g + \ddot{\eta})\sin\phi = 0\,. \tag{8.15}$$

This is a damped and parametrically driven system. Its rotational equilibrium points are $\phi_1 = 0$ and $\phi_2 = \pi$.

3. $\ddot{\xi} \neq 0$ and $\ddot{\eta} \neq 0$. This choice corresponds to the nail exhibiting oscillation in both horizontal and vertical directions. In this case also the equilibrium points $\phi_1 = 0$ and $\phi_2 = \pi$ of the unexcited system disappear.

8.2.2 *Loose Contact*

In the case of loose contact, the model is the inhomogeneous circular ring Γ of a mass (representing the propeller) rolling without slipping in a fixed plane on the outer edge of a circular disk D (representing the nail). Let R and r ($r > R$) are the radii of the disk and the ring, respectively. A and C are the centres of the disk and the ring, respectively. G is the CoM of the ring. ϕ is the angle between the segment AC and the downward vertical strainght-line passing through A. α is the angle between CG and the straight-line passing through the points A and C and $a = CG$.

The angular velocity $\omega_{\rm r}$ of the ring is given by

$$\omega_{\rm r} = \frac{r-R}{r}\dot{\phi}\,. \tag{8.16}$$

Oxy is the inertial frame of reference. The centre A of the disk moves according to the law of the form

$$A - O = \xi(t)\mathbf{e}_x + \eta(t)\mathbf{e}_y\,. \tag{8.17}$$

The Lagrangian of the system is

$$\begin{aligned} L = {} & \frac{1}{2}m\left[(r-R)^2 + a^2 + 2(r-R)a\cos\alpha\right]\dot{\phi}^2 + \frac{1}{2}I\frac{(r-R)^2}{r^2}\dot{\phi}^2 \\ & + m\dot{\xi}\dot{\phi}\left[(r-R)\cos\phi + a\cos(\phi+\alpha)\right] \\ & + m\dot{\eta}\dot{\phi}\left[(r-R)\sin\phi + a\sin(\phi+\alpha)\right] \\ & + mg\left[(r-R)\cos\phi + a\cos(\phi+\alpha)\right]\,. \end{aligned} \tag{8.18}$$

The Lagrange equation of motion is

$$\begin{aligned} & m\left[(r-R)^2 + a^2 + 2(r-R)a\cos\alpha\right]\ddot{\phi} + I\frac{(r-R)^2}{r^2}\dot{\phi} \\ & + m\ddot{\xi}\left[(r-R)\cos\phi + a\cos(\phi+\alpha)\right] \\ & + m\ddot{\eta}\left[(r-R)\sin\phi + a\sin(\phi+\alpha)\right] \\ & + mg\left[(r-R)\sin\phi + a\sin(\phi+\alpha)\right] = \beta\dot{\phi}\,, \end{aligned} \tag{8.19}$$

where I is the moment of inertia of the ring relative to the point G. This is a linearly damped and parametrically driven system.

For $\ddot{\xi} = 0$ and $\ddot{\eta} = 0$ (unexcited system) corresponding to the fixed nail, that is, the disk is at rest the rotational equlibrium points obtained by setting $\dot{\phi} = 0$ and $\ddot{\phi} = 0$ are the roots of the equation

$$(r-R)\sin\phi + a\sin(\phi+\alpha) = 0\,, \tag{8.20}$$

where $\ddot{\xi} = 0$ and $\ddot{\eta} \neq 0$, which corresponds to purely vertical motion of the disk, the rotational equilibria are again the roots of the Eq. (8.20). Small amplitude oscillations occurring around the stable equilibrium point can be due to the parametric resonance of the system. It is easy to note that for $\ddot{\eta} = 0$ and $\ddot{\xi} \neq 0$ the equilibrium points of the unexcited system disappear. The growth of the amplitude of oscillation can be attributed to the occurrence of ordinary resonance. For $\ddot{\xi} \neq 0$ and $\ddot{\eta} \neq 0$ there are no rotational equilibria and both ordinary and parametric resonances occur.

8.3 Gravity-Spinning Magnet

Ring magnets behave fascinatingly when inserted into a metal rod that is magnetic. For example, when a ring magnet moves down a cycle spoke, a paper spinner spins magically.

Take a long cycle spoke. Pass a wooden or a plastic bead through the spoke's screw end and allow it rest on the bent end. Prepare a small paper spinner of circular, rectangular, star or any other symmetrical shape, punch a hole at its centre, insert it into the spoke and make it rest on the bead. Insert a small ring magnet into the spoke and at the start keep it near the screw end. Keep the screw end at the top and hold the spoke vertically. Give the magnet enough twirl with a slight downward push so that it can start moving down the spoke. As the magnet moves downward its inner boundary hits the spoke and induces the spoke to vibrate. This vibration leads to rotational motion of the spinner. If the magnet is set to rotate in the clockwise (counter clockwise) direction, then the spinner also rotate in the clockwise (counter clockwise) direction.

8.4 Conclusion

Notched stick is a mechanical device which converts the oscillatory motion of the stick into rotational motion of the propeller. It is not quite easy to explain how and why linear motion along the notches induces a circular motion of the propeller and what defines the direction of the motion. Earlier theories suggested that the rotation of the propeller was due to an elliptical motion of the nail at the end of the stick. They assumed that the nail having an elliptical motion was always in contact with the propeller. They also proposed that the speed of the propeller was determined by the magnitude of the areal velocity of the nail. But recent experimental and theoretical studies have established that elliptical movement of the nail is not necessary for starting and maintaining the rotation of the propeller. Also, the speed of the propeller is not necessarily related to the areal movement of the nail.

It has been proposed and verified experimentally that the nail collides with the propeller and impart angular momentum to the propeller to rotate it. It has been found both theoretically and experimentally that no rotation takes place whenever the propeller is tightly connected to the stick nail. Also, it has been found that no rotation will take place if the CoM of the propeller lies on the symmetry axis of the nail. The device succeeds in converting oscillations into rotations only if the nail-propeller coupling is not tight or the propeller is not completely axisymmetrical relative to the nail axis. Elliptical motion of the nail is found to be unnecessary as even a pure linear oscillation of the nail along any fixed direction is enough to convert the oscillatory motion of the stick into rotational motion of the propeller.

8.5 Bibliography

[1] R.W. Leonard, Am. J. Phys. **5**, 175 (1937).
[2] J.S. Miller, Am. J. Phys. **23**, 176 (1955).
[3] E.R. Laird, Am. J. Phys. **23**, 472 (1955).
[4] G.J. Aubrecht II, The Phys. Teach. **20**, 614 (1982).
[5] http://www.mugwumps.com/whammy.htm.
[6] J. Satonobu, S. Ueha and K. Nakamura, Jpn. J. Appl. Phys. **34**, 2745 (1995).
[7] J.F. Wilson, Int. J. Non-Linear Mechanics **33**, 189 (1998).
[8] R. Sedigh, Young Scientist Res. **2**, 9 (2018).
[9] M. Broseghini, C. ceccolini, C.D. Volpe and S. Siboni, PLOS ONE **14**, e0218666 (2019).
[10] G. David Scott, Am. J. Phys. **24**, 464 (1956).
[11] S.S. Welch, The Phys. Teach. **11**, 303 (1973).
[12] M. Marek, M. Badin and M. Plesch, Sci. Rep. **8**, 3718 (2018).
[13] S. Rajasekar and M.A.F. Sanjuan, *Nonlinear Resonances* (Springer, New York, 2016).

8.6 Exercises

8.1 For the notched stick system determine the equilibrium points and analyse their stability for $a = 0$.

8.2 Show that for $\ddot{\eta} \neq 0$ and $\ddot{\xi} = 0$ we get the equilibrium points $\phi_1 = 0$ and $\phi_2 = \pi$.

8.3 Show that for the case (i) $\ddot{\eta} = 0$ and $\ddot{\xi} \neq 0$ and the case (ii) $\ddot{\eta} \neq 0$ and $\ddot{\xi} \neq 0$ the equilibrium points $\phi_1 = 0$ and $\phi_2 = \pi$ disappear.

8.4 In Ref. [12], it is proposed that for larger speeds of the propeller, its motion can be described by free fall motion and short collision with the nail. Assume that the collision of the propeller and the nail takes place at an angle ϕ as shown in Fig. 8.4 at point C for a short duration. Let the impact normal momentum be $\int F dt$ and the tangential frictional

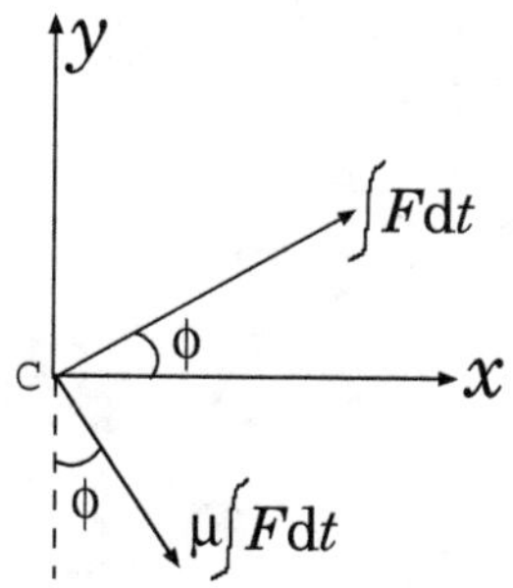

Fig. 8.4: Collision of the propeller and nail at C.

momentum be $\mu \int F\mathrm{d}t$ where μ is the frictional coefficient. Find the change in linear momentum ΔP_x and ΔP_y due to the collision. Show that the angular momentum imparted to the propeller due to this collision is given by $\Delta L = r(\Delta P_x \sin\phi - \Delta P_y \cos\phi)$ where r is the radius of the centre hole.

8.5 In the model discussed in the exercise 8.4 the propeller has a free fall motion when there is no collision. Find ΔP_x, ΔP_y and ΔL for the free fall motion for a duration Δt.

Chapter 9

Gyroscope

A gyroscope is a spinning disc or wheel wherein the spin axis becomes free to take any orientation by itself. When a gyroscope is spun at a high speed, it would be able to stand up vertically without falling over. Moreover, as it rotates any tilting or rotation of the mounting will not affect the orientation of the spin axis. That is, the orientation of an object can be stabilized by gyroscopes. This is due to the conservation of the angular momentum. Actually, a spinning gyroscope produces the so-called *gyroscopic force* that keeps it from falling over and defy gravity. A gyroscope is developed for maintaining or measuring orientation and angular velocity. The name gyroscope is derived from the ancient Greek words *guros* (circle) and *skopeo* (to look). The behaviours of spinning tops, earth, the wheels of bicycles and motorcycles are some kind of gyroscopic motion. Gyroscopes are used to demonstrate rotational dynamics, particularly, angular momentum and torque.

In 1743 John Serson invented an instrument similar to the gyroscope [1]. Another gyroscope-like apparatus was made by Johann Bohnenberger in 1817. A similar device based on a rotating disc was introduced by Walter R. Johnson in 1832. In 1852 Leon Foucault used a gyroscope in his investigation of rotation of the earth [2]. Well before 1900, gyroscopes were in use in the Vanderbilt University, Maroetta College in Ohio and Cornell University. For various forms of gyroscope, one may refer [3].

There are number of papers devoted to the analysis of behaviours of the gyroscope with certain approximations [4-14]. A simple relation between the angle of dip of a gyroscope and its precession velocity is derived in [15] with the gyroscope being a heavy thin disc on a massless axle. For a mathematical model for motions of a gyroscope suspended from flexible cord, one may refer to [16]. In [17] physics of gyroscope nutation is explained. Pseudo torques have been calculated [18].

The ability of a gyroscope to resist changes in its orientations has several uses. Applications of gyroscopes are found in inertial navigation systems (for example in the Hubble Telescope and submerged submarines), gyrotheodolites to maintain direction in tunnel mining, gyrocompasses (in spacecraft, ships and aircraft) and assisting stability in ships, bicycles and motorcycles. Commercially diffused gyroscope technologies include mechanical gyroscopes, silicon micro-electromechanical systems (MEMS) gyroscopes, ring laser gyroscopes, optical gyroscopes and fiber optic gyroscopes and extremely sensitive quantum gyroscope [19]. Gyroscope cages with sensors and electrical contact provide details to the pilot when a plane rolls or pitches. This allows the pilot to know the current relative orientation of a plane in space.

From 1917, the Chandler company of Indianapolis (which was purchased by TEDCO Inc., in 1982) manufactured a toy gyroscope called *Chandler gyroscope*. The Chandler gyroscope toys are commercially available with a pull string and a pedestal. In this chapter, first we present tricks with the Chandler gyroscope. Next, we describe elementary analysis of the gyroscope with a simple model. Using conservation of angular momentum, we setup a relation between the precession velocity and the dip angle of the gyroscope released from rest. Then, performing the stability analysis, we show that the spinning motion of the gyroscope about its symmetry axis is stable.

9.1 Tricks with the Chandler Gyroscope

Figure 9.1 shows the Chandler gyroscope. Its various parts are indicated. The rotor can spin in clockwise or counter-clockwise direction. In order to have a successful operation of the gyroscope, the rotor (the flywheel) has to be set to rotate with a sufficient high speed. For a smooth and easy spinning of the rotor, leave a drop of oil on the spindle points. When we make the gyroscope to stand on a table or floor or on the top of the pedestal, it will fall down due to gravity. However, when the rotor is set to spin with a high speed, the gyroscope exhibits fascinating and mesmerizing behaviours.

Hold the gyroscope firmly between thumb and forefingers of the left-hand. Take a string through the small hole on the spindle, rotate the rotor and tie the string around the spindle from the hole to the central part of the rotor and back again as tight and as smooth as possible with the right-hand.

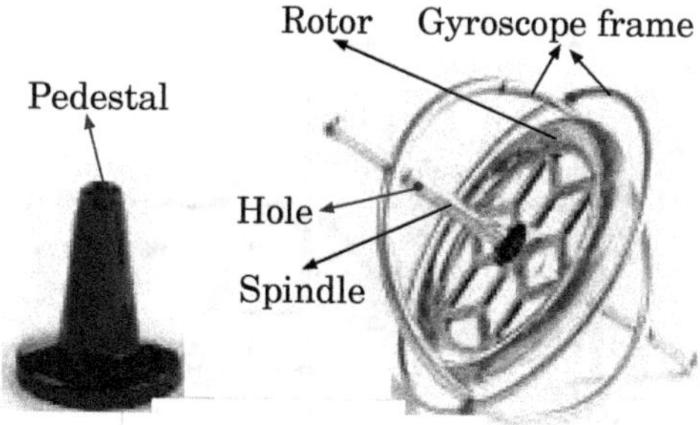

Fig. 9.1: A commercially available Chandler gyroscope set.

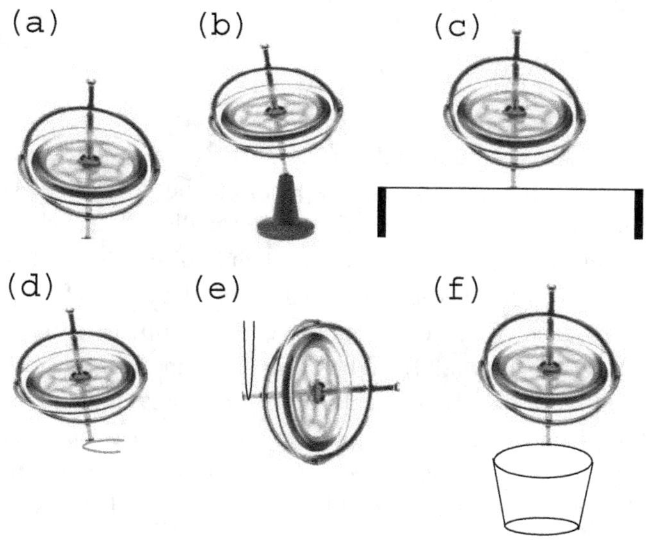

Fig. 9.2: Some of the tricks with the Chandler gyroscope. For details see the text.

Don't wind the string above the side. The hands should not touch the rotor. Drawing the string-off with a strong steady pull makes the rotor to spin with a high speed. When you gently leave the gyroscope to stand on a table, it stands vertically (Fig. 9.2a). The force created by the spinning rotor keeps the gyroscope to operate. The toy precesses in the clockwise or counter-clockwise direction about the vertical. When the spinning of the rotor ceases, the force disappears and the gyroscope falls over.

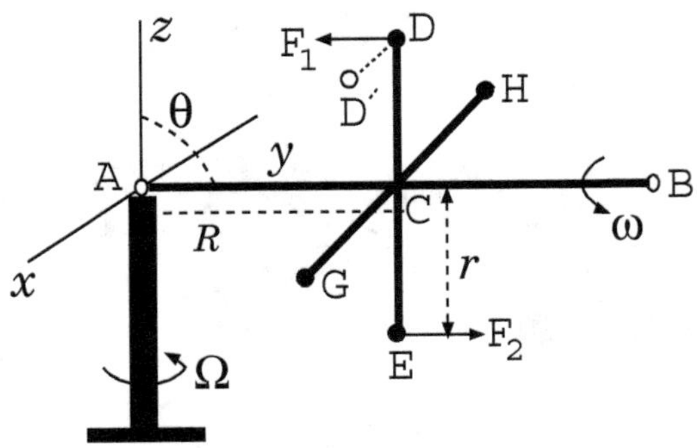

Fig. 9.3: An approximate model of the gyroscope [7]. (Reproduced from Am. J. Phys. **28**, 808 (1960), with permission of American Association of Physics Teachers. Copyright 1960 E.F. Barker, licensed under a Creative Commons Attribution License.)

In another trick, to our high surprise, the force is able to support the gyroscope as in Fig. 9.2b. The toy appears to be levitating. While doing so, the gyroscope rotates about the pedestal illustrating the rotation of the globe and also demonstrates the principle of equilibrium of bodies in motion. Several other astounding tricks can also be performed with the spinning gyroscope. It can even dance on a tight rope (Fig. 9.2c) or spin on the tip of a finger (Fig. 9.2d). When you suspend the gyroscope in a loop made by doubling the string, then it will spin at any angle placed above or below the horizontal (9.2e). The gyroscope can be set to spin on the curved top edge of a glass (Fig. 9.2f). Recorded videos of these tricks are available in YouTube [20-23]. These incredible tricks are difficult to explain.

9.2 A Four-Mass Gyroscope Model

To describe the working of a gyroscope, approximate the rotor as an evenly spaced four point masses D, E, G and H, each with mass m, about the centre of rotation [7]. They are mounted symmetrically as shown in Fig. 9.3 by means of two perpendicular cross-bars. They pass through the axis AB at its centre of mass (CoM) C. We neglect the friction. The gyroscope is assumed to be rotating about the axis AB with the angular velocity ω. The CoM has precessional motion and is rotating about the pedestal with the angular velocity Ω. The associated angular momentum is along the positive vertical z-axis. The plane with the four masses is considered as the rotor

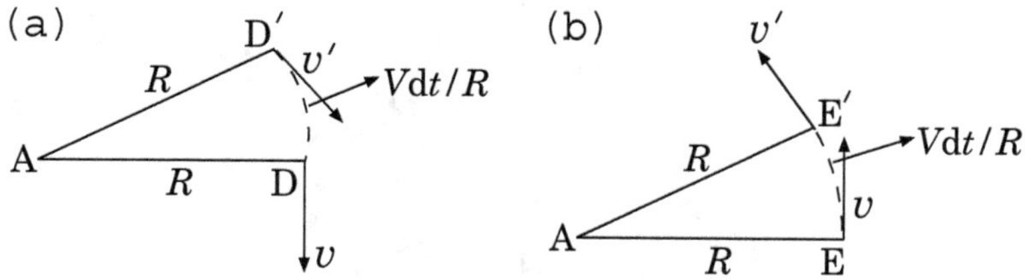

Fig. 9.4: Change in the direction of the rotational velocity v of (a) D and (b) E due to precession. (Reproduced from Am. J. Phys. **28**, 808 (1960), with permission of American Association of Physics Teachers. Copyright 1960 E.F. Barker, licensed under a Creative Commons Attribution License.)

plane and is at a distance R from the point of support A. We discuss the effect of rotation and precession on the motion of the point masses.

9.2.1 *Forces on Rotating Masses Due to Precession*

Let $V = \Omega R$ is the linear velocity of the CoM due to precession, v and v' are the initial rotational velocity of D and its rotational velocity after an infinitesimal precession, respectively. The difference between v and v' is, say, $\mathrm{d}v$. The resulting momentum per unit time is in the direction of $\mathrm{d}v$. This is because of the force acts on the mass D by the framework consisting of the cross-bars and the bar AB in Fig. 9.3. In general the forces acting on the masses accelerate them and their paths continuously vary in direction. The rotational force associated with each mass is proportional to its instantaneous acceleration and opposite in direction. For example, for the mass point D, the reaction force pushes the framework at D in the negative y-direction.

During the infinitesimal precession, the rotational velocity of D changes its direction but the magnitude remains the same. This reaction force F_1 (refer Fig. 9.3) at D pushes the framework radially inward. The force F_1 can be determined [7,12]. As shown in Fig. 9.4a [7], in time $\mathrm{d}t$ due to the precession the linear displacement of v is $V\mathrm{d}t$ while the change in angle is $V\mathrm{d}t/R$. Now, we write

$$\frac{V\mathrm{d}t}{R} = \frac{\mathrm{d}v}{v}, \tag{9.1}$$

which gives

$$\frac{\mathrm{d}v}{\mathrm{d}t} = \frac{Vv}{R} = a_1 . \tag{9.2}$$

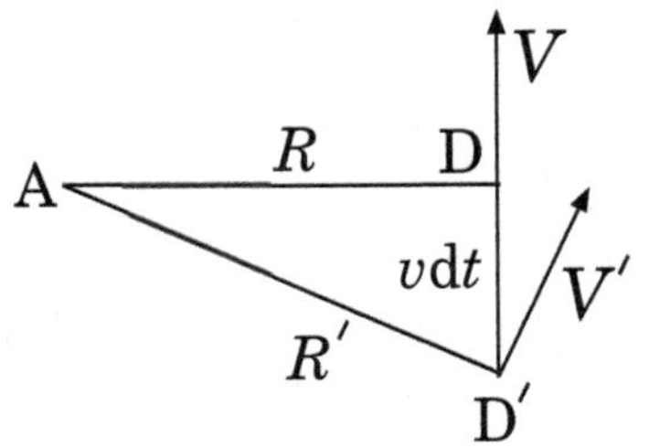

Fig. 9.5: Change in the direction of V due to rotation. (Reproduced from Am. J. Phys. **28**, 808 (1960), with permission of American Association of Physics Teachers. Copyright 1960 E.F. Barker, licensed under a Creative Commons Attribution License.)

Then,

$$F_1 = ma_1 = \frac{Vv}{R}m\,. \tag{9.3}$$

For the mass E we refer Fig. 9.4b and similar analysis gives the reaction force F_2 as

$$F_2 = ma_2 = \frac{Vv}{R}m\,, \tag{9.4}$$

but the direction of it is opposite to that of F_1. We note that $|F_1| = |F_2|$.

Next, consider the masses G and H. At these points, the rotational motion has v and is directed downward. Due to the precession v changes to v' with the same magnitude of v and as well as in direction. Thus, the resulting change in momentum is zero and the total force exerted by the mass G (and H) on the framework is zero.

9.2.2 *Force on Precessing Masses Due to Rotation*

Assume that as shown in Fig. 9.3 DE rotates through a small angle in time dt. D is moved to D′, R changed into R' and V changed into V'. As shown in Fig. 9.5 [7], V is changed through the angle $\mathrm{d}V/V = DD'/R = v\mathrm{d}t/R$. This gives the acceleration a_{DV} as

$$a_{\mathrm{DV}} = \frac{\mathrm{d}V}{\mathrm{d}t} = \frac{Vv}{R} = a_1 \tag{9.5}$$

and

$$F_{\mathrm{DV}} = ma_{\mathrm{DV}} = m\frac{Vv}{R} = F_1\,. \tag{9.6}$$

Note that $a_{\text{DV}} = a_1$ and is in the same direction. This is the effect of rotation due to precession.

Similarly,

$$F_{\text{EV}} = ma_{\text{EV}} = m\frac{Vv}{R} = F_2 = F_1\,, \tag{9.7}$$

but a_{EV} is in the opposite direction. Thus, there are four accelerations: two are due to rotation and another two are due to precession and all are present simultaneously.

The total force at D (E) acting on the framework is $2F_1$ $(2F_2)$. The total torque about A due to rotation of the masses is

$$\tau = 2rF_1 + 2rF_2 = 4rF_1 = 4mr\frac{Vv}{R}\,. \tag{9.8}$$

With the total mass $M = 4m$, $\Omega = V/R$ and $\omega = v/r$, we can rewrite the above equation as

$$\tau = Mr^2\Omega\omega\,. \tag{9.9}$$

A steady motion can be realized when this torque is equal to MgR, the gravitational torque. That is, for the known magnitude and direction of ω the magnitude and direction of Ω can be determined for steady motion.

The mass particles of the rotor are moving and their velocities are continuously changing in direction. Accelerations result due to these changes. Internal torques are thus exerted on the moving system by the masses. When the gravitational torque is balanced by the internal torques steady motion occurs. In other words, due to the nature of the kinematics, the masses in the rotor experience accelerations in such a way that the gravitational force is able to maintain the angle θ of the gyroscope as it precesses. *What will happen if the spinning rotor has no precession?*

9.3 Dipping of Gyroscope and its Precession Motion

In a gyroscope a heavy spinning flywheel or a disc has its axis along the horizontal direction and the axis is pivoted on a vertical pedestal. The horizontal axis can precess about the vertical plane. When a large spin is given to the wheel, then the horizontal axis of the wheel precesses about the vertical axis. When the wheel is given an angular velocity ω, an angular momentum L acts along the direction of the horizontal axis. *How does then the gyroscope get the angular velocity Ω about the vertical axis and precesses about the vertical axis? How does an angular momentum L given*

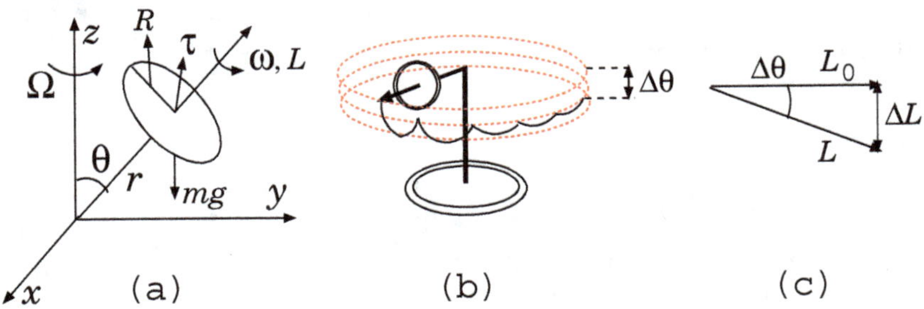

Fig. 9.6: (a) The basic parameters of the gyroscope. (b) Precession and decaying nutation. Here $\Delta\theta$ is the dip angle. (c) Representation of the initial and final spin angular momenta after the decay of the nutation [15]. (Reproduced from The Phys. Teach. **49**, 216 (2011), with permission of American Association of Physics Teachers. Copyright 2011 S. Kostov and D. Hammer, licensed under a Creative Commons Attribution License.)

in the horizontal direction generates an angular momentum along the vertical direction? Is not angular momentum conserved in gyroscope motion? Richard Feynman has discussed this apparent paradox in *The Feynman Lectures on Physics* [24]. He explained clearly in this topic that the spin axis has to drop vertically (dip), such that the gyroscope can get a vertical angular momentum so that the total angular momentum of the system is conserved. We discuss below a simple model given in [15] for the dipping gyroscope.

Consider a disc of radius R and mass m having the horizontal axis x at a distance r from the vertical z-axis as shown in Fig. 9.6a. The torque acting on the disc is $\tau = \mathbf{r} \times m\mathbf{g} = rmg$. Note that τ is in the x-y plane. The angular velocity Ω of the constant precession about the vertical axis z due to the large angular velocity ω given to the disc has been derived [25] as

$$\Omega = \frac{mgr}{I_3\omega}, \quad I_3 = \frac{1}{2}mR^2. \tag{9.10}$$

I_3 is the moment of inertia of the disc about the horizontal axis.

Actually, the disc will precess along with a nutation which will damp out due to friction leaving a constant precession and with a dip angle $\Delta\theta$ of the horizontal axis as shown in Fig. 9.6b. Due to this dip $\Delta\theta$, the angular momentum given initially to the disc has changed from L_0 to L along the rotation axis as shown in Fig. 9.6c. As this ΔL is in the negative

z-direction an angular momentum $-\Delta L$ will be induced in the positive z-direction. This induced angular momentum forces the axis to precess about z-axis with the precession angular velocity Ω. We find that

$$\Delta L = L_0 \Delta\theta. \tag{9.11}$$

As the angular momentum is given by the moment of inertia times the angular velocity, we get $\Delta L = \Omega I_1$ where I_1 is the moment of inertia of the disc about the z-axis and $L_0 = I_3\omega$ where I_3 is the moment of inertia of the disc about its centre axis. We get from (9.11) $\Omega I_1 = I_3\omega\Delta\theta$. Using Eq. (9.10), we obtain

$$\Delta\theta = \frac{I_1\Omega^2}{mgr}. \tag{9.12}$$

For large ω the dip angle is proportional to the square of the precession angular velocity Ω.

9.4 Stability Analysis of the Gyroscopic Motion

In a gyroscope, the spinning flywheel retains its orientation as the angular momentum is conserved. An analysis for this stability of the rotational motion of a gyroscope around its symmetric axis due to small perturbation is given in [14] using Euler's equation for a rotating body. The Euler's equation for a body with an external torque $\mathbf{T} = (T_x, T_y, T_z)$ is given by

$$I_1\dot{\omega}_x + (I_3 - I_2)\,\omega_z\omega_y = T_x, \tag{9.13a}$$

$$I_2\dot{\omega}_y + (I_1 - I_3)\,\omega_x\omega_z = T_y, \tag{9.13b}$$

$$I_3\dot{\omega}_z + (I_2 - I_1)\,\omega_y\omega_x = T_z, \tag{9.13c}$$

where I_1, I_2 and I_3 are the principal moments of inertia of the rotating body and $\boldsymbol{\omega} = (\omega_x, \omega_y, \omega_z)$ is its angular velocity. (x, y, z) is the rotating body fixed reference frame.

As a gyroscope is an axis symmetric body, $I_1 = I_2 = I_\perp$ and $I_3 = I_\parallel$. As no external torque acts on a gyroscope we get from Eqs. (9.13)

$$I_\perp\dot{\omega}_x + \left(I_\parallel - I_\perp\right)\omega_z\omega_y = 0, \tag{9.14a}$$

$$I_\perp\dot{\omega}_y + \left(I_\perp - I_\parallel\right)\omega_x\omega_z = 0, \tag{9.14b}$$

$$I_\parallel\dot{\omega}_z = 0. \tag{9.14c}$$

Solved Problem 1:

Using Eqs. (9.14), for a constant ω_z, find the solutions of ω_x and ω_y.

The first two subequations of Eq. (9.14) give

$$\dot{\omega}_x = -\frac{I_{\|} - I_{\perp}}{I_{\perp}} \omega_z \omega_y, \tag{9.15a}$$

$$\dot{\omega}_y = \frac{I_{\|} - I_{\perp}}{I_{\perp}} \omega_z \omega_x. \tag{9.15b}$$

Differentiation of Eq. (9.15a) and substitution of Eq. (9.15b) lead to

$$\ddot{\omega}_x + \nu^2 \omega_x = 0, \quad \nu = \left(\frac{I_{\|} - I_{\perp}}{I_{\perp}}\right) \omega_z. \tag{9.16}$$

The solution of this equation is

$$\omega_x = A\cos(\nu t + \phi) \tag{9.17}$$

where A and ϕ are integration constants. Equations (9.15a) and (9.17) give

$$\omega_y = -\dot{\omega}_x/\nu = A\sin(\nu t + \phi). \tag{9.18}$$

Note that both ω_x and ω_y oscillate with angular frequency ν.

Suppose the flywheel rotates around its principal inertia axis (z-axis) at an angular velocity $\boldsymbol{\omega} = (0, 0, \omega_z^0)$. Let us assume a small initial perturbation $\boldsymbol{\delta}^0 = (\delta_x^0, \delta_y^0, \delta_z^0)$. Then, $\boldsymbol{\omega} = (\delta_x^0, \delta_y^0, \omega_z^0 + \delta_z^0)$. Substitution of $\omega_x = \omega_x + \delta_x$, $\omega_y = \omega_y + \delta_y$ and $\omega_z = \omega_z^0 + \delta_z^0$ in Eqs. (9.14) and negleting the higher powers of δ's, we obtain the evolution equations for the perturbations as

$$I_{\perp}\dot{\delta}_x + \left(I_{\|} - I_{\perp}\right) \omega_z^0 \delta_y = 0, \tag{9.19a}$$

$$I_{\perp}\dot{\delta}_y + \left(I_{\perp} - I_{\|}\right) \omega_z^0 \delta_x = 0, \tag{9.19b}$$

$$I_{\|}\dot{\delta}_z = 0. \tag{9.19c}$$

Equation (9.19c) gives δ_z = a constant. As Eqs. (9.19a) and (9.19b) do not contain δ_z any change in ω_z does not affect the motion of the system. *How does the change in δ_x and δ_y affects the motion?* Differentiating Eq. (9.19a) and eliminating $\dot{\delta}_y$ from the Eq. (9.19b) we obtain

$$\ddot{\delta}_x + \left(\frac{I_{\|} - I_{\perp}}{I_{\perp}}\right)^2 \left(\omega_z^0\right)^2 \delta_x = 0. \tag{9.20}$$

This equation describes a harmonic oscillation of δ_x. A similar equation can be derived for δ_y also. So, both δ_x and δ_y do not grow in time but oscillate around zero for any initial perturbations δ_x^0 and δ_y^0. Thus, we have found that any change in ω_z does not affect the motion of the system. This analysis establishes that the rotational motion of the gyroscope about its symmetry axis (z-axis) is stable.

9.5 Conclusion

Gyroscope demonstration in a physics lecture can help to teach students about rotational dynamics, angular momentum and torque. When a gyroscope is demonstrated to the students, it never fails to command attention and to arouse keen interest in them. They will be quite surprised to observe that a spin angular momentum given to the disc in the horizontal axis leads the horizontal axis to precess about the vertical axis. They will wonder how does the gyroscope get its vertical angular momentum. Though it appears to violate the conservation of angular momentum, they will understand that a dip of the horizontal axis leads to the vertical angular momentum, thus conserving the total angular momentum of the system. Furthermore, in spite that there is initially a vertical oscillation of the horizontal axis (nutation), it damps out fairly rapidly in toy gyroscopes. So, it will be very difficult to observe nutation in toy gyroscope. Only the uniform precession about the vertical axis can be demonstrated to the students. For discussion on misconceptions about gyroscopic stabilization, one may refer to [26].

9.6 Bibliography

[1] https://en.wikipedia.org/wiki/Gyroscope.

[2] L. Foucault, Comptes rendus hebdomadaires des séances de l'Académie des Sciences (Paris) **35**, 424 (1852).

[3] http://physics.kenyon.edu/EarlyApparatus/Mechanics/Gyroscope/-Gyroscope.html.

[4] G.D. Scott, Am. J. Phys. **25**, 80 (1957).

[5] A.E. Benfield, Am. J. Phys. **26**, 396 (1958).

[6] E. Pettersen, Am. J. Phys. **27**, 429 (1959).

[7] E.F. Barker, Am. J. Phys. **28**, 808 (1960).

[8] J.L. Snider, Am. J. Phys. **33**, 847 (1965).

[9] H.L. Armstrong, Am. J. Phys. **35**, 883 (1967).

[10] W. Case, Am. J. Phys. **45**, 1107 (1977).

[11] H. Soodak and M.S. Tiersten, Am. J. Phys. **62**, 687 (1994).

[12] H. Kaplan and A. Hirsch, The Phys. Teach. **52**, 30 (2014).

[13] W. Rueckner, Am. J Phys. **85**, 228 (2017).

[14] P. Muller, A. Sack and T. Poschel, Am. J. Phys. **88**, 175 (2020).

[15] S. Kostov and D. Hammer, The Phys. Teach. **49**, 216 (2011).

[16] R. Usubamatov, Cogent Eng. **3**, 1245901 (2016).

[17] R. Usubamatov, AIP Advances **9**, 105101 (2019).

[18] R.H. Price, Am. J. Phys. **88**, 1145 (2020).
[19] V.M.N. Passaro, A. Cuccovillo, L. Vaiani, M. De Carlo and C.E. Campanella, Sensors **17**, 2284 (2017) and references therein.
[20] https://www.youtube.com/watch?v=-NgCc42PTis.
[21] https://www.youtube.com/watch?v=BzEbKE3ND8Y.
[22] https://www.youtube.com/watch?v=_9Q0VH2CEE4.
[23] https://www.youtube.com/watch?v=X1DvMKzWreo.
[24] R. Feynman, R. Leighton and M. Sands, *The Feynman Lectures on Physics* (CIT, California, 1963) Vol-I.
[25] R.A. Serway and J.W. Jewett, *Physics for Scientists and Engineers* (Brooks/Cole Cengage Learning, Boston, 2014) 9th edition, pp.351.
[26] P. Muller, A. Sack and T. Poschel, Am. J. Phys. **88**, 175 (2020).

9.7 Exercises

9.1 Consider a gyroscope supported at one end. State the associated conserved quantities assuming that the pivot point is frictionless.

9.2 Assume that a gyroscope is started with its axis in a horizontal position and at rest. The gyroscope is then released. Denote the angle between the axis of the gyroscope and the horizontal by θ. Obtain the total energy of the gyroscope.

9.3 Consider the four-point mass model of the gyroscope. Suppose the rotor is supported at both A and B and is spinning at a fairly high speed but without precession. What will happen if the support at A is removed? [7].

9.4 Assume that the precession angular velocity $\mathbf{\Omega} = \Omega\mathbf{k}$ and the spin angular velocity ω is in the spin axis direction $\mathbf{s}$. For a steady Ω and ω, the rate of change of the axes orientation is given by $\mathbf{s}' = \mathbf{\Omega} \times \mathbf{s}$. Find the torque due to this change in orientation of the symmetry axis and equate with the gravitational torque to prove $\Omega = mgr/(I_3\omega)$ where I_3 is the moment of inertia about the horizontal axis and r is the distance of the centre of the disc from the origin.

9.5 Consider a gyroscope having a disc of mass 0.2 kg, radius $R = 0.015$ m and $r = 0.15$ m. If the disc is given an angular velocity ω of 400 rpm find Ω and the dip angle $\delta\theta$.

Chapter 10

The Buzzer

A whirligig is one of the simplest toys that spins or whirls. A simple version of it called a *button spinner* or a *buzzer* consists of a circular disc (rotator) with two holes on each side of its centre of gravity and a loop of thread or a string through the holes. It is very easy to make. Take a two feet length of thread or string and a button with two holes. Thread the string through the holes in the button and tie their free ends together. Figure 10.1 shows three buzzers with one being with a button. Keep the button in the middle of the loop and insert the index or middle finger of each hand into the loop at each end. The button spinner hangs in V-shape when the strings are loose. The disc (button) is like a pendulum. With one hand, swing the button in a circle to twist the strings and continue until it is wrapped around itself fully. Applying pulling and releasing alternately the tension of the string keeps the button spinning.

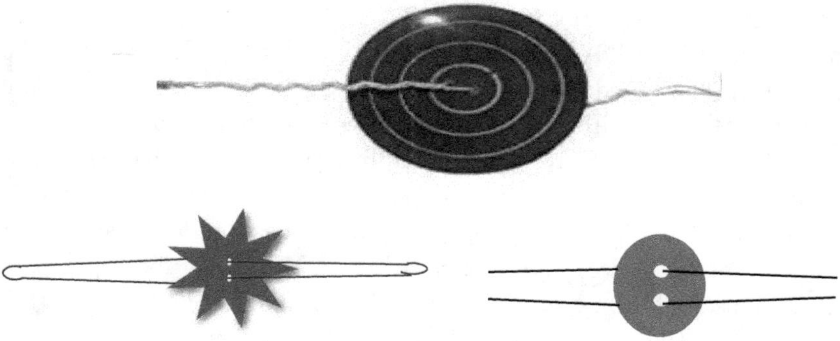

Fig. 10.1: A few buzzers.

The working of this toy has two successive process, namely, unwinding and winding. In the unwinding stage the axial tensile forces act at the ends of the string. This makes the string to unwind leading to spinning of the disc. When the string is completely unwound the rotation comes to a momentary halt. During the winding process the string is loosened. The disc begins to spin in the reverse direction. The string is rewound by the momentum of the whirling disc. It comes to a momentary halt when the number of turns reaches a maximum. By reapplying the stretching force, the process of unwinding/winding is restarted. Increasing the speed of the action and applying and releasing the tension in an appropriate rhythm makes the disc to produce a humming or whirring sound. The buzzer illustrates conversion of work to energy and potential energy to kinetic energy and vice-versa. When the string is twisted and stationary it possesses potential energy. As it is pulled the disc starts spinning and the potential energy starts to convert to kinetic energy. When the disc winds the string again, the kinetic energy is converted back to potential energy.

What is surprising with this toy is that if it is properly operated, *it can spin several thousands of revolutions per minute (rpm).* This is revealed by frames captured by a slow motion camera. Zhao and his collaborators noticed $200,000$ rpm [1]. The circular disc in the buzzer can be a button or cardboard disc, plastic lids, soda bottle tops (after flattening with a hammer and making two holes), old CDs and disc shaped candys.

The word whirligig means *to whirla top* and derived from the English words *whirlen* (to whirl) and *gigg* (top) [2]. The whirligig (buzzer) had been in use more than 5000 years ago [3]. In early 500 BC in India wood, stone or bone were used to construct buzzers. In China in 400 BC the bamboo butterfly or bamboo-copter was introduced [4]. String-operated buzzer were in use in Egypt by 100 BC. A book published in 1500 AD depicts the Christ child in the margin with a buzzer [4]. The buzzer is one kind of whirligig. The other kinds of whirligig include pinwheels, comic weathervanes, whirlybird and a plain whirly, a few to mention.

The buzzer systems have potential applications. The spinning motion of a buzzer has been utilized in a hand powered blood centrifuge to separate plasma from blood [5,6]. The string-disc system has found applications in developing novel devices and materials like carbon nanotube ropes [7,8], artificial muscles [9,10], centrifugation and microfluidics [11-14].

We outline the theoretical model of the buzzer toy. Particularly, we present the equation of motion for the rotation angle and input, drag and twisting torques.

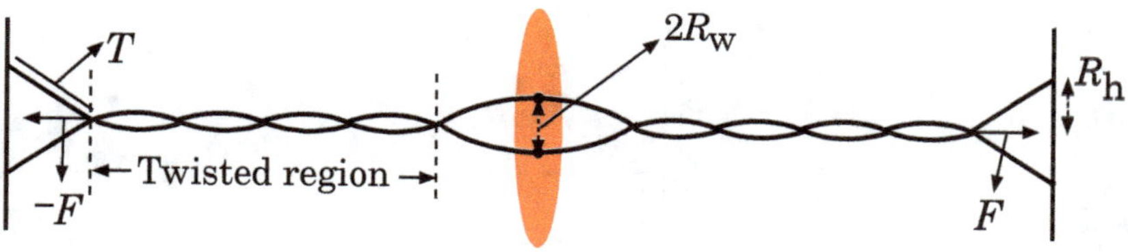

Fig. 10.2: The theoretical model of the buzzer.

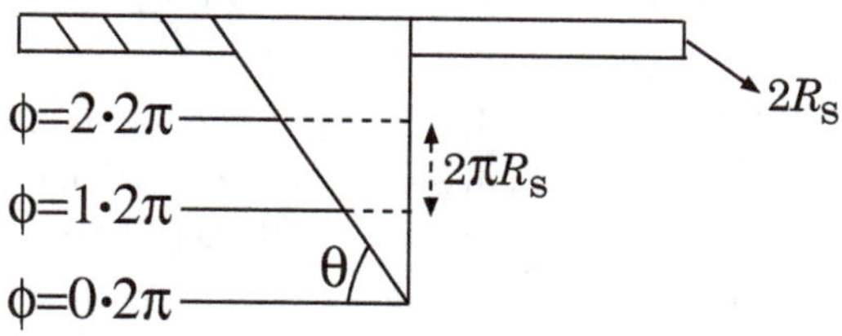

Fig. 10.3: Representation of the rotation angle ϕ [15]. (Reproduced with permission from Institute of Physics Publishing Ltd: H.J. Schlichting and W. Suhr, Eur. J. Phys. **31**, 501 (2010).)

10.1 Theoretical Model

Let us setup a physical model of the buzzer toy [6].

Figure 10.2 depicts the theoretical model of the buzzer. We neglect the weight of the disc and the string. Assume that a human hand supplies a periodic input force **F** which is translational and acts along the horizontal axis. This force is transferred to rotational motion of the disc. In the unwound state the string is, say, a thin cylinder with a constant cross-section. Also, neglect the elasticity of the string. R_s is the radius of the circular cross-section of the string. A twisting of a pair of strings can be regarded as a spiraling of central filament on a cylinder with radius R_s. Denote ϕ as the rotation angle or winding angle. One complete winding of the string increases the value of ϕ by 2π (refer Fig. 10.3). The speed of rotation is $\dot{\phi}$.

Solved Problem 1:

The force **F** has two components F_t (vertical) and F_a (axial). The forces **F**, F_t and **T** are represented in Fig. 10.4. Using the triangle shown in this figure determine F_t.

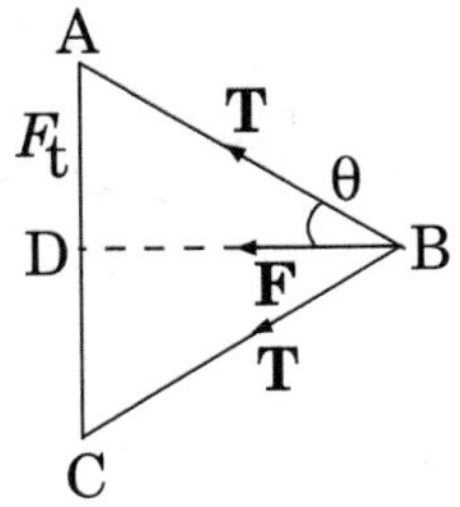

Fig. 10.4: Representation of the forces $\mathbf{F}$, F_t and $\mathbf{T}$.

From the triangle ABD, $F_t = T\sin\theta$ where θ is the helix angle. Further, $F_a = T\cos\theta$. As, another F_a will arise due to the tension on the string BC, we get $F = 2F_a = 2T\cos\theta$. That is, $T = F/(2\cos\theta)$. Substituting this T in $F_t = T\sin\theta$ we obtain $F_t = \frac{1}{2}F\tan\theta$.

10.1.1 *Energy Sources and Equation of Motion of ϕ*

The prime energy components of the buzzer toy are [6]:

1. $E_{\text{input}}(\phi)$ – Input source energy from human.
2. $E_{\text{drag}}(\dot{\phi})$ – Energy of air resisting the motion of the circular disc.
3. $E_{\text{twist}}(\phi)$ – Energy of the string twisting and compressing, resisting motion.
4. $E_{\text{KE}}(\dot{\phi})$ – Kinetic energy of the circular disc.

The source of energy from human is transferred into kinetic energy of the disc. This energy is dissipated into air and string twisting energy. Due to the continuous loss of energy the disc finally comes to a halt.

The Lagrangian of the system is

$$L = T - V = E_{\text{KE}}(\dot{\phi}) - \left(E_{\text{drag}}(\dot{\phi}) + E_{\text{twist}}(\phi) + E_{\text{input}}(\phi)\right). \quad (10.1)$$

The Lagrange equation of motion

$$\frac{\mathrm{d}}{\mathrm{d}t}\left(\frac{\partial L}{\partial \dot{\phi}}\right) = \frac{\partial L}{\partial \phi} \quad (10.2)$$

with $E_{\text{KE}} = I\dot{\phi}^2/2$ where I is the moment of inertia of the disc becomes

$$\begin{aligned} I\ddot{\phi} &= \frac{\mathrm{d}}{\mathrm{d}t}\frac{\partial}{\partial \dot{\phi}}E_{\text{drag}} - \frac{\partial}{\partial \phi}E_{\text{twist}} - \frac{\partial}{\partial \phi}E_{\text{input}} \\ &= \tau_{\text{drag}}(\dot{\phi}) + \tau_{\text{twist}}(\phi) + \tau_{\text{input}}(\phi), \end{aligned} \quad (10.3)$$

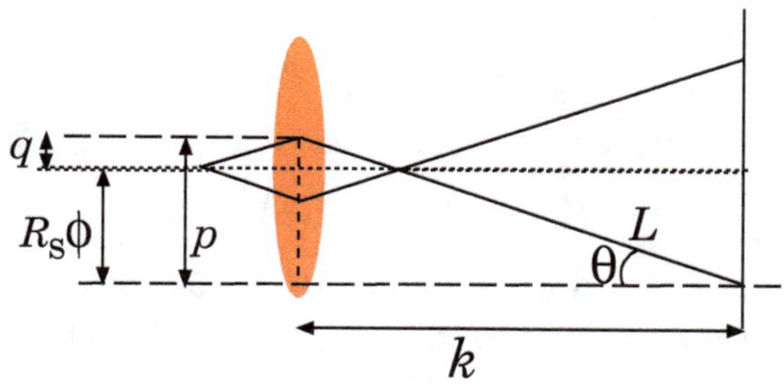

Fig. 10.5: Representation of q, p, L, θ and $R_s\phi$.

where τ_{input}, τ_{drag} and τ_{twist} are the input, air-drag and string-twisting contributions to the torque, respectively. For a circular disc with width w_d and radius R_d the moment of inertia is $I = \pi\rho w_d R_d^4/2$.

10.1.2 *Input Torque*

We apply a torque balance by the strings on the circular disc. We have $F_t = (F/2)\tan\theta$. Because there are four strings passing through the disc the torque on the disc is $\tau_{\text{input}} = -4R_sF_t = -2R_sF\tan\theta$.

Next, find the relation between θ and ϕ and then replace $\tan\theta$ in terms of ϕ. To account for the changing sign of ϕ, consider the geometry of the buzzer shown in Fig. 10.5 [15] from which we write

$$p = \text{sgn}(\phi)q + R_s\phi, \quad k = \sqrt{L^2 - p^2}, \quad \tan\theta = p/k\,. \tag{10.4}$$

Then,

$$\begin{aligned}\tau_{\text{input}} &= -\text{sgn}(\phi)2R_sF\tan\theta \\ &= -\text{sgn}(\phi)2R_sF\frac{p}{k} \\ &= -\text{sgn}(\phi)2R_sF\frac{\text{sgn}(\phi)q + R_s\phi}{\sqrt{L^2 - (\text{sgn}(\phi)q + R_s\phi)^2}}\,.\end{aligned} \tag{10.5}$$

An appropriate form of $F(t)$ is [15]

$$F(t) = A|\sin(\omega t + \alpha)|^3 + F_{\text{min}}\,, \tag{10.6}$$

where α is the initial phase and F_{min} is a minimum value of F.

10.1.3 *Drag Torque*

When an object rotates in a fluid at an angular velocity $\dot{\phi}$ a boundary layer is developed. The drag force is approximated as [6]

$$F_{\mathrm{D}} = -A_{\mathrm{R}} S_{\mathrm{d}} \left(R_{\mathrm{d}} \dot{\phi} \right)^2 , \tag{10.7}$$

where R_{d}, S_{d} and $R_{\mathrm{d}}\dot{\phi}$ are the radius, the surface area, the velocity v, respectively, of the disc. A_{R} is the drag coefficient. The v for a small surface element on the disc is a function of radius along the disc. Integration over the surface of the disc with radius R_{d} and width w_{d} gives τ_{drag} as [6]

$$\tau_{\mathrm{drag}} = -A_{\mathrm{R}} \mathrm{sgn}(\dot{\phi}) \left(\frac{4\pi}{5} R_{\mathrm{d}}^5 + 2\pi w_{\mathrm{d}} R_{\mathrm{d}}^4 \right) \dot{\phi}^2 . \tag{10.8}$$

10.2 Twisting Torque

We can easily observe three cases associated with strings in the buzzer. At $\phi = 0$, there is no torque on the disc and hence $\tau_{\mathrm{twist}}(\phi = 0) = 0$. Next, at a critical value of $\theta (= \theta_{\mathrm{crit}})$ (the corresponding ϕ as ϕ_{crit}), in the twisting region the string lays against itself. That is, θ_{crit} is the value of θ beyond which the string will be unable to twist further without compression. θ_{crit} is given by $\cos\theta_{\mathrm{crit}} = 2/\pi$. In this case the torque for $\phi = \phi_{\mathrm{crit}}$ can be approximated as $\frac{\partial}{\partial\phi}\tau_{\mathrm{twist}}(\phi)|_{\phi=\phi_{\mathrm{crit}}} = \kappa$ where κ is the spring constant and is set to unity. In the third case, the twisted region indicated in Fig. 10.2a fully extends between the handle and the disc. The corresponding value of ϕ is denoted as ϕ_{max} and ϕ cannot exceed beyond ϕ_{max}. In this case $\tau_{\mathrm{twist}}(\phi_{\mathrm{max}}) = \infty$.

A form of $\tau_{\mathrm{twist}}(\phi)$ satisfying the above mentioned three boundary conditions is [6]

$$\tau_{\mathrm{twist}} = -\mathrm{sgn}(\phi) \frac{1}{\gamma} \left(\phi_{\mathrm{max}} - \phi_{\mathrm{crit}} \right)^{\gamma+1} \left[\frac{1}{\left(\phi_{\mathrm{max}} - |\phi| \right)^{\gamma}} - \frac{1}{\phi_{\mathrm{max}}^{\gamma}} \right] , \tag{10.9}$$

where γ is a parameter to be determined by comparing the theoretical model with experimental data. We have

$$\sin\theta = \frac{|\phi| R_{\mathrm{s}} + R_{\mathrm{h}} + R_{\mathrm{w}}}{L} . \tag{10.10}$$

ϕ_{crit} can be calculated from the above equation by substituting $\theta = \theta_{\mathrm{crit}} = \cos^{-1}(2/\pi)$ and $\phi = \phi_{\mathrm{crit}}$. Next, to calculate ϕ_{max}, we use the fact that when $\theta = \theta_{\mathrm{crit}} = \cos^{-1}(2/\pi)$ with ϕ being ϕ_{max} and $R_{\mathrm{w}} = R_{\mathrm{h}} = 0$. Then, Eq. (10.9) gives

$$|\phi_{\mathrm{max}}| = \frac{L}{R_{\mathrm{s}}} \sin\theta_{\mathrm{crit}} . \tag{10.11}$$

As $\sin^2\theta = 1 - \cos^2\theta$, for $\theta_{\text{crit}} = \cos^{-1}(2/\pi)$ we find

$$\sin^2\theta_{\text{crit}} = 1 - \cos\theta_{\text{crit}}\cos\theta_{\text{crit}} = 1 - (2/\pi)^2\,. \tag{10.12}$$

That is, $\sin\theta_{\text{crit}} = \sqrt{\pi^2 - 2^2}/\pi$.

Solved Problem 2:

Find the values of τ_{twist} for $\phi = 0$ and $\phi = \phi_{\text{max}}$. What do they signify?

Consider the expression for τ_{twist} given by Eq. (10.9). When $\phi = 0$ we obtain $\tau_{\text{twist}} = 0$. As the strings are not rotated at all, they are not twisted. So, no torque acts on the disc. When $\phi = \phi_{\text{max}}$, τ_{twist} becomes infinity. So, this model predicts that the twist of the two strings will not be uniform after a maximum value of ϕ. A further increase of ϕ will make the twist nonuniform and compress the string one over the other.

10.3 Conclusion

A buzzer transforms translation motion into high-speed rotational motion. Buzzer is one of the oldest toys which has been found in all most all ancient civilizations. Though it is a simple toy, its physical analysis is found to be highly complex. The behaviour of the buzzer depends heavily on the system parameters and the physical properties of the string. So, finding a universal model describing a buzzer is not yet achieved. Different models have been used and usally the model parameters are found out by matching them with the experimental results.

The most important application of this ancient toy is found in centrifugation. The design of such a low cost light weight human powered centrifuge made out of paper is discussed in [6]. A very high rotation upto 200,000 rpm has been obtained by a manually operated device.

10.4 Bibliography

[1] Z.L. Zhao, S. Zhou, S. Xu, X.Q. Feng and Y.M. Xie, Sci. Rep. **7**, 3111 (2017).
[2] https://www.merriam-webster.com/dictionary/whirligig.
[3] G.W. van Beek, Bull. Am. Schools Orient. Res. 53 (1989).
[4] https://en.wikipedia.org/w/index.php?title=Whirligig&oldid=903704710.
[5] S. Iqbal, Science Reporter, May 2017, p49-50.

[6] M. Saad Bhamla, B. Benson, C. Chai, G. Katsikis, A. Johri and M. Prakash, Nature Biomed. Eng. **1**, 0009 (2017) and supplementary information.
[7] J.P. Salvetat, G.A.D. Briggs, J.M. Bonard, R.R. Bacsa, A.J. Kulik, T. Stockli, N.A. Burnham and L. Forro, Phys. Rev. Lett. **82**, 944 (1999).
[8] Z.L. Zhao, H.P. Zhao, J.S. Wang, Z. Zhang and X.Q. Feng, J. Mech. Phys. Solids **71**, 64 (2014).
[9] J. Foroughi, G.M. Spinks, G.G. Wallace, J. Oh, M.E. Kozlov, S. Fang, T. Mirfakhrai, J.D.W. Madden, M.K. Shin, S.J. Kim, R.H. Baughman, Science **334**, 494 (2011).
[10] S.H. Kim, M.D. Lima, M.E. Kozlov, C.S. Haines, G.M. Spinks, S. Aziz, C. Choi, H.J. Sim, X. Wang, H. Lu, D. Qian, J.D.W. Madden, R.H. Baughman and S.J. Kim, Energy Env. Sci. **8**, 3336 (2015).
[11] J. Ducree, S. Haeberle, S. Lutz, S. Pausch, F. von Stetten and R. Zengerle, J. Micromech. Microeng. **17**, S103 (2007).
[12] R. Gorkin, J. Park, J. Siegrist, M. Amasia, B.S. Lee, J.M. Park, J. Kim, H. Kim, M. Madouab and Y.K Cho, Lab. Chip **10**, 1758 (2010).
[13] A.W. Martinez, S.T. Phillips, G.W. Whitesides and E. Carrilho, Anal. Chem. **82**, 3 (2010).
[14] S.O. Sundberg, C.T. Wittwer, C. Gao and R.K. Gale, Anal. Chem. **82**, 1546 (2010).
[15] H.J. Schlichting and W. Suhr, Eur. J. Phys. **31**, 501 (2010).

10.5 Exercises

10.1 Assume that in the buzzer system, the dominant force is the human input. Neglecting the other forces solve the equation of motion for ϕ and determine $\dot{\phi}_{\text{max}}$ and the relative centrifugal force (RCF).

10.2 A button buzzer with 2 cm diameter experiences a centripetal acceleration $100,000\,\text{m/s}^2$. Determine the tangential velocity and number of rotations per minute (rpm).

10.3 A buzzer has the parameters $L = 0.3\,\text{m}$, $R_{\text{s}} = 0.25\,\text{mm}$, $R_{\text{w}} = 0.5\,\text{mm}$ and $R_{\text{h}} = 20\,\text{mm}$. Find ϕ_{max}.

10.4 Using the parameters of a buzzer given in the exercise 10.3 find ϕ_{crit}.

10.5 Given $\phi_{\text{max}} = 862.18\,\text{rad}$ and $\phi_{\text{crit}} = 843.42\,\text{rad}$ find τ_{twist} for $\gamma = 1$ and $\gamma = 3$ for $\phi/\phi_{\text{max}} = 0.99$.

Chapter 11

Yo-Yo

A yo-yo is a traditional toy consisting of two identical disks connected by a cylindrical axle at their centres and a length of string is attached to and looped around the axle. The radius of the axle is much smaller than the radius of the disks. In a typical play the free end of the string is held by a finger and the toy is released. The string unwinds and the yo-yo moves downward with spinning. Once the string is completely unwound a small jerk given to the yo-yo will make it to climb up towards the hand with rewinding of the string. Yo-yos are popular for their rises and falls [1-6]. Figure 11.1 shows the front and side views of a simple yo-yo. When a yo-yo is spinning at the end of the string (at the bottom stage), then it is said to be in a *sleeping state*. More tricks can be performed if a yo-yo can sleep a longer period. The sleeping time depends on the design of a yo-yo.

Who invented yo-yos? Though we do not have a clear evidence, however, a boy playing with the toy yo-yo appeared in a Greek Vast painting made around 500 BC. There is an evidence for yo-yo played in Berlin around

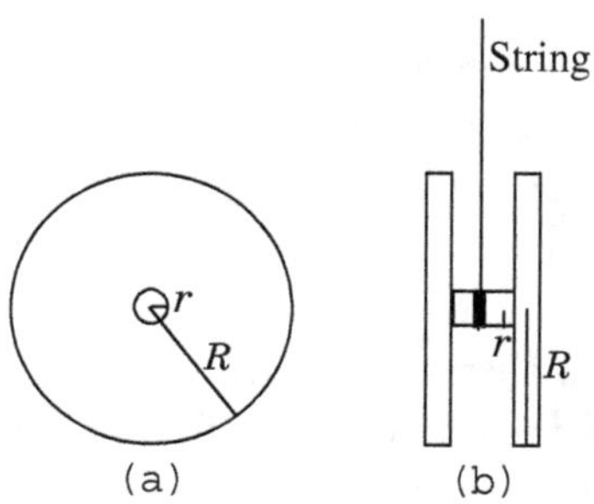

Fig. 11.1: Front view (a) and side view (b) of a yo-yo.

40 BC. A portrait of a lady with a yo-yo was found in Rajasthan, India in 1770. On 20, November 1866 a patent for a yo-yo had been granted to James L. Haven and Charles Hettrick, USA. In 1928 Pedro Flores started the yo-yo manufacturing company in Santa Barbara, California. Sam Dubner in Vancouver, Canada, had first registered the yo-yo trademark in 1932. The Duncan Toys Company in 1946 began a yo-yo factory in Luck, Wisconsin and its yo-yo was inducted into the National Toy Hall of Fame in 1999 at The Strong in Rochester, New York. Nowadays, yo-yos are manufactured in a variety of forms.

Yo-yo competitions are conducted in several countries. The first world yo-yo competition was held in 1932 in London and the winner was Harvey Lowe. There are a variety of yo-yo competitions. In one type, a player with a long sleeping yo-yo does string tricks. In another format, a player with two yo-yos perform simultaneously reciprocating or looping tricks.

We start with reviewing the linear and rotational motions of an object. We explain the basic working principle of a yo-yo. Next, we point out the forces acting on it and obtain expressions for the acceleration, moment of inertia and tension force. Then, we obtain a formula for the time required for the yo-yo to completely unwind. Finally, we setup the equation for the vertical displacement.

11.1 Linear and Rotational Motions

The displacement of an object refers to the vector connecting the position of the object with respect to an origin. Linear motion involves an object moving from one point to another in a straight-line. The rate at which the displacement changes with respect to time is the velocity of the object. If h is the displacement, then the velocity is $\mathbf{v} = \mathrm{d}\mathbf{h}/\mathrm{d}t$. The rate at which the velocity changes is called the *acceleration* of the body $\mathbf{a} = \mathrm{d}\mathbf{v}/\mathrm{d}t = \mathrm{d}^2\mathbf{h}/\mathrm{d}t^2$. As the linear motion is a motion along a straight-line, it can be described by one-dimensional Newton's equation of motion for a body of mass m and acted upon by a force as $m\mathrm{d}^2h/\mathrm{d}t^2 = F$. If the body is subjected to a constant force, then $F = ma$ is a constant. So, $\mathrm{d}^2h/\mathrm{d}t^2 = a$ where a is a constant.

If the body moves with the initial position h_0 and initial velocity v_0, then integrating $\mathrm{d}^2h/\mathrm{d}t^2 = a$ twice and applying the initial conditions, we get the position of the body as a function of time as $h(t) = h_0 + v_0t + \frac{1}{2}at^2$.

Pure rotational motion occurs when every particle of the body moves in a circle about a single line called the *axis of rotation*. In a rotational motion, the displacement vector rotates and changes its angular position.

If θ is the angle of rotation, then the *angular velocity* ω is defined as the rate of change of the angular displacement θ: $\omega = \mathrm{d}\theta/\mathrm{d}t$. If r is the radius of the circle in which the particle moves due to rotation, then the linear displacement will be $r\mathrm{d}\theta$. So, the velocity of the rotating particle will be $v = r\mathrm{d}\theta/\mathrm{d}t = r\omega$. Further, $\mathbf{v} = \boldsymbol{\omega}\times\mathbf{r}$ as $\boldsymbol{\omega}$ is in the direction of the rotation axis. The rate of change of the angular velocity gives the angular acceleration $\alpha = \mathrm{d}\omega/\mathrm{d}t = \mathrm{d}^2\theta/\mathrm{d}t^2$. Since $v = r\omega$, the translational acceleration is $a = \mathrm{d}v/\mathrm{d}t = r\mathrm{d}\omega/\mathrm{d}t = r\alpha$. Further, since ω and α are perpendicular to $\mathbf{r}$, then $\mathbf{a} = \boldsymbol{\alpha}\times\mathbf{r}$. It is also called the tangential component of the acceleration. It can be proved that the particle has a radial acceleration v^2/r.

If the angular acceleration is a constant, then integrating the equation $\mathrm{d}^2\theta/\mathrm{d}t^2 = \alpha$ twice we get $\theta = \theta_0 + \omega_0 t + \frac{1}{2}\alpha t^2$ where θ_0 is the initial angle and ω_0 is the initial angular velocity. The angular acceleration is caused by the torque $\boldsymbol{\tau}$, which is defined as the moment of the force and is given by

$$\begin{aligned}\boldsymbol{\tau} &= \mathbf{r}\times\mathbf{F}\\ &= \mathbf{r}\times m\mathbf{a}\\ &= \mathbf{r}\times m(\boldsymbol{\alpha}\times\mathbf{r})\\ &= m\left((\mathbf{r}\cdot\mathbf{r})\alpha - (\mathbf{r}\cdot\boldsymbol{\alpha})\mathbf{r}\right)\\ &= mr^2\boldsymbol{\alpha} \text{ as } \mathbf{r}\cdot\boldsymbol{\alpha} = 0.\end{aligned} \tag{11.1}$$

mr^2 is called the *moment of inertia* I of the particle with respect to the axis of rotation. Then, $\boldsymbol{\tau} = I\boldsymbol{\alpha}$.

For a system of N particles we can write

$$\begin{aligned}\boldsymbol{\tau} &= \sum_{i=1}^{N}\mathbf{r}_i\times m_i(\boldsymbol{\alpha}\times\mathbf{r}_i)\\ &= \sum_{i=1}^{N} m_i\left[(\mathbf{r}_i\cdot\mathbf{r}_i)\,\boldsymbol{\alpha} - (\mathbf{r}_i\cdot\boldsymbol{\alpha})\,\mathbf{r}_i\right]\\ &= \sum_{i=1}^{N} m_i r_i^2\boldsymbol{\alpha}\\ &= I\boldsymbol{\alpha},\end{aligned} \tag{11.2}$$

where the moment of inertia of the system about the axis of rotation is $I = \sum_{i=1}^{N} m_i r_i^2$. If the body is a rigid body, then

$$I = \int_V \rho(x,y,z)r^2\mathrm{d}x\mathrm{d}y\mathrm{d}z, \tag{11.3}$$

where ρ is the mass density.

The *angular momentum* **L** is defined as the moment of linear momentum as

$$\begin{aligned}\mathbf{L} &= \mathbf{r}\times\mathbf{p}\\ &= \mathbf{r}\times m\mathbf{v}\\ &= \mathbf{r}\times m(\boldsymbol{\omega}\times\mathbf{r})\\ &= m(\mathbf{r}\cdot\mathbf{r})\omega - m(\mathbf{r}\cdot\boldsymbol{\omega})\mathbf{r}\\ &= mr^2\omega \text{ as } \mathbf{r}\cdot\boldsymbol{\omega}=0\,.\\ &= I\boldsymbol{\omega}\,. \end{aligned} \tag{11.4}$$

Then, the rate of change of **L** is $\mathrm{d}\mathbf{L}/\mathrm{d}t = I\mathrm{d}\boldsymbol{\omega}/\mathrm{d}t = I\boldsymbol{\alpha} = \boldsymbol{\tau}$. The rate of change of angular momentum gives the torque. This is the Newton's equation of motion for rotational motion. Yo-yo is a system which has both linear and rotational motions.

11.2 Physical Mechanism of Working of Yo-Yo

A yo-yo can be thrown in any direction. Let us consider a simple case of released downward as shown in Fig. 11.2. Initially, you hold a yo-yo in your hand. At this initial stage, it possesses a potential energy because it is at a height from the floor. When a yo-yo is released, the gravitational force acting on its centre of mass (CoM) pulls it down. Further, the yo-yo is made to rotate while it drops since the string of the yo-yo is wrapped around the axle of the yo-yo and also because one end of the string is hold at your finger. Obviously, the yo-yo would not move downward if it is not rotated. It begins its motion with gravitational potential energy. When it falls down the gravitational potential energy is gradually transformed into kinetic energy. It gains both linear kinetic energy, because of its downward motion, and rotational kinetic energy (angular momentum), because of its spinning. The acceleration of the yo-yo will be relatively smaller than that if it were freely falling. *Why?*

The rate of rotation and the rate of falling down of the yo-yo increases with time due to the gravitational attraction. These rates become maximum at the bottom when the string is completely unwound. At the bottom stage, the yo-yo can continue its rotation because the string is not tightly tied around the axle. The rotation rate cannot increase further since the yo-yo cannot fall down further. Due to the friction between the string and the axle the angular momentum starts to dissipate.

If the yo-yo is left at the bottom, then it will eventually come to halts its rotation. When the yo-yo arrive the bottom, give a tug (a short and

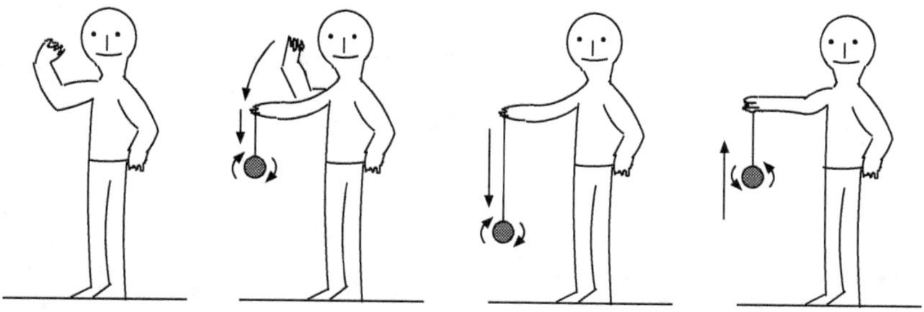

Fig. 11.2: Falling and rising of a yo-yo.

a sharp upward pull) on the string so that the yo-yo starts to climb above because of its spinning motion. The applied pull supplies an additional kinetic energy to compensate the energy loss due to friction. As the yo-yo climbs up, the string's kinetic energy is converted to gravitational potential energy due to the rising of the height of the CoM of the yo-yo.

After a first few upward rotational climbings, the string keeps it away from slipping. Hence, even after stopping the tugging of the string, the climbing up of the yo-yo continues. The rate of rotation steadily decreases. The physical process happening when the yo-yo climbs upward from the bottom is exactly the reverse process of it dropped from the finger. Because of the loss of energy due to the friction, the yo-yo would climb up to a height shorter than the finger position. Then, it begins to fall down. By means of giving tugs when ever the yo-yo arrives at the bottom, you can make it in rising and falling motion indefinitely.

11.3 Motion of the Unwinding Yo-Yo

It is possible to describe the downward (unwinding) of the yo-yo by applying the laws of classical mechanics [2,6].

Consider a yo-yo with mass m with axle radius r and two circular disks with equal radii R. The position of the hand holding the free end of the string at time $t = 0$ at which yo-yo will be released is set as the origin of the coordinate system. The initial height of the yo-yo at $t = 0$ is h_0, as shown in Fig. 11.3. P_1 be a point near the rim of the disk at time $t = 0$ and is making an angle ϕ_0 with the horizontal axis. P_2 is the point of CoM. Denote h, v and a as the vertical height, velocity and acceleration of the CoM point P_2.

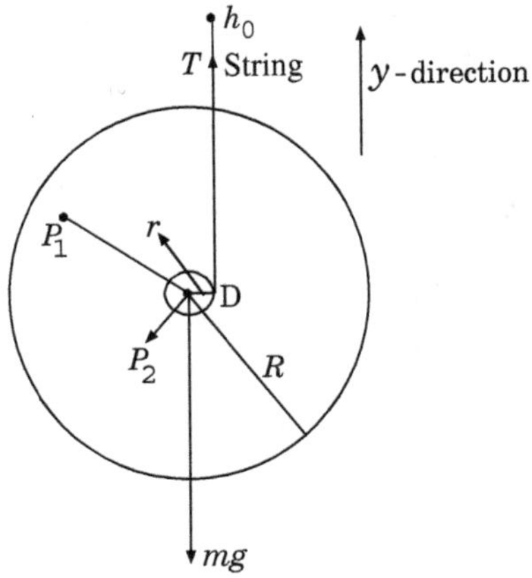

Fig. 11.3: Force for a yo-yo.

11.3.1 *Forces and an Expression for Acceleration*

What are the forces acting on the yo-yo when it moves downward? The forces are the gravity acting in the downward direction and the tensional force ($F_{\mathrm{T}} = T$) of the string acting in the upward direction [2,6]. If we choose the downward direction as the negative direction, then the total force F is given by $F = T - mg$. According to the Newton's law

$$T - mg = ma\,. \tag{11.5}$$

This is for the linear downward (translational) motion. It is the dynamical equation for motion of the CoM. For the rotational motion with respect to the CoM, the torque τ (moment of force) about P_2 which enables the yo-yo to rotate is

$$\tau = \mathbf{r} \times \mathbf{F}_{\mathrm{T}} = rT, \tag{11.6}$$

since it depends only on T. With L being the angular momentum and ω being the angular frequency of rotation we write

$$\tau = \frac{\mathrm{d}L}{\mathrm{d}t} = I\frac{\mathrm{d}\omega}{\mathrm{d}t} = I\alpha\,, \tag{11.7}$$

where α is the angular velocity and I is the moment of inertia. From $v = -r\omega$ we have $a = \mathrm{d}v/\mathrm{d}t = -r\mathrm{d}\omega/\mathrm{d}t = -r\alpha$ or

$$\alpha = -\frac{a}{r}\,. \tag{11.8}$$

From Eqs. (11.6)-(11.8), we find

$$T = \frac{I\alpha}{r} = -\frac{Ia}{r^2}\,. \tag{11.9}$$

Now, we are ready to obtain an expression for a from the Eqs. (11.5) and (11.9). The result is

$$a = -\frac{g}{1 + \dfrac{I}{mr^2}}\,. \tag{11.10}$$

a and T are constants.

11.3.2 *Expression for I*

What is the expression for I? Note that here I is the sum of I of each of the identical disks with mass m_1 and the I of the axle with mass m_2. Further, $m = 2m_1 + m_2$. Then,

$$\begin{aligned} I &= 2I_{\text{disks}} + I_{\text{axle}} \\ &= 2\left(\frac{1}{2}m_1R^2\right) + \frac{1}{2}m_2r^2\,. \end{aligned} \tag{11.11}$$

Solved Problem 1:

Consider a yo-yo having two identical disks of mass m_1 each and radius R connected by a cylindrical axle of mass m_2 and radius r. Find the moment of inertia of the yo-yo about its axis of rotation.

Consider the disk of mass m_1 and radius R with the axis passing through its centre O perpendicular to the plane. Then, consider an elemental ring of radius x and thickness $\mathrm{d}x$ as shown in Fig. 11.4a. The mass per unit area of the disk is $\rho = m_1/(\pi R^2)$. The mass of the elemental ring is $\mathrm{d}m = \rho 2\pi x \mathrm{d}x$. The moment of inertia of the ring is $\mathrm{d}I = \mathrm{d}mx^2$. The moment of inertia of the disk is

$$I_{\text{disk}} = \int_0^R \mathrm{d}I = \int_0^R 2\pi\rho x^3 \mathrm{d}x = \frac{1}{2}\pi\rho R^4\,. \tag{11.12}$$

To find the moment of inertial of the axle consider a disk of thickness $\mathrm{d}x$ at position x from an end as shown in Fig. 11.4b. The density of the axle is $= m_2/V = m_2/(\pi r^2 L)$. The volume of the elemental disk is $\pi r^2 \mathrm{d}x$. Then,

$$\mathrm{d}m = \frac{m_2}{\pi r^2 L}\pi r^2 \mathrm{d}x = \frac{m_2}{L}\mathrm{d}x\,. \tag{11.13}$$

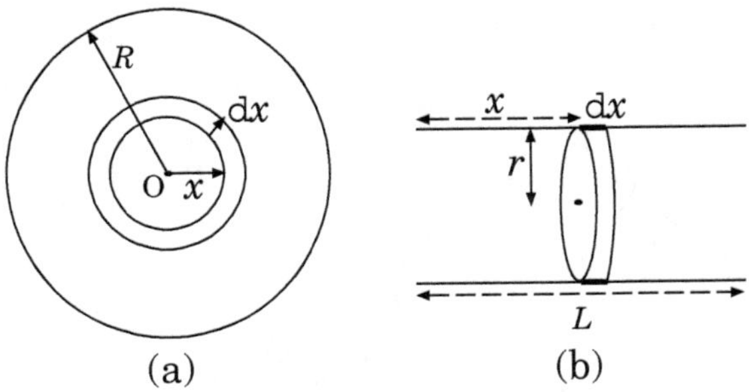

Fig. 11.4: (a) An elemental ring of radius x and thickness dx of the yo-yo disk with radius R. (b) A disk of thickness dx at position x from an end.

Next, the moment of inertia of the elemental disk of radius r is

$$\mathrm{d}I = \frac{1}{2}r^2\mathrm{d}m = \frac{m_2r^2}{2L}\mathrm{d}x\,. \tag{11.14}$$

The moment of inertia of the axle about the rotation axis is

$$I_{\text{axle}} = \frac{m_2}{2L}r^2\int_0^L \mathrm{d}x = \frac{1}{2}m_2r^2\,. \tag{11.15}$$

The total moment of inertia of the yo-yo is

$$\begin{aligned} I &= 2I_{\text{disk}} + I_{\text{axle}} \\ &= 2\frac{1}{2}m_1R^2 + \frac{1}{2}m_2r^2 \\ &= m_1R^2 + \frac{1}{2}m_2r^2\,. \end{aligned} \tag{11.16}$$

11.3.3 *Expression for T*

From the Eqs. (11.5) and (11.10) we find

$$\begin{aligned} T &= mg + ma \\ &= mg - \frac{mg}{1+\dfrac{I}{mr^2}} \\ &= \frac{mgI/(mr^2)}{1+I/(mr^2)} \\ &= \frac{mg}{1+(mr^2/I)} < mg\,. \end{aligned} \tag{11.17}$$

The tension force is $< mg$.

11.3.4 *Energy of Yo-Yo*

What about the energy of the yo-yo? At rest it has potential energy. When it is in the downward or upward motion, it possesses potential energy and two kinds of kinetic energies - translational $(mv^2/2)$ and rotational $(I\omega^2/2)$. With E_{Tot} is the total energy of the yo-yo, then if the friction is neglected then $E_{\text{Tot}} = E_{\text{Top}} - E_{\text{Bottom}}$ and is also equal to its total energy in motion.

11.3.5 *Expression for h and Time Required for Complete Unwinding*

We have

$$h = h_0 + v_0 t + \frac{1}{2} a t^2 . \tag{11.18}$$

Setting $h_0 = 0$ and the initial velocity of the yo-yo as zero gives

$$h = \frac{1}{2} a t^2 . \tag{11.19}$$

Note that both a and h are negative in the unwinding motion. *What is the time required for the yo-yo to completely unwind?* The length of the string is, say, H. Suppose t_{H} is the time taken for complete unwinding. At $t = t_{\text{H}}$ the yo-yo went downward through the distance (height) $h = H$. Then, from Eq. (11.19) we write

$$t_H = \sqrt{\frac{2H}{a}} = \sqrt{\frac{2|H|\,(1 + I/mr^2)}{g}} . \tag{11.20}$$

The upward motion of the yo-yo can be treated by changing the signs in the relevant formulas.

11.3.6 *Equation for the Vertical Displacement*

Consider Eq. (11.11). Since $m_2 \ll m_1$ and $r \ll R$ the second term in the right-side of Eq. (11.11) can be neglected compared to the first term. Then, $I = m_1 R^2$. Then, a given by Eq. (11.10) becomes, with $m = 2m_1 + m_2 \approx 2m_1$

$$a = -\frac{g}{1 + \frac{1}{2}(R/r)^2} . \tag{11.21}$$

Consider a point P_1 on the disk at the distance R' from the centre o. To determine the vertical displacement of P_1 consider the Fig. 11.5. In this figure the angle between P_1OD is ϕ and hence the angle between P_1OB is

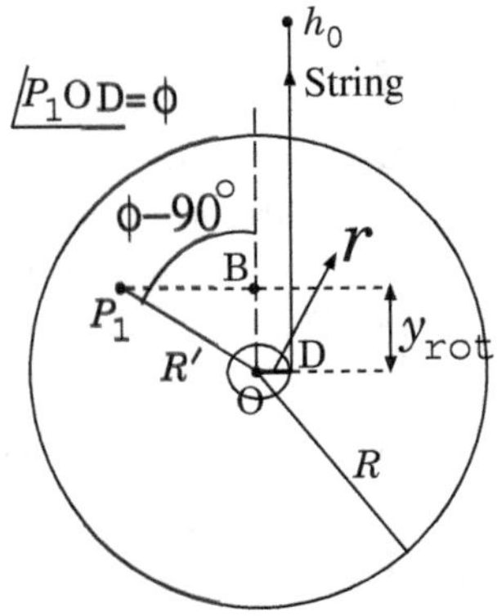

Fig. 11.5: Representation of the point P_1 on the disk of yo-yo at a distance R' from the centre O.

$\phi - 90°$. So, the projection of the point P_1 along the vertical direction is given by

$$y_{\text{rot}} = R' \cos(\phi - 90^o) = R' \sin \phi \,. \tag{11.22}$$

The vertical displacement y of the point P_1 is a superposition of the vertical h of the point P_2 (that is the point O in Fig. 11.5) and the vertical projection y_{rot} of the rotation about the axle, that is, $y = h + y_{\text{rot}}$ where y_{rot} is given by Eq. (11.22). Setting the initial velocity of the yo-yo as zero and the initial height $h_0 = 0$ and defining ϕ_0 as the angle between P_1OD at the initial position of the yo-yo, we have

$$\phi = -\frac{h}{r} + \phi_0 = -\frac{1}{2r}at^2 + \phi_0 \,, \tag{11.23}$$

Then,

$$y = h + y_{\text{rot}} = \frac{1}{2}at^2 + R' \sin\left(-\frac{1}{2r}at^2 + \phi_0\right) . \tag{11.24}$$

11.4 Conclusion

The study of motion of particles is called *mechanics*. The knowledge of linear and rotation motions is very important for the students to learn mechanics. The study of yo-yo will help the students to learn in practice these two fundamental motions. Also, the yo-yo is a system which possesses both linear and rotational motions simultaneously. The students can see in a yo-yo experiment how the energy in linear motion is transformed into rotational energy. They can also understand that the combined energies of linear and rotational motions are always conserved in absence of friction. For a report on the study of a modern yo-yo with a centrifugal clutch allowing free falling one may refer to [7]. In [8] using smartphone sensors both theoretically and experimentally the dynamics of the yo-yo have been investigated. Near the lowest position the string has a jerk and is due to the change from a linear motion to a pure rotational motion. This is studied in Ref. [9].

11.5 Bibliography

[1] https://science.howstuffworks.com/yo-yo.htm.

[2] A. Heck and P. Uylings, *Yoyo Joy*; https://www.researchgate.net/publication/25490159_yoyo_Joy.

[3] https://en.wikidedia.org/wiki/Yo-yo.

[4] http://www.explainthatstuff.com/yoyos.html.

[5] W. Davis, *How do yo-yos work? Why do they come back after being thrown down?*; http://www.physlink.com/education/askexperts/ae18.cfm.

[6] https:/wiki.brown.edu/confluence/download/attachments/.../Yo-yo.pdf.

[7] C. de Izarra, Eur. J. Phys. **32**, 1097 (2011).

[8] I. Salinas, M. Monteiro, A.C. Marti and J.A. Monsoriu, The Phys. Teacher **58**, 569 (2020).

[9] L. Minkin and D. Sikes, The Phys. Teach. **58**, 656 (2020).

11.6 Exercises

11.1 Explain that a spinning yo-yo behaves like a gyroscope.

11.2 Consider a yo-yo with $m_1 = 0.05\,\text{kg}$, $m_2 = 0.005\,\text{kg}$, $r = 0.004\,\text{m}$, $R = 0.03\,\text{m}$ and string length $0.6\,\text{m}$ released with zero initial velocity. Calculate the time taken for complete unwinding of the string.

11.3 Consider the origin of the coordinate system as the position of the hand at $t = 0$ at which a yo-yo is released with zero initial velocity. Assume that $m_1 = 0.05\,\text{kg}$, $m_2 = 0.005\,\text{kg}$, $r = 0.004\,\text{m}$, $R = 0.03\,\text{m}$ and string length $0.6\,\text{m}$.

(a) Calculate the initial total energy (total kinetic energy plus the potential energy).

(b) Show that the total kinetic energy after it has moved downward through 0.3 m is 0.3087 J (assume that the total energy is conserved).
(c) Calculate the speed of the CoM and the angular speed of the yo-yo when it reached downward 0.3 m.

11.4 A yo-yo with $m_1 = 0.075\,\text{kg}$, $m_2 = 0.005\,\text{kg}$, $r = 0.004\,\text{m}$, $R = 0.03\,\text{m}$ is in a sleeping state.
(a) Calculate its moment of inertia.
(b) How much rotational energy has the yo-yo to make 5 revolutions per second?

11.5 Consider a yo-yo with $m_1 = 0.075\,\text{kg}$, $m_2 = 0.005\,\text{kg}$, $r = 0.004\,\text{m}$, $R = 0.03\,\text{m}$ and a string of 0.8 m. It is released down with zero initial velocity. Calculate the amount of work to be done by the force of tension to make one revolution of the yo-yo.

Chapter 12

Astrojax

The toy astrojax consists of three balls of equally sized masses on a string. Two balls are fixed to the ends of the string. The centre ball is free to slide along the string and can go back and forth between the end balls. An astrojax is shown in Fig. 12.1 [1]. An astrojax is an orbiting skill toy and can be made swing in any direction. The motion of the balls on a string are easily controllable. Hold one end ball in one hand and the other end ball in the other hand. Leave the middle ball down so that the astrojax is in 'V' shape. Bring the left-hand ball near to the right-hand ball and drop it in such a way that the third ball begins to rotate about the middle ball vertically. By means of giving a proper push-pull of the right-hand holding the first ball, the third ball can be made to rotate about the middle ball either in the clockwise or anticlockwise direction. Horizontal rotation is also possible. It is possible to do a variety of tricks with number of patterns. Simultaneously throwing two balls and catching them simultaneously by the two hands in a fixed position is a skill with the astrojax.

Fig. 12.1: An Astrojax.

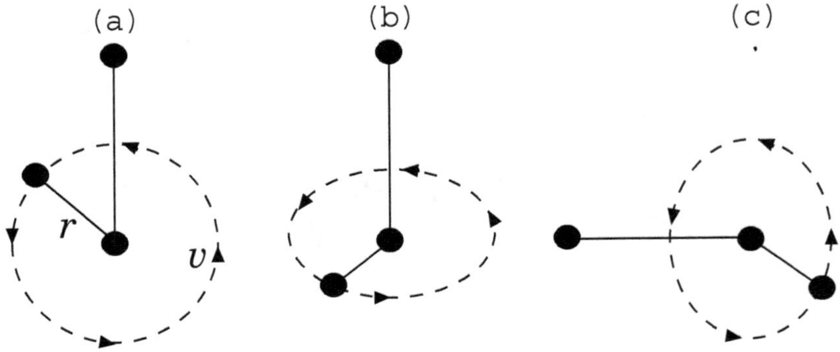

Fig. 12.2: Three types of orbital motion of one of the free end balls.

The toy astrojax was invented by Larry Shaw in 1986 when he was a physics graduate student in Cornell University. First, this toy was made with hex nuts and dental floss [2]. In 1994 it was sold in the market as a polyurethane foam version. Nowadays, astrojaxs are coming with balls in many colours, variety of materials and design and balls and string with lighting effect.

The rhythm and beautiful movement of the astrojax makes one want to keep doing it. Once you watch the videos of the various tricks with the astrojax available in the internet, you would like to get training to do certain incredible tricks. This toy is sold around, won awards and taken into outer space by NASA as part of its Science Education Programme. In the Guinness Book of World Records database, there are three astrojax categories.

Astrojax can be used to teach physics in school and college levels. Research has been performed on exploring its complex dynamics and is treated as a generalized double pendulum [3,4].

12.1 Orbital Motions of a Ball of an Astrojax

When performing various tricks with astrojax, its string does not tangle. This is because the balls in the astrojax are specially designed [5]. Each ball has a metal weight at the middle. This metal weight of the middle ball allows it to rotate fastly in response to the forces exerted by the string. Further, the moment of inertia is lowered by the central metal weight. The forces acting on the balls are gravity, spring forces and damping forces.

In a vertical orbital motion as shown in Fig. 12.2a the outer balls orbits around the middle ball. The orbiting ball has to pass the string at the top of its orbit while the middle ball has to rotate by 180° in about $1/20$ sec in order to make the string not to snag around the middle ball. This happens due to the reduced moment of inertia caused by the central weight and the shape of the bore.

In the case of the horizontal orbital motion depicted in Fig. 12.2b the outer ball traces out a circular orbit with the middle ball being the centre of the orbit. The middle and the outer balls orbit in circles. The string attached to the outer orbiting ball exerts a force on it that pulls toward the centre of the orbit. This is like the earth and moon system.

In another trick we can notice the conservation of angular momentum. Let us keep one of the end balls in one hand and the middle ball in the other hand. Spin the free end ball so that it traces a circular motion with radius r and velocity v as depicted in Fig. 12.2c. Suppose you pull the free end ball, which is spinning, by increasing the distance between your hands. This speed up the rotating ball. *What is happening?* The angular momentum is proportional to rv and is a constant. When you increase the distance between the balls in your hands, the radius r gets reduced. If r becomes, say, $r/2$, then to conserve the angular momentum, v changes into $2v$ [5].

Solved Problem 1:

Assume that an astrojax is a generalized double pendulum as described in the exercise 12.2. Write the Newton's equations of motion for the motion of the middle and the free end third ball considering the forces of action are gravity, force of the string and the drag force.

Denote g as the acceleration due to gravity, a_{ij} as the acceleration of the ith mass in jth direction, $F_{\mathrm{s}_{\alpha\beta}}$ as the force of the string on mass α in the β-direction and $F_{\mathrm{d}_{\alpha\beta}}$ is the drag force. The Newton's equations of motion are

$$m_2 a_{2x} = F_{\mathrm{s}_{1x}} - F_{\mathrm{s}_{2x}} + F_{\mathrm{d}_{2x}}\,, \quad (12.1a)$$
$$m_2 a_{2y} = F_{\mathrm{s}_{1y}} - F_{\mathrm{s}_{2y}} + F_{\mathrm{d}_{2y}}\,, \quad (12.1b)$$
$$m_2 a_{2z} = -m_2 g + F_{\mathrm{s}_{1z}} - F_{\mathrm{s}_{2z}} + F_{\mathrm{d}_{2z}}\,, \quad (12.1c)$$
$$m_3 a_{3x} = F_{\mathrm{s}_{2x}} + F_{\mathrm{d}_{3x}}\,, \quad (12.1d)$$
$$m_3 a_{3y} = F_{\mathrm{s}_{2y}} + F_{\mathrm{d}_{3y}}\,, \quad (12.1e)$$
$$m_3 a_{3z} = -m_3 g + F_{\mathrm{s}_{2z}} + F_{\mathrm{d}_{3z}}\,. \quad (12.1f)$$

12.2 Some Other Tricks

There is a variety of astrojax tricks that one may specialize. The demonstration of them can be found in excellent video clips. Let us mention some of them [6,7].

1. **Juggling:** Tug a bit harder on the balls such that the distant one comes up to hand height. Now, leave the ball in the hand and catch that one.
2. **Skyscraper:** Perform a few orbital motion and pull on the ball in the hand to make a big orbit.
3. **String Trick:** Wrap the astrojax around the various parts of the body. Temporarily allow the string to go and catch it once it has swung arround the finger, leg, etc.
4. **Butterfly:** Perform some orbits and then pull the ball in the hand left and right. The balls move in a helical shape.
5. **Rebound Trick:** Create a smooth pattern by bouncing a ball off the floor or kicking a ball lightly.
6. **Venus:** Do a vertical orbital motion. Pull hardly the bottom ball making it travels over your arm, followed by the middle ball.
7. **Switch:** Perform the vertical orbital motion. When the orbiting ball comes nearer to the hand holding the ball, release the ball to catch the orbiting one.

12.3 Conclusion

Astrojax is a fascinating toy that can illustrate a wide variety of basic physics problems in rotational dynamics and orbital motions. Through the astrojax toy, the students can understand that by concentrating the mass at the centre of an object, it lowers its moment of inertia. When energy is conserved, by lowering the moment of inertia, it increases its angular velocity. That is the reason why a diver in a swimming pool rolls his body into a circle for having more number of rotations before falling into water. In the horizontal motion of the astrojax, the tension on the string pulls the outer ball towards the middle ball which is balanced by the centrifugal force mv^2/r arises due to circular motion of the outer ball. The students can also be taught the conservation of angular momentum using astrojax as discussed in the text.

12.4 Bibliography

[1] https://www.majaskogstad/fr/nature-decouvertes-astrojax-jonglages/.

[2] https://astrojax.com/pages/the-history-of-astrojax.

[3] D. Dichter and K. Maschan, *Modeling and Simulation of Astrojax*, 2013; http://testtestestes423432435645.weebly.com/uploads/2/4/4-/5/24451800/astrojax_final_report.pdf.

[4] A. Karsai, S. Harrington and C. Campbell, *The Astrojax Pendulum*, 2014; https://www.semanticscholar.org/paper/The-Astrojax-Pendulum-Karsai-Harrington/1d65c87ac85170ae87652acd02da036a-1750d7f1.

[5] https://astrojax.com/pages/the-science.

[6] http://www.jugglingworld.biz/tricks/prop-manipulation-tricks/ast-rojax-tricks/.

[7] https://en.wikipedia.org/wiki/Astrojax.

12.5 Exercises

12.1 How many spring forces are there in an astrojax?

12.2 Astrojax can be considered as a general type of a double pendulum [4]. The traditional double pendulum has two degrees of freedom with θ_1 and θ_2 being the two azimuthal angles of the pendulum axis from their respective pivots. Sketch the astrojax as a double pendulum and mark the dynamical variables.

12.3 Consider the previous exercise. How many degrees of freedom the system has?

12.4 Treating an astrojax as a generalized double pendulum as described in the exercise 12.2 setup its Lagrangian.

12.5 Consider a system of a rod of negligible mass and length 0.20 m carrying a sphere of radius 0.01 m at both ends suspended from the centre of the rod. A rotational energy E is given to this system by an external torque. Find its angular velocity. If both spheres are brought to the centre of the rod, touching each other, how much does its angular velocity increase now?

Chapter 13

Hula Hoop

A hula hoop is a popular sports equipment composed of a thin walled hoop that goes around the waist, limbs or neck of an athlete. Hula hooping is a complex skill of keeping an unstable hoop or a perfect ring in its circular shape in steady oscillation parallel to the ground or other planes by means of coordinate oscillations of various parts of the body such as hips, arms, waist, neck, legs, knees etc. A hula hoop [1] and a hula hooping with three rings are shown in Fig. 13.1. A hula hoop dancer waves the ring around his/her waist, shoulders, neck, arms, also roll it like a wheel over the body, toss it to the air or floor and jump through the hoop. During hula hooping a stimulation creativity takes place making beautiful flow of movements for enjoyment.

Hula hoopers apply different strategies to maintain hoop oscillation [2,3]:

1. Use of extensor movement of the knee for making the rotational motion.
2. Use of ankle-hip movement in a more diagonal elliptical motion.

Fig. 13.1: A beautiful hula hoop and a hula hooping.

3. Moving the hula hoop to legs, waist and sides in same proportion in order to achieve a balanced strategy toward two directions.

The term hula hoop came from British sailors who had noticed hula dancing in the Hawaiian Islands. Nowadays, hula hoops are made by a plastic ring. In the early years, hula hoops were made by a circle of willow, grapevines and stiff grasses.

Since centuries ago people practiced hula hooping to keep their good health. The ancient Greeks used hoops as a tool for exercise. A vase in the Louvre (500-490 BC) displays Ganymede rolling a hoop. Back in 1000 BC Egyptian children used to play with hoops made out of dried grapevines. They rolled hoops with sticks or whirling around waist. In 1400s, in Native American Indians hoop dancing is a way of story telling.

In 1958 Wham-O Company in Los Angeles manufactured polythylene hoops. More than 100 million hoops were sold in the first year. In 1960s hoops were introduced in circus world in Russia and China. In the year 1980 World Hula Hoop Championship was conducted in more than 2000 cities. In primary school classes hula hooping is used to explain friction and periodic motion.

13.1 Hula Hoop is a Sport/Game

Hula hooping is a sport/game because [3] of the following:

1. It enhances the body metabolism and tones belly and thighs.
2. It gives ways for improved posture, balance and proprioception.
3. It has three of the four characteristics of sports or games.

The four characteristics of a sport/game are, ilnex, mimicry, alea and agon [4]. Ilnex is a pursuit of vertigo and inducing shock/panic. In other words, it is about producing a state of dizziness and disorder through a falling movement or a rapid whirling. In the hula hooping a slight dizziness occurs due to the spinning movement and rotation of the body together with the hoop. Mimicry refers to pretend to another or convince others one is another. Many hula hoopers perform the actions with fancy dresses, beautiful and imaginative costumes as a part of creativity. Alea is a Latin word meaning chance or randomness. This is happening in hula hooping because the falling down and change of direction of the hoop occurs in unexpected ways. The fourth characteristic, namely, agon (competition) does not happen in hula hooping, because it is at individual level.

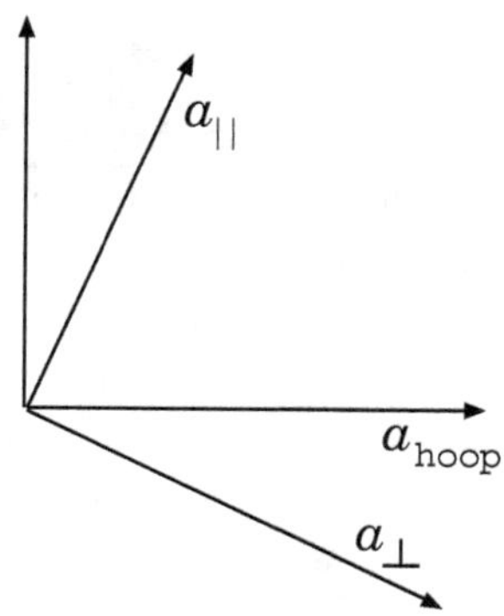

Fig. 13.2: Components of acceleration.

13.2 Physical Basis of the Hula Hooping Skill

What are the forces acting on a hula hoop? The forces concerned are:

1. **Force of gravity:** This force pulls the hula hoop down.
2. **Centripetal force:** When a hooper moves his/her body to sling the hoop around him/her, exert a torque, a turning force. This force is parallel, outward and necessary to keep a rotating object moving in a circle.
3. **Friction:** The friction between the hula hoop and the clothes of the hooper.

We wish to understand the spinning and upward movement of a hula hoop. *What are the sources for these movements?* First, the friction of the hips causes the rotational motion of the hula hoop about its axis. Further, the hips exert a force almost perpendicular to the velocity of the hoop. This keeps the hoop to totate about the hips. Next, consider the upward motion of the hoop. When the body of a hooper is in contact with the hoop ,then it makes an angle inward at the point of contact. Because the hoop is moving in a circle, it possesses an acceleration ($\mathbf{a}$) directly toward the centre of rotation. We can split the acceleration into two components $\mathbf{a}_{\perp}$ and $\mathbf{a}_{||}$ as shown in Fig. 13.2. $\mathbf{a}_{||}$ can have both horizontal and vertical (upward) components. When the upward component becomes greater than the downward component of the hoop (due to gravity) then the hoop will move upward, otherwise fall down.

13.3 Theoretical Study

For a rotating hula hoop the waist of the performer (the athlete) executes periodic motions in the horizontal plane leading to stable rotations. Caughey [5] setup equations of motion for the hula hoop motion and considered the periodic motion of the waist of the athlete along one axis. The equivalent problem becomes a pendulum with a vibrating suspension point and the gravity is assumed absent. One of the possible solutions is associated with the mechanism by which the hula hoop may undergo rotation by an oscillating motion of the point of support.

Bogolyubov and Mitropolsky applied the method of average and investigated the stability condition [6]. Seyranian and Belyakov [7,8] determined the exact solutions for the case of a circular motion of the waist centre and found approximate results for an elliptical motion. A condition for maintaining contact with the waist during twirling has been obtained. In the following we follow [8].

13.3.1 *Model Equations*

Assume that the waist of the athlete is represented by a circle with the centre O' and its moves along an elliptical path described by

$$x = a\sin\omega t\,, \quad y = b\cos\omega t\,, \tag{13.1}$$

where a and b are semi axes and ω is the excitation frequency as shown in Fig. 13.3. $b > 0$ and < 0 refer to the clock-wise and anti-clockwise movements of the centre of the waist, respectively. $b = 0$ corresponds to a strainght-line. Define θ-the angle of rotation around the centre of mass C, R-the radius of the hula hoop, m-mass of the hula hoop, r-radius of the waist, ϕ-the angle between the axis X and the radius CO', $I_c(= mR^2)$-the moment of inertia of the hula hoop, k-friction coefficient, F_T-tangential friction force, N-normal reaction force of the hula hoop to the waist.

As the hula hoop has a radius R and rotation angle θ about the centre we can write its torque as $I_c\ddot{\theta}$. The damping torque is given by $k\dot{\theta}$. F_T is the frictional force acting tangentially at the point of contact of waist and N is the normal force acting radially at the point of contact. The torque due to F_T is RF_T as $\mathbf{R}$ and $\mathbf{F}_T$ are perpendicular. So, we can write the torque equation of motion of the hoop as

$$I_c\ddot{\theta} + k\dot{\theta} = -F_T R. \tag{13.2}$$

As the point O' moves in an eliptical path the force on O' will be $\mathbf{F} = \mathbf{i}m\ddot{x} + \mathbf{j}\ddot{y}$. Since the angle between the x-axis and the normal force is

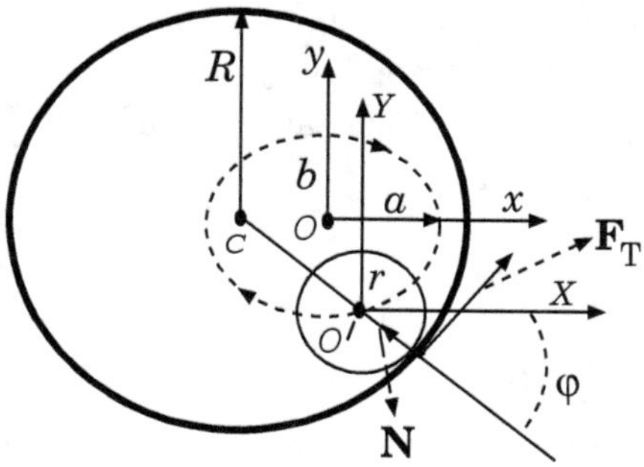

Fig. 13.3: Representation of the motion of a hula hoop and athlete's waist in a coordinate system [8]. (Reproduced from Am. J. Phys. **79**, 712 (2011), with permission of American Association of Physics Teachers. Copyright 2011 A.P. Seyranian and A.O. Belyakov, licensed under a Creative Commons Attribution License.)

ϕ, the angle between F_{T} and the x-axis will be $(90° - \phi)$. So, the component of force acting in the N-direction will be

$$F_{\text{normal}} = m\ddot{x}\cos\phi - m\ddot{y}\sin\phi. \tag{13.3}$$

The force acting in the direction of F_{T} will be

$$F_{\text{tangential}} = m\ddot{x}\sin\phi + m\ddot{y}\cos\phi. \tag{13.4}$$

The tangential force acting in the direction of F_{T} is given by $m(R-r)\ddot{\phi}$. The force equation along the tangential direction can be written as

$$m(R-r)\ddot{\phi} = m\left(\ddot{x}\sin\phi + \ddot{y}\cos\phi\right) + F_{\mathrm{T}}\,. \tag{13.5}$$

The centripetal force acting in the normal direction is given by $m(R-r)\dot{\phi}^2$. The force equation along the normal direction is given by

$$m(R-r)\dot{\phi}^2 = N + m\left(\ddot{x}\cos\phi - \ddot{y}\sin\phi\right)\,. \tag{13.6}$$

Taking into the account of the viscous friction, in the waist fixed coordinate system the equations of motion of the hula hoop are thus written as

$$I_c\ddot{\theta} + k\dot{\theta} = -F_{\mathrm{T}}R\,, \tag{13.7a}$$

$$m(R-r)\ddot{\phi} = m\left(\ddot{x}\sin\phi + \ddot{y}\cos\phi\right) + F_{\mathrm{T}}\,, \tag{13.7b}$$

$$m(R-r)\dot{\phi}^2 = N + m\left(\ddot{x}\cos\phi - \ddot{y}\sin\phi\right)\,. \tag{13.7c}$$

Equation (13.7a) governs the variations in the angular momentum arising as a result of the linear viscous damping with k and F_{T}. The Eqs. (13.7b) and (13.7c) dictate the motion of the hoop in longitudinal and transverse directions to the radius CO', respectively.

Slipping is assumed to be absent at the point of contact. Then $(R-r)\dot{\phi} = R\dot{\theta}$. Next, introducing

$$\gamma = \frac{k}{2mR^2\omega}\,, \quad \alpha = \frac{a}{2(R-r)}\,, \quad \beta = \frac{b}{2(R-r)}\,, \quad \tau = \omega t\,, \quad N > 0 \qquad (13.8)$$

we obtain

$$\ddot{\phi} + \gamma\dot{\phi} + \alpha \sin\tau \sin\phi + \beta \cos\tau \cos\phi = 0\,, \qquad (13.9)$$

$$\dot{\phi}^2 + 2\,(\alpha \sin\tau \cos\phi - \beta \cos\tau \sin\phi) > 0\,. \qquad (13.10)$$

In the above, $\alpha = \beta$ corresponds to a circular trajectory of the centre of waist while $\beta = 0$ means the trajectory is a line.

For $\alpha = \beta$ and $|\gamma| \leq |\alpha|$, Eq. (13.9) admits the solution

$$\phi_{\pm} = \tau + \psi_{\pm}\,, \quad \psi_{\pm} = \pm\cos^{-1}(-\gamma/\alpha) + 2n\pi,\ n = 0, 1, 2, \ldots. \qquad (13.11)$$

These solutions describe the rotation of the hula hoop wherein the angular velocity is equal to the constant excitation frequency ω. In the next section, we explain the stability of these solutions.

Solved Problem 1:

Show that $\alpha = \beta$ corresponds to a circular trajectory of the centre of the waist and $\beta = 0$ means that the trajectory is a line.

As $\alpha = \dfrac{a}{2(R-r)}$ and $\beta = \dfrac{a}{2(R-r)}$ the choice $\alpha = \beta$ gives $a = b$. So, from Eq. (13.1) $x = a\sin\omega t$, $y = a\cos\omega t$ and $x^2+y^2 = a^2(\sin^2\omega t+\cos^2\omega t) = a^2$. The trajectory is thus a circle. When $\beta = 0$ we have $b = 0$ so $y = 0$ and $x = a\sin\omega t$. Therefore, the motion is only along the x-line.

13.4 Stability Analysis

The stabilities of the solutions (13.11) are discussed in [8]. Let us consider the choice $\alpha = \beta$ for a circular trajectory. Equation (13.9) becomes

$$\ddot{\phi} + \gamma\dot{\phi} + \alpha\cos(\phi - \tau) = 0. \qquad (13.12)$$

The angle ϕ is assumed to be in the form $\phi = \tau + \psi + \eta(\tau)$, where $\eta(\tau)$ is very small. Substituting this expression for ϕ in Eq. (13.12) and linearizing it, we obtain (see the exercise 13.5 at the end of this chapter)

$$\ddot{\eta} + \gamma\dot{\eta} - (\alpha\sin\psi)\eta = 0. \qquad (13.13)$$

Without loss of generality, we can take $\alpha > 0$ since $\alpha < 0$ will be reduced to the same Eq. (13.13) by the transformation $\tau' = \tau + \pi$.

To perform the stability analysis described in sec. 1.8 of chapter 1 in Part I, we rewrite Eq. (13.13) as

$$\dot{\eta} = \zeta = P(\eta, \zeta), \quad \dot{\zeta} = -\gamma\zeta + (\alpha \sin\psi)\eta = Q(\eta, \zeta). \tag{13.14}$$

The equilibrium point, that is the steady state solution, is obtained by substituting $\dot{\eta} = \dot{\zeta} = 0$. We obtain $(\eta^*, \zeta^*) = (0, 0)$. The stability determining eigenvalues are the roots of the equation

$$\begin{vmatrix} \partial P/\partial\eta - \lambda & \partial P/\partial\zeta \\ \partial Q/\partial\eta & \partial Q/\partial\zeta - \lambda \end{vmatrix}_{\eta=\eta^*,\, \zeta=\zeta^*} = \begin{vmatrix} -\lambda & 1 \\ \alpha\sin\psi & -\gamma - \lambda \end{vmatrix} = 0. \tag{13.15}$$

Expanding the determinant we obtain $\lambda^2 + \gamma\lambda - \alpha \sin\psi = 0$. Its roots are

$$\lambda_{\pm} = \frac{1}{2}\left[-\gamma \pm \sqrt{\gamma^2 + 4\alpha \sin\psi}\right]. \tag{13.16}$$

For a steady state to be stable, the real part of all the eigenvalues must be negative. For $\lambda_{\pm}$ given by Eq. (13.16) this stability condition is satisfied for $\gamma > 0$ and $\sin\psi < 0$. There are two possible steady state values of ψ, namely, $\psi_{\pm}$ given by Eq. (13.11) for $\alpha = \beta$ and $|\gamma| \le |\alpha|$. We find that for $0 < \gamma < \alpha$ the solution $\phi = \tau + \psi_-$ is stable while $\phi = \tau + \psi_+$ is unstable. The stable solution ϕ_- for $0 < \gamma < \alpha$ with $-\pi \le \psi_- \le -\pi/2$ describes stable twirling of the hoop with constant ω without loosing the contact with the athlete's waist. The allowed phase ψ_- describes how to twirl a hula hoop.

13.5 Conclusion

Rotation of a hula hoop is based on several physical laws such as gravity, rotation, torques, friction and inertia. The most important concept that can be learned from the motion of the hula hoop is the centripetal force, which is a noninertial force arising due to the rotation of the frame of reference. Increase in the friction can be used to prevent the hula hoop from falling and this is experimentally described [9].

Hula hooping is a very useful fitness activity. The effect of hula hoop on the strength of core muscle groups has been experimentally analysed [10]. The analysis has shown an enhancement of core muscles particularly thighs, gluteus, abs and lower back muscles. Note that core muscle strength is importance for all human movements. The push pull activity makes the hoop up leading to stretch of the muscles. This gives movement to spine and assists bring nutrients to the intervertebral discs. Further, close kinematic movement is found to happen over the spine, pelvic, hip and back extensor muscle groups when the weighed hula hoop is made to swirl around the waist. Continuous resistance has been provided to these muscles resulting in co-contraction of the back extensors and abdominals and enhancing their strength.

13.6 Bibliography

[1] https://www.hamleys.com/frozen-hula-hoop.
[2] T. Cluff, D. Robertson and R. Balasubramanian, Human Movement Science **27**, 622 (2008).
[3] A.G. Falgueras, Psychology **7**, 1503 (2016).
[4] K. Salen and E. Zimmerman, *The Game Design Reader: Rules of Play Anthrology* (The MIT Press, Massachusetts, 2006).
[5] T.K. Caughey, Am. J. Phys. **28**, 104 (1960).
[6] N.N. Bogolyubov and Yu.A. Mitropolsky, *Asymptotic Methods in the Theory of Nonlinear Oscillations* (Gordonam Breach, New York, 1961).
[7] A.O. Belyakov and A.P. Seyranian, Doklady Phys. **55** (2), 99, 2010.
[8] A.P. Seyranian and A.O. Belyakov, Am. J. Phys. **79**, 712 (2011).
[9] R. Cross, Phys. Edu. **56**, 033001 (2021).
[10] M.S. Raorane, K. Rao, S. Bhalerao, S. Redij, R. Kamthe, S. Gattani, R. Chaudhari and K. Kataria, Int. J. Appl. Res. **3**, 578 (2017).

13.7 Exercises

13.1 Consider a hula hoop rotating about frictionless hips. Refer Fig. 13.2 with θ is the angle between $a_{\parallel}$ and $a_{\perp}$. Calculate the force to be exerted on the hoop to make it to go upward.

13.2 If the hula hoop weighs 0.5 kg and a hooper's body makes an angle 30° with the y-axis, find the force to be provided to the hoop to keep it from falling towards the ground.

13.3 Consider a hula hoop rolling on the ground. What is the linear speed of the hoop if its mass is 0.5 kg and kinetic energy is 0.104 J?

13.4 Prove that for $\alpha = \beta$ and $|\gamma| \leq |\alpha|$, Eq. (13.9) admits the solutions given by Eq. (13.11).

13.5 Show that the equation $\ddot{\phi} + \gamma\dot{\phi} + \alpha\cos(\phi - \tau) = 0$ reduces to $\ddot{\eta} + \gamma\dot{\eta} - (\alpha\sin\psi)\eta = 0$ for $\phi = \tau + \psi + \eta(\tau)$ where η is very small and the initial phase is given by $\cos\psi = -\gamma/\mu$.

Part IV: Flying Toys

Chapter 1

Balsa Gliders

Humans are fascinated with flying. Fascination with flight led enthusiasts to build toys which are able to fly with the help of wind or propulsion. Balsa gliders and paper planes are popular among these types of toys. A glider is a fixed-wing aircraft without an engine which flight is supported by the dynamic reaction of the air against its lifting surfaces. Balsa gliders are made of balsa wood. They are usually lighter because they have no engines and the balsa wood is light. Other woods with lighter weight and stronger can also be used as a construction material of a glider. The art of construction of paper planes and wood gliders is referred to as *aerogami* after origami, the Japanese art of paper folding [1].

A toy glider, for example the one shown in Fig. 1.1, essentially consists of three parts: fuselage, wing and tail. Each of these parts has its major role in the glider for a successful flight. Fuselage is the main part that houses the passangers and cargo in a aircraft. In a glider there is simply a two-dimensional fuselage, having most of the weight of the glider.

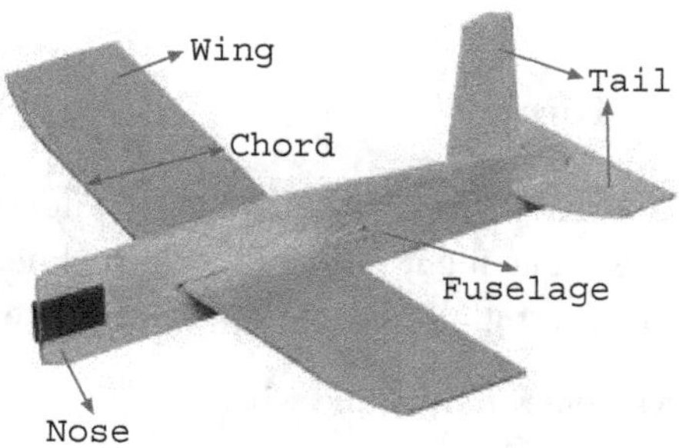

Fig. 1.1: A balsa glider

For a smooth flow and to avoid drag the front part of the fuselage is made into a round shape and is called *nose*. Paper clips or coins are added on the nose in order to assist the glider for a better penetration in a windy weather, to speed it and also to reduce the spiraling. The most essential part of the glider is the wing. Its upper part is curved while the bottom part is flat. This is for generating an upward lift force that balances the weight of the glider and make it to fly.

Usually a slot is cut in the fuselage and the wing is attached to it. This keeps the wing firmly attached. The tail is set at the rear end of the glider. It consists of two parts: horizontal and vertical stabilizers. The main purpose is to control the vertical up and down motions of the nose and directional turning. For details of a proper construction of gliders one may refer to [2,3]. A properly designed balsa glider is able to fly for hours by balancing the forces of gravity, lift and drag.

The ninth century poet Abbas Ibn Firnas near Cordoba, Spain made an attempt at flight [4]. Eilmer of Malmesbury, England glided about 200 metres some time between 1000 and 1010 AD [5]. Around 1630-1632 Hezarfen Ahmed Celebi had flown a glider in Istanbul [6]. More than 2000 successful flights were made by Lilienthal. In 1905 Daniel Maloney launched a balloon-launched glider from 4000 feet [7]. Then, the Wright brothers developed a series of gliders. During the World War II many countries developed military gliders for landing troops.

Nowadays, gliders are used as a sport and recreation. They are used to illustrate the balance of forces and the features of fluid mechanics such as Bernoulli's theorem. High School students can be asked to construct a glider that will fly through a target and a glider competition can also be conducted.

In this chapter, we bring out the aerodynamic forces associated with a glider and describe its flight.

1.1 Aerodynamic Forces

Consider a flight at a constant velocity $\mathbf{v}$ and at an angle θ to the horizontal as shown in Fig. 1.2 [8]. The flight path is a straight-line. Essentially, there are three forces acting on a glider in its flight. They are:

1. Upward lift force generated by the wing.
2. Backward drag force (air resistance).
3. The downward gravity due to the weight of the glider.

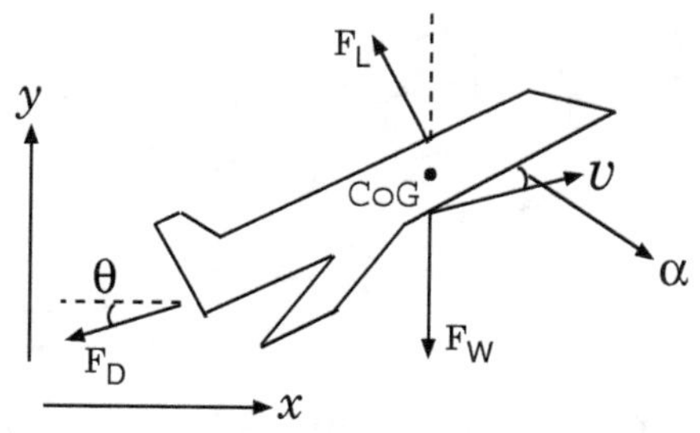

Fig. 1.2: Force diagram of a balsa glider. CoG is the centre of gravity where the weight of the glider acts. θ and α are the glide angle and the angle of attack, respectively.

The lift force [8-11] arising from the force of the air on the bottom of the wing is

$$F_{\rm L} = \frac{1}{2}\rho v^2 C_{\rm L} A\,, \tag{1.1}$$

where ρ is the air density, $C_{\rm L}$ is the lift coefficient and A is the planform area of the wing. The lift force acts in the direction normal to the direction of flight. When the glider motion is horizontal, the lift force exactly opposes gravity. The drag force pointing opposite to the velocity is [9,10]

$$F_{\rm D} = \frac{1}{2}\rho v^2 C_{\rm D} A\,, \tag{1.2}$$

where $C_{\rm D}$ is the drag coefficient. The weight $F_{\rm W}$ points in the downward direction as shown in Fig. 1.2. The angle α in Fig. 1.2 is called *angle of attack* and can be related to $C_{\rm L}$ and $C_{\rm D}$.

Lift is controlled by A and the shape of the wing. The higher the value of A, the higher the pushing of the wing by the air. From Eqs. (1.1) and (1.2), we note that for a given ρ, v and A we need to find $C_{\rm L}$ and $C_{\rm D}$. They can be determined experimentally and depend on the angle of attack α. Increasing the angle of attack leads to increase in the quantum of air incident on the bottom resulting in more lifting of the wings.

1.2 How Does a Glider Fly?

The wing is usually flat on the bottom and curved on the top. As it moves through the air, the air splits into two parts at the leading edge. One part moves along the curved top portion of the wing while the other part moves

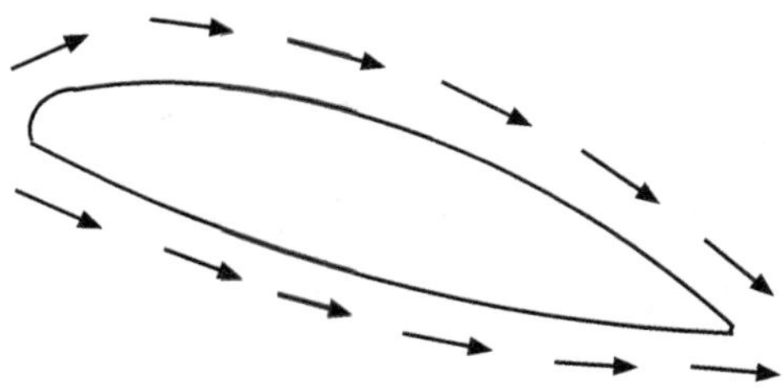

Fig. 1.3: Splitting of air in the leading edge and meeting again at the trailing edge of the wing.

along the bottom as shown in Fig. 1.3. These parts meet again at the trailing edge. The air in the top must move relatively faster than the air in the bottom since it has to travel a larger distance because of curved shape of the top of the wing. This causes, according to the Bernoulli's principle, lower pressure on the top. The higher air pressure below the wing is able to push the glider up.

For a glider to fly the lift force must be higher than the gravity. This is the case when the glider is thrown to fly or moves upward. When it is parallel to the ground, then the lift and the gravity are balanced. In its descent motion towards ground, the lift force is overcome by the gravity. A glider always descends to the air in which it is flying. As it travels from a higher altitude to a lower altitude, a difference in the potential energy occurs. This difference is converted into kinetic energy, that is velocity. In this way a glider generates the velocity required for fly.

Under certain circumstances, a glider can stay aloft for hours. A glider can be designed to descend very slowly. Suppose in the region where a glider is flying there exists a pocket of air moving up with a rate higher than the descend rate of the glider. Such pockets of air can be found where a wind at a hill or a mountain need to rise to move over it. When the ground has dark land masses, it can absorb more heat from the sun and in turn heats the surrounding air. This leads to rising of pockets of hot air. These pockets of air can cause the glider to gain altitude and stay aloft for hours even if it descends constantly.

What is the significance of the angle of attack α? It is the angle the wind makes with the wing. Experiments have shown that an increase of α produces more lift upto a certain value of it. Beyond that critical value of α smooth flow of air over the wing is destroyed, the loss of lift occurs and the weight of the glider dominates causing it to descend towards ground.

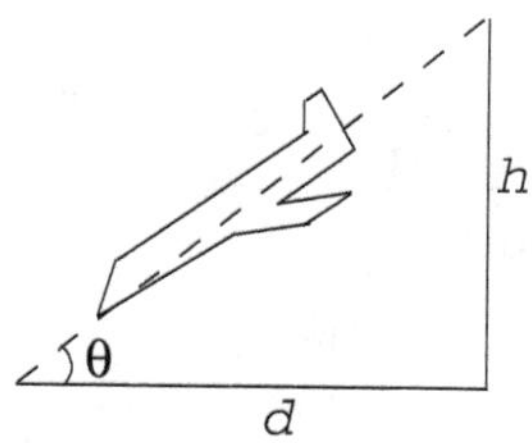

Fig. 1.4: Representation of glide angle, distance flown and the change in the altitude.

1.3 Significant Factors of Bulsa Gliders

Assume that a glider moves downward without acceleration and with a constant velocity. Its path is a straight-line as shown in Fig. 1.4. It loses altitude as it travels towards the ground. From Fig. 1.4 we write

$$\tan\theta = h/d\,. \tag{1.3}$$

From this equation we infer the following:

1. If we throw a glider at a glide angle θ, then the distance it would fly for a given change in altitude h is $d = h/\tan\theta$.
2. For a small θ, $\tan\theta$ is also small. In this case, even for a small change in h, the glider can fly a long distance d.
3. For a large θ, $\tan\theta \gg 1$. The glider can fly over a short distance only for a given h.
4. The higher the releasing point of the glider, the longer the distance traveled.

1.4 Determination of Glide Angle and Lift and Drag Forces

The balance of the forces give two equations, the vertical force equation

$$F_{\mathrm{L}}\cos\theta + F_{\mathrm{D}}\sin\theta = F_{\mathrm{W}}\,, \tag{1.4}$$

and the horizontal force equation

$$F_{\mathrm{L}}\sin\theta = F_{\mathrm{D}}\cos\theta\,. \tag{1.5}$$

How do we determine θ? We can determine θ experimentally. Release a glider at a height h from the ground and measure the distance d traveled by it along the ground. Calculate θ from Eq. (1.3). Repeat this experiment for several values of h and compute the average value of θ. Once the weight of the glider is measured, we now have two equations, Eqs. (1.4) and (1.5) for the two unknowns F_L and F_D.

From the Eqs. (1.3) and (1.5) we write

$$R = \frac{F_L}{F_D} = \frac{\cos\theta}{\sin\theta} = \frac{1}{\tan\theta} = \frac{d}{h}, \tag{1.6}$$

where R is called the *glider ratio*. Alternately, from the Eqs. (1.1) and (1.2)

$$R = \frac{F_L}{F_D} = \frac{C_L}{C_D}. \tag{1.7}$$

R is considered as an *efficiency factor*. R is high if the lift is large or the drag or the glide angle is small. A glider with a high R can fly over a long distance by reducing the glide angle. These clearly brings out the significance of a good wing design.

Solved Problem 1:

It has been found experimentally that the lift and drag coefficients depend on the angle of attack α as $C_L = a\alpha + b$ and $C_D = c\alpha^2 + d$ where a, b, c and d are constants that depend on the characteristics of the wing. Find the angle of attack which gives the maximum efficiency factor R in terms of these constants.

We have $R = C_L/C_D = (a\alpha + b)/(c\alpha^2 + d)$. For maximum R, the condition $\mathrm{d}R/\mathrm{d}\alpha = 0$ gives

$$\frac{a\left(c\alpha^2 + d\right) - (a\alpha + b)2c\alpha}{\left(c\alpha^2 + d\right)^2} = 0 \tag{1.8}$$

or $ac\alpha^2 + 2bc\alpha - ad = 0$. Its roots are

$$\alpha = \frac{-bc \pm \sqrt{b^2c^2 + a^2cd}}{ac}. \tag{1.9}$$

The positive root will give the answer.

1.5 Maximum Time Aloft

It is easy to find an expression for maximum time aloft for a fixed height h. For the glider traveling with a constant velocity writing $d = vt$, Eq. (1.3) gives $t = h/(v\tan\theta)$. Using Eq. (1.6) for $\tan\theta$ gives

$$t = \frac{h}{v}\frac{F_L}{F_D}, \tag{1.10}$$

where v from the Eq. (1.1) is given by

$$v = \left(\frac{2F_{\mathrm{L}}}{\rho C_{\mathrm{L}} A}\right)^{1/2} . \tag{1.11}$$

For small θ Eq. (1.4) gives $F_{\mathrm{L}} = F_{\mathrm{W}}$. Substitutions of this and $F_{\mathrm{L}}/F_{\mathrm{D}} = C_{\mathrm{L}}/C_{\mathrm{D}}$ in Eq. (1.10) give

$$t = h\left(\frac{C_{\mathrm{L}}}{C_{\mathrm{D}}}\right)\left(\frac{\rho C_{\mathrm{L}} A}{2F_{\mathrm{W}}}\right)^{1/2} = h\left(\frac{\rho}{2}\right)^{1/2}\left(\frac{A}{F_{\mathrm{W}}}\right)^{1/2}\left(\frac{C_{\mathrm{L}}^{3/2}}{C_{\mathrm{D}}}\right) . \tag{1.12}$$

t can be maximized by maximizing the quantity $C_{\mathrm{L}}^{3/2}/C_{\mathrm{D}}$ and minimizing $(F_{\mathrm{W}}/A)^{1/2}$ which is the wing load. Equations (1.6) and (1.12) clearly indicate the significance of $F_{\mathrm{L}}/F_{\mathrm{D}}$ or $C_{\mathrm{L}}/C_{\mathrm{D}}$ in the design of the wings of the glider in achieving a large d and a maximum time aloft.

1.6 Conclusion

Balsa glider can be used as a valuable tool in physics education as it teaches the basis of the applied Aviation Physics. It is a special kind of aircraft that has no engine. So, no thrust acts on it. Hence, its motion is decided by the lift, drag and gravitational forces only. As a glider does not have an engine, it generates its velocity needed for its flight, by reducing its altitude. Reducing the altitude decreases its potential energy and hence its kinetic energy increases. So, it gains velocity, and as a consequence it increases its lift force. The design of wings decides how long it will keep flying. The students can be asked to design gliders with different shapes of the wings and asked to study their flights.

1.7 Bibliography

[1] https://www/pragyan/16/home/events/chill_pill/aerogami/.
[2] https://www.accurateessays.com/samples/balsa-wood-glider/.
[3] C. Waltham, Am. J. Phys. **67**, 620 (1999).
[4] Lynn Townsend White, Technology and Cultures **2**, 97 (1961).
[5] Lynn Townsend White, *Eilmer of Malmesbury, an Eleventh Century Aviator. Medieval Religion and Technology* (University of California Press, Los Angeles, 1978).
[6] http://www.privatetour.net/hezarfen-ahmet-celebi-the-first-man-to-fly.

[7] S.C. Harwood and B.G. Fogel, *Quest for Flight: John J. Montgomery and the Dawn of Aviation in the West* (University of Oklahoma Press, 2012).
[8] W. Brian Lane, The Phys. Teach. **51**, 242 (2013).
[9] N. Dreska and L. Weisenthal, *Physics for Aviation* (Jeppesen, 1992).
[10] J.D. Anderson, *Introduction to Flight* (McGraw-Hill, New York, 1999).
[11] https://www.mansfieldct.org/Schools/MMS/staff/hand/flightglider.-htm.

1.8 Exercises

1.1 Solve the Eqs. (1.4) and (1.5) for F_L and F_D.

1.2 Why do gliders have very long wings?

1.3 State the advantages of thin wing profile over a thick wing profile.

1.4 Consider an airplane in flight moving in the $x-y$ plane as shown in Fig. 1.5 with an acceleration $\mathbf{a}$. $\mathbf{F}_T$ is the thrust given by the flight engine, which acts at an angle ϕ to the horizontal direction. $\mathbf{v}$ is the velocity of the flight making an angle θ with the horizontal direction. $\mathbf{F}_L$ is the lift force and $\mathbf{F}_D$ is the drag force. $\mathbf{F}_W$ is the weight of the flight. Using Newton's second law, setup the equations of motions along x and y directions. Prove that for a horizontal uniform motion of the flight, $F_T = (C_D/C_L)F_W$ and velocity $v = \sqrt{2F_W/(\rho A C_L)}$.

1.5 Consider the previous exercise with $C_D = 0.05$, $C_L = 0.5$, $g = 9.8\,\text{m/s}^2$, $A = 15\,\text{m}^2$, $F_W = 5000\,\text{N}$ and $\rho = 1.2\,\text{kg/m}^3$. Calculate the thrust force for maintaining uniform horizontal motion and the corresponding velocity.

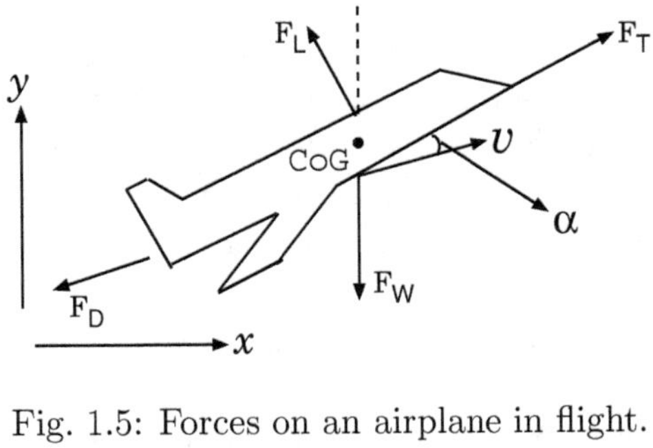

Fig. 1.5: Forces on an airplane in flight.

Chapter 2

Magnus Glider

Stand on a cliff and drop, say, a basket ball. The ball will land directly underneath along a straight-line path to a point towards which it is dropped. Suppose drop the ball giving a little spin to it. The ball will not follow a straight-line path instead it will take a curved path. See the video in [1] for viewing this unexpected phenomenon. The spinning of the ball actually disturbs the streamlines of air around it. The air pressure will be different on the sides of the ball. A pressure difference will be created and this will provide a lift to the ball to deviate its trajectory from the straight-line. Essentially, apart from the drag force the spinning of the ball (or an object) gives rise a force and is perpendicular to the velocity [2]. This is known as *Magnus effect*.

In 1852 Heinrich Gustav Magnus conducted an experiment consisting of a brass cylinder kept between two conical bearings. The cylinder was mounded on a freely rotatable arm. Using a air blower, a current of air was directed towards the cylinder. A strong lateral deviation was observed when the cylinder got rotated. The rotating cylinder tended to deflect toward the side of the rotor which was moving in the same direction as the wind coming from the air blower. This was due to the *Magnus effect* [2,3].

A Magnus glider is a simple flying machine constructed using two paper cups that works on the Magnus effect. It is an alternative to paper airplanes. Its construction is very simple [4-6]. Take two paper cups and join the bottom of them. Wrap up the joining by many layers of a tap. This is the glider and is shown in Fig. 2.1. The open mouths of the cups are facing opposite sides. Take a few rubber bands and couple them to make a chain of them. To couple them hold one rubber band horizontally and the second one vertically inside the first. Next, push the one end of the horizontal band through the second band. Pull that end with one hand and the top

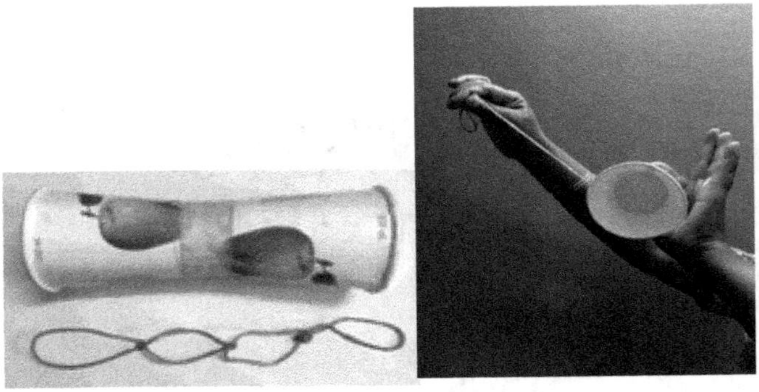

Fig. 2.1: A Magnus glider, a chain of rubber band and throwing it in air.

of the second band with the other hand in opposite directions. With this two rubber bands joined add the others one by one in the above manner.

Place one end of the chain of the rubber band on the joining (over the tap) of the paper cups, stretch the rubber band and wrap the chain over the cup tighter and tighter and a couple of inch of the chain is left. The free end of the chain is the *launcher*. Hold the part of the cup, over which the rubber band chain is wrapped, between the thumb and the index finger of the left-hand. Hold the free end of the rubber band chain with your other hand away from your body. Make sure that the launcher is coming out from under the bottom of the glider and pointing slightly upward as shown in Fig. 2.1. Release the glider from your left-hand. The glider will spin and fly up into the air along a *curved path* and then slowly glide to the ground. The curved path of the flight is due to the Magnus effect. The Magnus effect can be seen when a rotating object moves through a fluid, for example, air. Essentially the trajectory of the rotating object gets deflected in a manner which is not taking place when the object is not rotating. This deflection can be accounted by considering the pressure difference of the fluid on opposite sides of the rotating object.

The Magnus effect is seen in many ball sports as curve-ball or curving cross, bending of a soccer ball and back spinning of table tennis ball. Using a light Styrofoam ball with a rough surface, certain interesting features of Magnus effect can be noticed [7]. To enhance the Magnus effect remove many small pieces from the surface of the ball. To clearly see the rotation of the ball, colour half of it. When the ball is thrown straight ahead with giving backspin, it is pushed upward. The Magnus effect is important in the study of spinning of guided missiles.

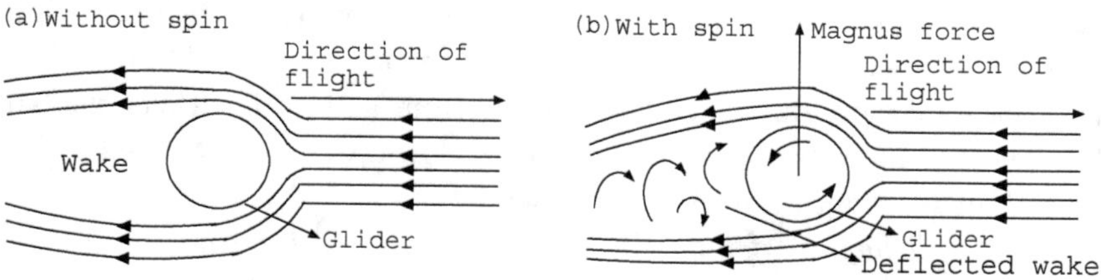

Fig. 2.2: Air flow around the ball moving in the air (a) without spin and (b) with spin.

In 1671 Isacc Newton reported the motion of spinning tennis ball [8]. In the same year Walker described the Magnus force [9]. In 1877 Lord John Rayleigh explained the curved path of a ball through Magnus effect [10]. By the Magnus effect B. Robins accounted the trajectories of musket balls [11,12]. In 1959 Lyman J Briggs reported the Magnus effect on base balls [13]. The Magnus effect has been used in rotor ships for propelling the ships. Several rotor airplanes have been built in the past [3]. For a detailed review on the Magnus effect one may refer to [3,14,15].

In this chapter, first we briefly bring out the physics of the Magnus effect. We briefly present the vector potential for incompressible, irrotational flow of fluids. Next, we develop the theoretical model for the Magnus glider. Particularly, treating the Magnus glider as a rotating cylinder, we calculate its velocity and obtain an expression for the lift (Magnus) force. Then, setup an equation of motion for the flight of the Magnus glider and compute the trajectory of the glider.

2.1 Magnus Force

The Magnus force takes place when spinning objects move through the air with the flight velocity vector making an angle with the axis of rotation [13]. This force acts in the direction perpendicular to the axis of rotation and the direction of flight. The magnitude of the force depends on the viscous coefficient, density, geometry of the object, spinning rate and flight velocity. The Magnus force is much greater in magnitude than the wing lifting force [3].

Consider a ball traveling horizontally in the air without spinning. The surface of the ball interacts with the thin layer of air around the ball forming the boundary layer. The air in the thin boundary layer come-off from the surface of the ball. This leads to a low-pressure part called a *wake* behind the ball as shown in Fig. 2.2a [16]. A backward force on the ball is developed due to the pressure difference between the front and back sides of the ball. The forward motion of the ball is slowed down. This is the well known normal air resistance (air drag).

What will happen if the ball is spinning as it is traveling through the air? Assume that the ball is spinning away from the flight direction. A thin layer of air around the surface of the ball is set in motion due to the spinning of the ball. The air meets the surface head-on, however, does not leave at the opposite side. The air flow around the ball is changed as shown in Fig. 2.2b [16]. In this figure, the air flow is deflected downwards and gains momentum downwards. Consequently, the ball gains momentum in the upwards direction. The spinning of the ball accelerates the flow of air on the upper side of the ball and slows down the speed of air in the bottom. Bernoulli's law states that "for an inviscid flow, an increase in the speed of the fluid occurs simultaneously with a decrease in pressure or a decrease in the fluid's potential energy" [17]. As the air is flowing relatively faster on the upper side of the ball than the lower side of it, the air pressure in the upper side is relatively lower than that in the lower side. That is, there is a pressure difference. This causes the ball to move upwards. The associated force arising due to spin is known as the *Magnus force*. The upstream of the air clings to the ball and hence there is a slight deflection of the air as shown in Fig. 2.2b. The downward stream turns backward. Near the surface of the ball the pressure is lower. The pressure away from the ball is higher. Therefore, the air flow follows a curved path and hence the ball.

The underlying Magnus force acts at right angles to the direction of the ball and the axis of rotation. The magnitude of the force is directly proportional to the flight velocity and the rate of spin of the ball where as the direction of the force depends on the spin direction. *What will happen if the ball is spinning in the flight direction?*

2.2 Vector Potential for Incompressible Irrotational Flow

Potential flow theory of fluid dynamics helps to solve many fluid mechanics problems. In this section, we give an elementary introduction to vector potential under the assumption that the fluid is incompressible and it has an irrotational flow. If $\mathbf{v} = v_x\mathbf{i} + v_y\mathbf{j} + v_z\mathbf{k}$ is the velocity then the vorticity is deformed as $\nabla \times \mathbf{v}$. For irrotational flow this is equal to zero at every point in the flow. That is, for irrotational flow, we must find the $\mathbf{v}$ which

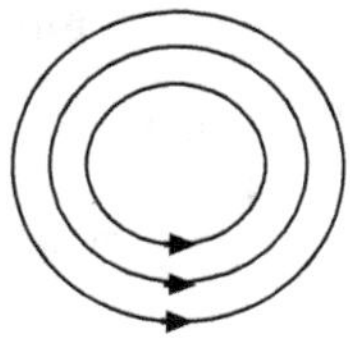

Fig. 2.3: Irrotation vortex with flow in cicular around the centre.

satisfies the equation

$$\nabla \times \mathbf{v} = 0. \tag{2.1}$$

As the curl of the gradient of a scalar function ϕ is always zero, we have

$$\nabla \times \nabla\phi = 0. \tag{2.2}$$

For an irrotational flow, we find from Eq. (2.1) and (2.2),

$$\mathbf{v} = \nabla\phi, \tag{2.3}$$

where ϕ is called the *velocity vector potential*. As mass is always conserved, the mass conservation in the flow of fluid gives the equation of continuity

$$\frac{\partial\rho}{\partial t} + \nabla\cdot(\rho\mathbf{v}) = 0\,, \tag{2.4}$$

where ρ is the density of the fluid. For an incompressible fluid, ρ is constant. So, $\partial\rho/\partial t = 0$ and from Eq. (2.4)

$$\nabla\cdot\mathbf{v} = 0\,. \tag{2.5}$$

Substituting for $\mathbf{v}$ from Eq. (2.3) we get

$$\nabla\cdot\nabla\phi = \nabla^2\phi = 0\,. \tag{2.6}$$

The velocity vector potential for incompressible, irrotational flow of a fluid satisfies the Laplace Eq. (2.6).

Next, we consider an irrotational vortex as shown in Fig. 2.3 in which the particles move with circular paths around a central point such that the velocity distribution still satisfies the irrotational condition. The fluid particles do not rotate themselves, but they simply move on a circular path. In the irrotational vortex flow, the individual particles do not rotate about the centre and the radial velocity $v_r = 0$. For a two-dimensional vortex $v_z = 0$. If we consider a cylindrical coordinate system

$$v_r = \partial\phi/\partial r = 0 \;\text{ and }\; v_\theta = (1/r)\partial\phi/\partial\theta\,, \tag{2.7}$$

where ϕ is the velocity potential for the two-dimensional irrotational vortex.

Because the flow is considered irrotational, the velocity $\nabla \times \mathbf{v} = 0$. It is given in cylindrical coordinates as

$$\left(\frac{1}{r}\frac{\partial v_z}{\partial \theta} - \frac{\partial v_\theta}{\partial z}\right)\mathbf{e}_r + \left(\frac{\partial v_r}{\partial z} - \frac{\partial v_z}{\partial r}\right)\mathbf{e}_\theta + \left(\frac{1}{r}\frac{\partial r v_\theta}{\partial r} - \frac{1}{r}\frac{\partial v_r}{\partial \theta}\right)\mathbf{e}_z = 0. \tag{2.8}$$

As the vortex is two-dimensional, v_z and all derivatives with respect to z are zero. So, Eq. (2.8) gives

$$\frac{\partial(r v_\theta)}{\partial r} - \frac{\partial v_r}{\partial \theta} = 0. \tag{2.9}$$

Further, as $v_r = 0$, we obtain $v_\theta = k/r$. By convention, k is set equal to $\tau/(2\pi)$ where τ is called *circulation*. Then from $v_\theta = (1/r)\partial\phi/\partial\theta$ we find

$$\phi = \frac{\tau\theta}{2\pi}, \tag{2.10}$$

where the physically insignificant constant has been neglected. Equation (2.10) gives the vector potential for an irrotational vortex flow. The circulation can be found mathematically as the line integral of the tangential component of the velocity taken about a closed surface C as $\tau = \oint_C \mathbf{V}_t \cdot \mathbf{ds}$ where $\mathbf{V}_t$ is the tangential velocity and $\mathbf{ds}$ is the differential length along the closed curve C.

2.3 Theoretical Model

The model of a Magnus glider made of two paper cups is shown in Fig. 2.4 with $L \gg a$, where L and a are the height and bottom radius, respectively, of the cup used. For theoretical analysis we assume that the glider is a long cylinder with radius a rotating in incompressible and irrotational air with centroid velocity V (translational velocity) and angular velocity ω. It is spinning in the anti-clockwise direction. We consider the motion in two-dimensions.

2.3.1 *Forces Acting on a Magnus Glider*

The following forces act on a Magnus glider.

1. **Weight** – It is the well known force of gravity ($-mg$) acting in the downward direction.

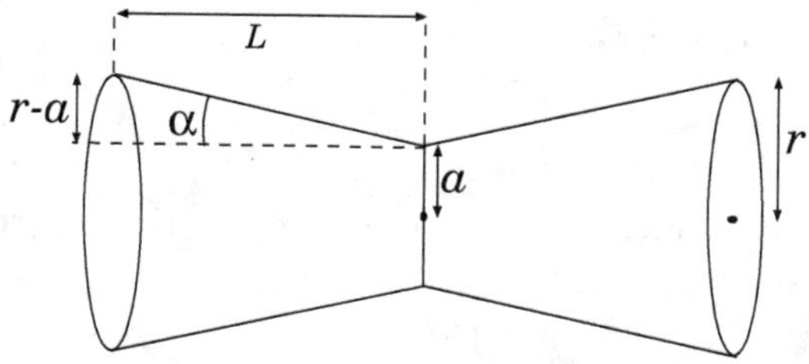

Fig. 2.4: The model of a Magnus glider.

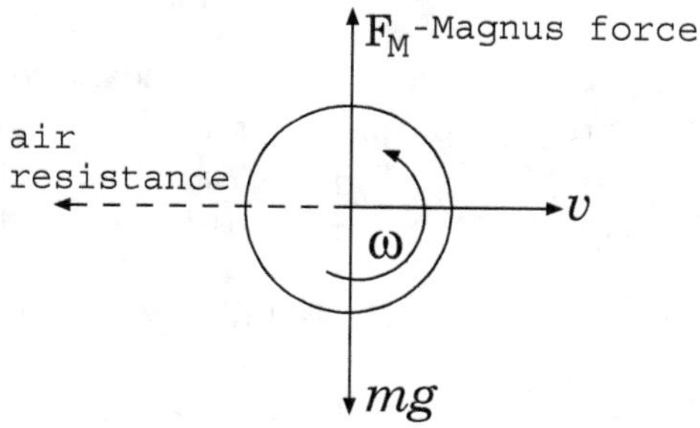

Fig. 2.5: Forces acting on a Magnus glider.

2. **Lift (Magnus force)** – This force is arising due to the difference in air pressure occurring in the top and bottom of the spinning glider. It acts at a right angle to the direction of flight of the glider.
3. **Drag** – It is due to the friction. It acts in the direction opposite to the flight direction.

Figure 2.5 depicts the representation of the above three forces.

2.3.2 *Velocity Potential*

Let us calculate the velocity potential (distribution) ϕ of the fluid outside the rotating cylinder [18,19]. For an incompressible fluid, ϕ satisfies the Laplace equation. Consider the continuity equation

$$\frac{\partial v_x}{\partial x} + \frac{\partial v_y}{\partial y} = 0. \tag{2.11}$$

As ϕ is the velocity potential we write

$$v_x = \frac{\partial \phi}{\partial x}, \quad v_y = \frac{\partial \phi}{\partial y}. \tag{2.12}$$

Substituting (2.12) in (2.11) we get

$$\frac{\partial^2\phi}{\partial x^2}+\frac{\partial^2\phi}{\partial y^2}=0. \tag{2.13}$$

As the glider is a cylinder, we consider ϕ in cylindrical coordinates. $\phi(r,\theta)$ satisfies the Laplace equation

$$\frac{1}{r}\frac{\partial}{\partial r}\left(r\frac{\partial}{\partial r}\right)\phi+\frac{1}{r^2}\phi_{\theta\theta}=0 \tag{2.14}$$

with the boundary conditions

$$\phi_r|_{r=a}=0,\quad \phi_r|_{r\to\infty}=V\cos\theta\,. \tag{2.15}$$

The solution of Eq. (2.14) subjected to the boundary conditions (2.15) is (see the exercise 2.1 at the end of this chapter)

$$\phi(r,\theta)=V\left(1+\frac{a^2}{r^2}\right)r\cos\theta+\frac{\tau\theta}{2\pi}\,, \tag{2.16a}$$

where the last term is the vector potential for a vortex and

$$\tau=\int_0^{2\pi}\omega a a d\theta=2\pi\omega a^2 \tag{2.16b}$$

is the anti-clockwise line vortex of circulation.

The velocities in the r and θ directions are

$$v_r=\phi_r=V\left(1-\frac{a^2}{r^2}\right)\cos\theta\,, \tag{2.17a}$$

$$v_\theta=\frac{1}{r}\phi_\theta=-V\left(1+\frac{a^2}{r^2}\right)\sin\theta+\frac{\tau}{2\pi r}\,. \tag{2.17b}$$

Solved Problem 1:

Show that the circulation τ for a cylinder of radius a rotating about the axis of the cylinder with angular velocity ω is equal to $2\pi\omega a^2$.

Consider Fig. 2.6. We have $\tau=\oint_C \mathbf{V}_\mathrm{t}\cdot\mathbf{ds}$. Since the cylinder is rotating with angular velocity ω, the radial velocity of the fluid at the surface will be ωa. As $\mathrm{d}s=a d\theta$, we get

$$\mathbf{V}_\mathrm{t}\cdot\mathbf{ds}=\omega a a d\theta. \tag{2.18}$$

Therefore,

$$\tau=\oint_C \omega a^2 \mathrm{d}\theta=2\pi a^2\omega. \tag{2.19}$$

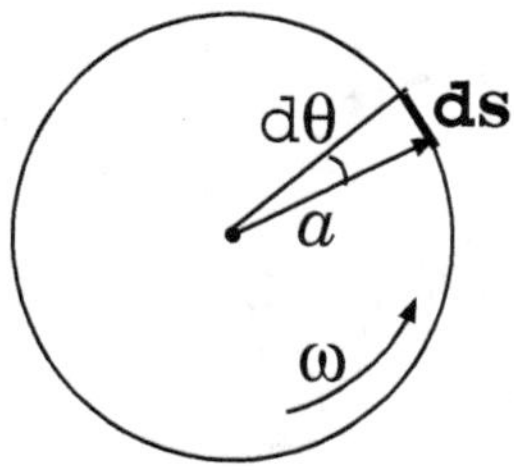

Fig. 2.6: Definition of dθ, **ds** and a of a rotating cylinder.

Solved Problem 2:

Determine the points on the cylindrical glider at which the fluid velocity $v = 0$ [20].

The point at which $v = 0$ is called *stagnation point*. Let us denote this point as (r^*, θ^*). We start from Eqs. (2.17). $v_r = 0$ and $v_\theta = 0$ give

(1) $r^* = a$, $\sin\theta^* = \dfrac{\tau}{4\pi a V}$.

(2) $r^* \neq a$, $\theta^* = \dfrac{\pi}{2}$ or $\dfrac{3\pi}{2}$, $-V\left(1 + \dfrac{a^2}{r^{*2}}\right)\sin\theta^* + \dfrac{\tau}{2\pi r^*} = 0$.

The case (1) gives the stagnation points on the glider. If $\tau/(4\pi a V) < 1$ then

$$\theta_1^* = \sin^{-1}\left(\frac{\tau}{4\pi a V}\right), \quad \theta_2^* = \pi - \theta_1^* . \tag{2.20}$$

Thus, for $\tau/(4\pi a V) < 1$ the stagnation points are $(r^*, \theta^*) = (a, \theta_1^*)$, (a, θ_2^*). When $\tau = 4\pi a V$ we find $(r^*, \theta^*) = (a, \pi/2)$.

For $\tau/(4\pi a V) > 1$ there is no stagnation point on the glider. The stagnation point will be outside the glider and can be calculated from the case (2).

2.3.3 *Lift (Magnus) Force*

Now, determine the lift (Magnus) force [18-21]. Consider the pressure at a point on the cylindrical glider. To find the pressure $P(r = a, \theta)$, consider two points, one on the glider and the other at far from the glider where the velocity is V and the pressure is P_∞. According to Bernoulli's principle

$$P_\infty + \frac{1}{2}\rho V^2 + \rho g h_1 = P(r = a, \theta) + \frac{1}{2}\rho v^2 + \rho g h_2 . \tag{2.21}$$

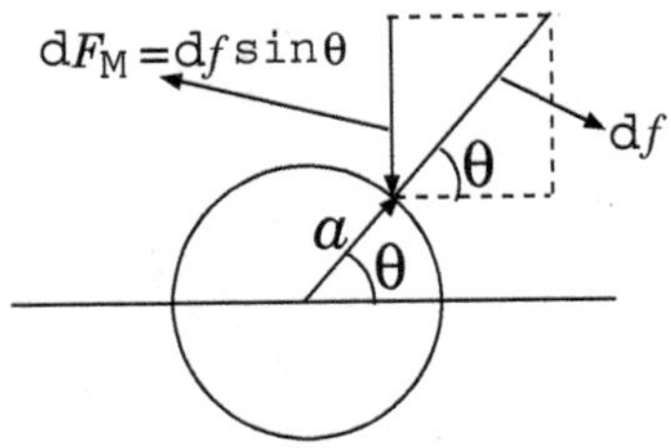

Fig. 2.7: Representation of the drag force $\mathrm{d}f$ and the Magnus force $\mathrm{d}F_{\mathrm{M}}$.

In the above equation in the both sides the first term is the pressure energy, second term is the kinetic energy per unit volume and the third term is the potential energy per unit volume. We choose $h_1 = h_2$. From Eqs. (2.17) we have

$$v^2 = v_r^2(r=a) + v_\theta^2(r=a) = \left(-2V\sin\theta + \frac{\tau}{2\pi a}\right)^2 . \tag{2.22}$$

Then Eq. (2.21) becomes

$$P(r=a,\theta) = P_\infty + \frac{1}{2}\rho\left[V^2 - \left(-2V\sin\theta + \frac{\tau}{2\pi a}\right)^2\right] . \tag{2.23}$$

From Fig. 2.7 we write

$$\mathrm{d}f = -P(r=a,\theta)a\mathrm{d}\theta\,, \quad \sin\theta = \frac{\mathrm{d}F_{\mathrm{M}}}{\mathrm{d}f} . \tag{2.24}$$

The total Magnus force is then

$$\begin{aligned} F_{\mathrm{M}} &= \int_0^{2\pi} \mathrm{d}F_{\mathrm{M}} \\ &= -\int_0^{2\pi} P(r=a,\theta)\sin\theta a\mathrm{d}\theta \\ &= -a\int_0^{2\pi}\left(P_\infty + \frac{1}{2}\rho\left[V^2 - \left(-2V\sin\theta + \frac{\tau}{2\pi a}\right)^2\right]\right)\sin\theta\mathrm{d}\theta \\ &= \frac{1}{2}a\rho\int_0^{2\pi}\left(-2V\sin\theta + \frac{\tau}{2\pi a}\right)^2\sin\theta\mathrm{d}\theta \\ &= -\frac{\rho V\tau}{\pi}\int_0^{2\pi}\sin^2\theta\mathrm{d}\theta \\ &= -\rho V\tau\,. \end{aligned} \tag{2.25}$$

The above equation indicates that the Magnus force is proportional to τ, the fluid circulation.

For the glider made of two paper cups the Magnus force is [19]

$$F_{\rm M} = \int_{-L}^{L} {\rm d}x \int_{0}^{2\pi} P\cos(\alpha) r \sin\theta {\rm d}\theta\,, \tag{2.26}$$

where

$$P = P_\infty + \frac{1}{2}\rho V^2 - \frac{1}{2}\rho\left[-V\left(1+\frac{a^2}{r^2}\right)\sin\theta + \frac{\tau}{2\pi r}\right]^2 \tag{2.27}$$

and α is the angle shown in Fig. 2.4. $F_{\rm M}$ can be written in a vector form as (for the proof see the exercise 2.3 at the end of this chapter)

$$\mathbf{F}_{\rm M} = k \cdot \boldsymbol{\omega} \times \mathbf{V}\,, \tag{2.28}$$

where k is to be determined from evaluating the integral.

The fluid (air) resistance is generally proportional to velocity and is taken as $\mathbf{f} = -\gamma\mathbf{v}$ where γ is the coefficient of resistance. More over, the spin of the glider drags the air and hence the angular velocity of the glider decays exponentially with time as $\omega = \omega_0 {\rm e}^{-\beta t}$.

2.3.4 *Equation of Motion*

Neglecting the air resistance and taking account of the gravity and the lift force the equation of motions for the horizontal distance x and the vertical height y are given by [19]

$$\ddot{x} = -\frac{k\omega}{m}\dot{y}, \quad \ddot{y} = -g + \frac{k\omega}{m}\dot{x}\,. \tag{2.29}$$

Inclusion of the drag force modifies the above equations as

$$\ddot{x} = -C_1\dot{x} - C_2{\rm e}^{-\beta t}\dot{y}\,, \tag{2.30a}$$

$$\ddot{y} = -g - C_1\dot{y} + C_2{\rm e}^{-\beta t}\dot{x}\,, \quad C_1 = \frac{\gamma}{m}, \; C_2 = \frac{k\omega_0}{m}\,. \tag{2.30b}$$

As Eqs. (2.30) are of variable coefficients equations, it is difficult to construct exact analytical solutions. They can be solved numerically for visualizing the trajectory of the glider in the $x-y$ plane. For this purpose, we fix $C_2 = 12$, $\beta = 1$, $x(0) = 0$, $\dot{x}(0) = 3$, $y(0) = 2$ and $\dot{y}(0) = 1$. Figure 2.8 shows the trajectories of the glider for $C_1 = 3$ and 8. We can clearly notice the effect of the Magnus force.

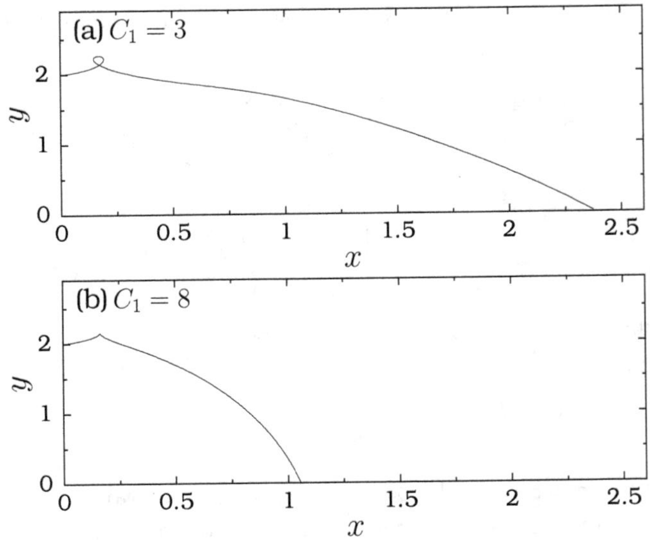

Fig. 2.8: Trajectories of the Magnus glider for two fixed vlues of the control parameter C_1.

2.4 Conclusion

In airplanes air moves over the top of the wings faster than it moves in the bottom. As faster moving air has lower pressure than the air moving in the bottom of the wings, a pressure difference is created. Therefore, there is a lift force pushing the wings upwards. The lift in the Magnus glider is generated in another way. A spherical or a cylindrical object spinning at the appropriate direction and orientation relative to its trajectory is able to generate a lift. This is due to the spinning surface imparting traction on the air. The speed of the air flowing under the spinning object will be reduced. The spinning continues because the air flow over the top of the object is increased leading to decrease in the pressure of that air. Magnus force is important in understanding of the spinning objects in the sports such as soccer, golf, baseball, cricket, tennis, badminton and shuttlecocks [23-35].

2.5 Bibliography

[1] https://www.youtube.com/watch?v=-svOj1Qgrbc.

[2] W. Johnson, Int. J. Mech. Sci. **28** 859 (1986).
[3] J. Seifert, Prog. Aerospace Sci. **55**, 17 (2012).
[4] http://www.instructables.com/id/Magnus-Glider/.
[5] http://www/cmhoustonblog.org/2010/04/26/make-your-own-magnus glider/.
[6] https://www.ncsciencefestival.org/sites/default/files/documents/Magnus20%Gliders.pdf.
[7] Y. Ogawara, The Phys. Teach. **47**, 555 (2009).
[8] I. Newton, Philos. Trans. R. Soc. London **6**, 3075 (1671).
[9] G.A. Benedetti, *Flight Dynamics of a Spinning Projectile Descending on a Parachute* (National Technical Information Service, 1989).
[10] L. Rayleigh, Messenger of Maths. **7**, 14 (1877).
[11] R.D. Lorenz, Adv. Space Res. **48**, 403 (2011).
[12] B.D. Steele, Tech. Cult. **35**, 348 (1994).
[13] L.J. Briggs, Am. J. Phys. **27**, 589 (1959).
[14] W.M. Swanson, J. Basic Eng. Trans. A SME **83**, 470 (1961).
[15] R.D. Mehta and J.D. Pallis, "Sports ball aerodynamics: Effects of velocity, spin and surface roughness" in *Materials and Science in Sports*, edited by F.H. Froes and S.J. Haake (TMS, Warrendale, 2001) pp.185-197.
[16] http://www.dept.aoe.vt.edu/~jschetz/fluidnature/unit03/unit3a.html.
[17] G.K. Batchelor, *An Introduction to Fluid Dynamics* (Cambridge University Press, Cambridge, 2000).
[18] Z. Wei, L. Ding, K. Wei, Z. Wang and R. Dai, arXiv:1610.02768v1[physics. flu.dyn] 10 Oct. 2016.
[19] T. Tau, D. Lu, Y. Deng, S. Wang and H. Zhou, "Magnus glider" in *International Young Physicists' Tournament . Problems & Solutions 2015*, edited by S. Wang and W. Gao (World Scientific, Singapore, 2018).
[20] https://ocw.mit.edu/courses/mechanical-engineering/2-25-advanced-fluid-mechanics-fall-2013/potential-flow-theory/MIT2_25F13_SolutionMagnus.pdf.
[21] Pardis Rafeii, "Magnus glider" in Persian Young Physicists' Tournament PYPT 2015. See https://iypt.ayimi.org/wp-content/uploads/sites/7/201 8/12/Rafeii-P33-37.pdf.
[22] R. Cross, The Phys. Teach. **45**, 334 (2007).
[23] J. Thomson, Nature **85**, 251 (1910).
[24] M. Peastrel, R. Lynch and A. Armenti, Am. J. Phys. **48**, 511 (1980).

[25] A. Stepanek, Am. J. Phys. **56**, 138 (1988).
[26] A.R. Penner, Rep. Prog. Phys. **66**, 131 (2002).
[27] A.M. Nathan, Am. J. Phys. **76**, 119 (2008).
[28] M. McBeath, A. Nathan, A. Bahill and D. Baldwin, Am. J. Phys. **76**, 723 (2008).
[29] J.E. Golf and M.J. Carre, Am. J. Phys. **77**, 1020 (2009).
[30] C. Frohlich, Am. J. Phys. **79**, 565 (2011).
[31] G. Dupeux, C. Cohen, A.L. Goff, D. Quere and C. Clanet, J. Fluid Struct. **27**, 659 (2011).
[32] A.M. Nathan, Proc. Eng. **34**, 116 (2012).
[33] C. Cohen, B.D. Texier, D. Quere and C. Clanet, New J. Phys. **17**, 063001 (2015).
[34] C. Clanet, Annual Rev. Fluid Mech. **47**, 455 (2015).
[35] J.E. Golf, J. Kelley, C.M. Hobson, K. Seo, T. Asai and S.B. Choppin, Eur. J. Phys. **38**, 044003 (2017).

2.6 Exercises

2.1 Construct the solution of the Laplace equation

$$\frac{1}{r}\frac{\partial}{\partial r}\left(r\frac{\partial}{\partial r}\right)\phi + \frac{1}{r^2}\phi_{\theta\theta} = 0 \tag{2.31}$$

associated with the cylindrical Magnus glider subjected to the boundary conditions $\phi_r|_{r=a} = 0$ and $\phi_r|_{r\to\infty} = V\cos\theta$, where a is the radius of the glider and the centroid velocity is V.

2.2 Prove the Eq. (2.28) and then find an expression for the value of k.

2.3 Show that the air cannot penetrate the glider radially.

2.4 (a) Calculate $v_r(r \to \infty)$ of the cylindrical glider.
(b) Show that the pressure drag force on a cylindrical glider in the case of non-viscous fluid is zero [20].

2.5 Drop a party balloon from a height. It falls straight down. To visualize the Magnus effect add a light weight strip of adhesive tape over the circumference of the balloon. This is to increase the rotational inertia and stabilize its rotation [22]. Now, spun the balloon (a) in the anti-clockwise direction and (b) clockwise direction and throw it vertically upward. Describe your observation. Make a rough sketch of the path of the balloon. Indicate the various forces acting on the balloon.

Chapter 3

Kite

A kite-flight is one of the mesmerizing joyful phenomena in everyday life throughout the world. A kite is a tethered heavier-than-air and flown by the lift induced by air in motion over its wing. A great skill is indeed required to make a kite to fly smoothly in changing wind patterns, display diving actions and cut nearby kites. For children, running with a kite is a fun. Kites can be prepared in a variety of shapes, however, the shapes are important in determining their flight dynamics. Figure 3.1 displays some kites. A delta shaped kite with single control line (string) is easy to construct and usually flies in different weathers.

Kites usually have a body (in different shapes and sizes), tethers (control lines), bridle (harness) and tails. The body of a kite is made with a framework and outer covering. The framework is prepared with a light-weight material like a wood or plastic and is stretched over it as a wing using a light-weight thin paper or silk fabric. The flight of the kite is controlled by the bridle and the control line. Setting the bridle is a nuance for a smooth flight of the kite. The flight of a kite is due to the lift created by the air

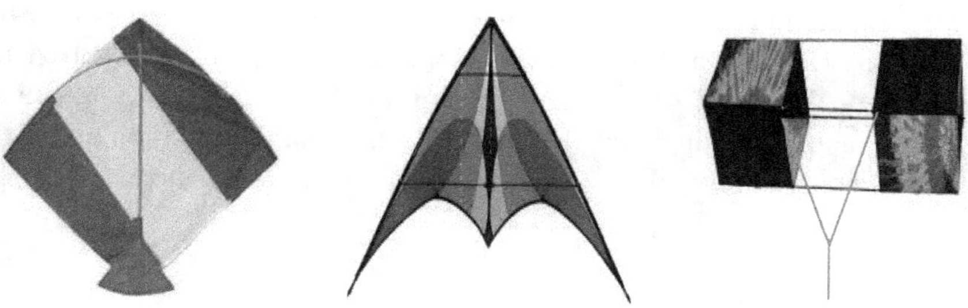

Fig. 3.1: Some kites.

moving around the surface of the kite. This air essentially produces high pressure below the wing and low pressure above it.

Kites are usually flown for recreation. There are sports kites as a part of a competition, power kites to generate large forces and for the purposes of kite surfing, kite fishing and snow kiting [1]. In certain countries, a game of cutting the control strings of other kites by a kite's string is played. To downing other kites the string is made of highly abrasive materials and is highly unsafe. In India several persons died due to the cutting of blood vessels in neck by such dangerous strings of downed kites. The highest altitude Guinness world record for a single kite is 4,879.54 m (16,009 ft). This was achieved by Robert Moore in Cobar, Australia on 23 September 2014.

A kite called Kaghati is found in the cave painting in Muna Island, Southeast, Indonesia dated from 9500-9000 BC [2]. Paper kites were flown from 549 AD. Dragon kites were described in Bellifortis about 1400 AD by Konrad Kyeser [1]. Interestingly, through the famous kite experiment in 1752 Benjamin Franklin proved that lightning was due to electricity. Wright brothers and many others used kites in their development of airplane in the late 1800s. Kite museums have been setup in Cambodia, Canada, China, England, Germany, India, Indonesia, Japan, Malaysia, Taiwan and USA.

During World War I, instability of kite-flight was studied through simplified models [3,4]. Stability of a kite with high lift/drag ratio was analysed [5]. The influence of the kite bridle and the existence of multiple equilibrium states were reported [6]. Structured design of a kite was considered in [7]. Physical models for a dual line kite [8] and tethered aerostats [9] have been proposed. The dynamics and control of a single line kite has been investigated through Lagrangian formulation [10,11]. Some interesting physics problems on kite strings were presented in [12]. Bifurcation theory was employed to a simplified kite model and multiple equilibrium states were found [13]. The problem of optimization was discussed in [14].

In this chapter, first we discuss about the various forces involved in kite-flight. We describe the four phases of kite-flight. Then, we present a physical model of a single-line kite and a Lagrangian formulation useful to analyse the dynamics and control of a kite. Next, we enumerate some of the physics of kite strings.

3.1 Forces Acting on a Kite

The prime forces determining the flight of a kite are the following [15-17]:

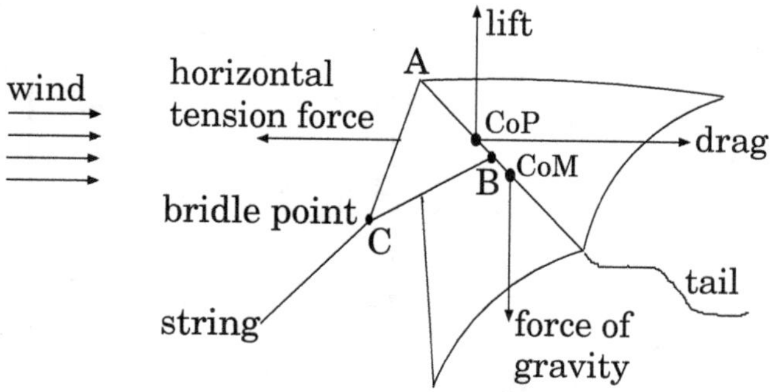

Fig. 3.2: Representation of the forces acting on a kite. A and B are the string attachment points for stability. CoM and CoP are the centre of mass and centre of pressure, respectively.

1. **Weight** – The gravitational downward force due to the gravitational attraction of the earth on the kite. This force pulls the kite downward towards the earth. The magnitude of this force is mg where m is the mass of the kite.
2. **Lift** – Pressure is created over the body of the kite by the wind moving across it. The shape and the horizontal angle of the kite are such that the air moving top of it has higher speed than over the bottom. According to Bernoulli, the pressure of a fluid decreases as the fluid speeds up. As a result, difference in air pressure develops an upward force called *lift* and it pushes the kite up.
3. **Tension** – A pulling force due to the holding of the string by a kite flyer. This force pulls the kite towards the flyer.
4. **Drag** – It is the force of the wind pushing the kite away. This is due to the air pressure difference between the front and back sides of the kite and the friction of the air moving the kite's surface.

The above four forces act simultaneously on the kite. However, they do not have the same centre of action. The directions of these forces are illustrated in Fig. 3.2. As the kite is tied to a string, the flyer can feel the force pulling the string. When a kite is set to fly interactions between the above mentioned, four forces take place mainly along three axes. These forces push or pull the kite forwards and backwards, up or down and side to side. The kite ascends when lift dominates weight and speeds away if the drag overcomes the tension.

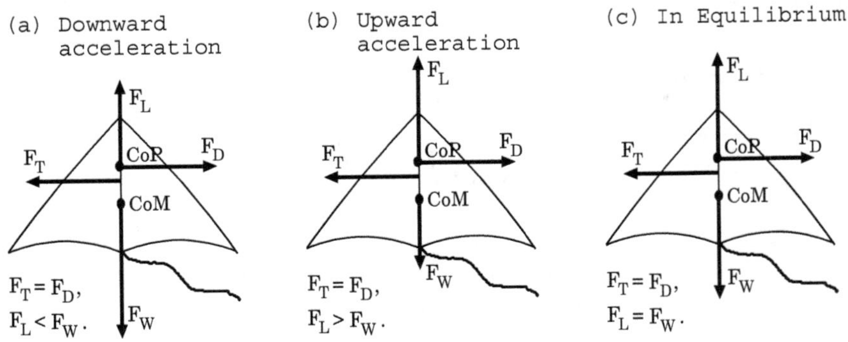

Fig. 3.3: Diagramatic representaion of forces of kite for three cases of its flight.

Solved Problem 1:

Denote the forces weight, lift, tension and drag as F_{W}, F_{L}, F_{T} and F_{D}, respectively, and the net force as F_{NET}. State the conditions on these forces for a kite to have (a) downward acceleration, (b) upward acceleration and (c) flying in equilibrium. Sketch the conditions diagramatically.

For downward acceleration of a kite

$$F_{\mathrm{T}} = F_{\mathrm{D}}, \quad F_{\mathrm{L}} < F_{\mathrm{W}} \quad \text{and} \quad F_{\mathrm{NET}} = F_{\mathrm{W}} - F_{\mathrm{L}}. \tag{3.1}$$

The conditions for upward acceleration is

$$F_{\mathrm{T}} = F_{\mathrm{D}}, \quad F_{\mathrm{L}} > F_{\mathrm{W}} \quad \text{and} \quad F_{\mathrm{NET}} = F_{\mathrm{L}} - F_{\mathrm{W}}. \tag{3.2}$$

The kite will be in equilibrium, that is, it will fly without acceleration in both the directions when

$$F_{\mathrm{T}} = F_{\mathrm{D}}, \quad F_{\mathrm{L}} = F_{\mathrm{W}} \quad \text{and} \quad F_{\mathrm{NET}} = 0. \tag{3.3}$$

These conditions are sketched in Fig. 3.3 [15].

3.2 The Four Phases of Flight

Airborne objects like kites, airplanes and rockets pass through four stages of flight, namely, *release, launch, climb* and *cruise* [17-19].

To release and launch, hold a kite up with its nose pointing straight-up and stand with your back to the direction of the wind. Gently release it.

The kite will rise up if the wind is strong enough. In case of less windy, run with the kite to generate sufficient lift. Alternately, ask your friend to hold the kite in two hands about a few meters away and gently throw above thereby giving enough lift. When it goes a little height release some string and pull (tug the string) it to gain altitude.

To make the kite climb and cruise pull the string occasionally. This will increase its speed and enhance the lift thereby make it to climb. When it rises up, release some length of string. Apply pull to add lift whenever the kite begins to descend. Add tension when it moves too far laterally. When you add the string a bit, due to the increased weight, the kite dips. To overcome it give a sufficient tug at the string. Remember that for a kite to cruise, all the forces acting on it have to be balanced. If the kite moves turbulently or dives often and fastly moves left and right, you need to adjust the attachment point B and the lengths AC and BC specified in Fig. 3.2 and add more tails. Long tails are usually attached to kites. Their fluttering develops beauty and fun. Further, adding drag at the kite's bottom will make the nose pointing to the sky and add stability also. If the kite is unstable, one can add tail and in case dragged back to the ground, reduce the length of the tail.

The amount of lift created can be adjusted by properly choosing the points A, B and the lengths AC and BC. The changes in these actually modify the *angle of attack* which is the angle that the kite leans into the wind. By properly adjusting these, a smooth flight of a kite can be achieved.

How does a kite stay in the air? Let us briefly explain it [20]. When a kite is near the ground, its tail can swing left and right similar to a pendulum. The kite has more weight below the point where the string is attached to it than above this point. The tail tries to hang straight downward. On the other hand, the nose points straight-upward. These make the kite climb quickly. When the kite is at a much height up, it is now also able to move left and right around the point of the string attached to it. But the end of the tail is not pointing straight-downward. The long and light-weight tail is able to make the nose of the kite not to point straight-upward, but to point straight into the wind.

Can a kite go backward to the person flying it? Why?

3.3 A Physical Model

We present the model considered by Sanchez [10]. Consider a kite as a tethered rigid and symmetrical body immersed in a homogeneous stream of fluid. The bridle lines are assumed as rigid solids.

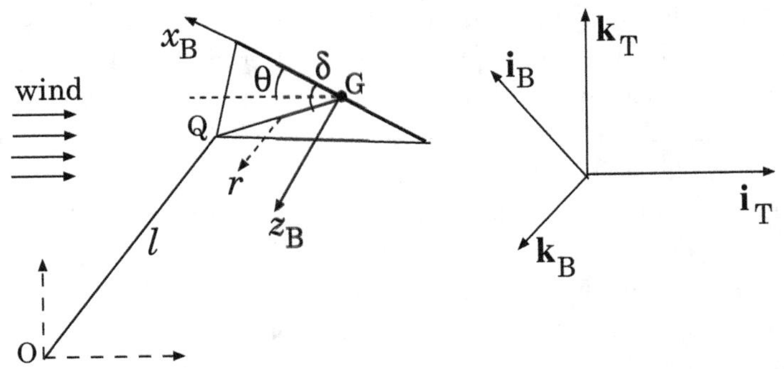

Fig. 3.4: Coordinate system of a kite. The point G is the CoM.

Figure 3.4 depicts the coordinate systems of a kite [10]. The inertial frame is $(x, y, z)_{\mathrm{T}}$ and $(x, y, z)_{\mathrm{B}}$ is a non-inertial frame and is body fixed attached to the kite with its CoM as the origin. The principal axes are body fixed. We restrict to longitudinal dynamics. The kite is restricted to move in a vertical plane with the mass centre coordinates $(x, 0, z)_{\mathrm{G}}$ and θ is the pitch angle defined in Fig. 3.4. OQ is the principal line (string) between the ground and the bridle and is massless and dragless rigid rod. The various parameters involved are defined in Fig. 3.4. $f(x_{\mathrm{G}}, z_{\mathrm{G}}, \theta, t) = 0$ is the holonomic constraint imposed by the bridle and the principal line. At O the horizontal and the vertical axes are x_{T} and z_{T}, respectively. Γ is the angle between l and the horizontal axis x_{T}. mg is the vertical downward force at the CoM.

3.3.1 *Lagrangian*

The mass centre position of the kite $\mathbf{r}_{\mathrm{G}}$ is given by

$$\mathbf{r}_{\mathrm{G}} = \begin{pmatrix} x_{\mathrm{G}} \\ y_{\mathrm{G}} \\ z_{\mathrm{G}} \end{pmatrix}_{\mathrm{T}} = \begin{pmatrix} l\cos\Gamma + r\cos(\delta - \theta) \\ 0 \\ l\sin\Gamma + r\sin(\delta - \theta) \end{pmatrix}_{\mathrm{T}} . \tag{3.4}$$

The velocity $\mathbf{v}_{\mathrm{G}}$ and the angular velocity $\boldsymbol{\omega}$ take the form

$$\mathbf{v}_{\mathrm{G}} = \begin{pmatrix} \dot{x}_{\mathrm{G}} \\ \dot{y}_{\mathrm{G}} \\ \dot{z}_{\mathrm{G}} \end{pmatrix} = \begin{pmatrix} -l\dot{\Gamma}\sin\Gamma + \dot{r}\cos(\delta - \theta) - r\sin(\delta - \theta)(\dot{\delta} - \dot{\theta}) \\ 0 \\ l\dot{\Gamma}\cos\Gamma + \dot{r}\sin(\delta - \theta) + r\cos(\delta - \theta)(\dot{\delta} - \dot{\theta}) \end{pmatrix}, \tag{3.5a}$$

$$\boldsymbol{\omega} = \dot{\theta}\mathbf{j}_0. \tag{3.5b}$$

Next, the kinetic and potential energies are

$$T = \frac{1}{2}m\mathbf{v}_{\mathrm{G}}^2 + \frac{1}{2}I_y\dot{\theta}^2, \tag{3.6a}$$

$$V = mgz_{\mathrm{G}}, \tag{3.6b}$$

where m is the mass of the kite. The Lagrangian L is

$$L = T - V = f(\dot{\Gamma}, \Gamma, \dot{\theta}, \theta, t). \tag{3.7}$$

3.3.2 *Generalized Forces*

We can express the aerodynamic force and moment as

$$\mathbf{F}_{\mathrm{a}} = -\frac{1}{2}\rho A\mathbf{v}_{\mathrm{a}}^2 C_{\mathrm{N}}\mathbf{k}_{\mathrm{B}}, \quad \mathbf{v}_{\mathrm{a}} = \mathbf{v}_{\mathrm{G}} - W_0\mathbf{i}_{\mathrm{T}}, \tag{3.8a}$$

$$\mathbf{M}_{\mathrm{a}} = \frac{1}{2}\rho A\mathbf{v}_{\mathrm{a}}^2 C_{\mathrm{N}} x_{\mathrm{CP}}\mathbf{j}_{\mathrm{B}}, \quad x_{\mathrm{CP}} = x_{\mathrm{G}} - x_{\mathrm{CoP}}. \tag{3.8b}$$

Here A-area of the kite, ρ-density of air, $\mathbf{v}_{\mathrm{a}}$-velocity of the kite, C_{N}-aerodynamic normal force coefficient, x_{CoP}-distance to centre of pressure (CoP) from leading edge, x_{CP}-distance to CoM from CoP and W_0-velocity of wind. The angle of attack (α) (in degree) is calculated as

$$\alpha = \theta + \tan^{-1}\left(\frac{\mathbf{v}_{\mathrm{a}} \cdot \mathbf{k}_{\mathrm{T}}}{\mathbf{v}_{\mathrm{a}} \cdot \mathbf{i}_{\mathrm{T}}}\right). \tag{3.9}$$

Due to the aerodynamic forces the terms to be added to the Lagrange equations are

$$Q_\Gamma = \mathbf{F}_{\mathrm{a}} \cdot \frac{\partial \mathbf{v}_{\mathrm{G}}}{\partial \dot{\Gamma}} + \mathbf{M}_{\mathrm{a}} \cdot \frac{\partial \omega}{\partial \dot{\Gamma}} = \frac{1}{2}\rho A\mathbf{v}_{\mathrm{a}}^2 C_{\mathrm{N}} l\cos(\Gamma + \theta), \tag{3.10a}$$

$$Q_\theta = \mathbf{F}_{\mathrm{a}} \cdot \frac{\partial \mathbf{v}_{\mathrm{G}}}{\partial \dot{\theta}} + \mathbf{M}_{\mathrm{a}} \cdot \frac{\partial \omega}{\partial \dot{\theta}} = \frac{1}{2}\rho A\mathbf{v}_{\mathrm{a}}^2 C_{\mathrm{N}}\left(x_{\mathrm{CP}} - r\cos\delta\right). \tag{3.10b}$$

For the derivations of Q_Γ and Q_θ see the exercises 3.4 and 3.5, respectively, at the end of the present chapter.

3.3.3 *Lagrange's Equations*

The Lagrange's equations are

$$\frac{\mathrm{d}}{\mathrm{d}t}\left(\frac{\partial L}{\partial \dot{\Gamma}}\right) - \frac{\partial L}{\partial \Gamma} = Q_\Gamma, \tag{3.11a}$$

$$\frac{\mathrm{d}}{\mathrm{d}t}\left(\frac{\partial L}{\partial \dot{\theta}}\right) - \frac{\partial L}{\partial \theta} = Q_\theta. \tag{3.11b}$$

For simplicity, we introduce the following change of variables and functions:

$$\widehat{x}_{\text{CP}} = \frac{x_{\text{CP}}}{l}, \quad \widehat{\mathbf{v}}_{\text{a}} = \frac{\mathbf{v}_{\text{a}}}{\sqrt{gl}}, \quad \widehat{r}_{\text{g}} = \frac{r_{\text{g}}}{l}, \quad \widehat{r}(t) = \frac{r}{l}, \quad \mu = \frac{\rho A l}{2m}, \tag{3.12a}$$

$$\tau = \sqrt{\frac{g}{l}}\,t, \quad f_1 = \sin(\delta - \theta - \Gamma), \quad f_2 = \cos(\delta - \theta - \Gamma), \tag{3.12b}$$

where r_{g} is the radius of gyration. The time evolution of the state vector $\mathbf{x} = (\dot{\Gamma}, \Gamma, \dot{\theta}, \theta)$ becomes

$$\frac{\mathrm{d}\mathbf{x}}{\mathrm{d}\tau} = \mathbf{h}_l(\mathbf{x}, t), \tag{3.13a}$$

where

$$\mathbf{h}_l = \begin{pmatrix} \frac{1}{|A|}\left[\left(\widehat{r}_{\text{g}}^2 + \widehat{r}^2\right) g_1 + \widehat{r} f_2 g_2\right] \\ \dot{\Gamma} \\ \frac{1}{|A|}\left[\widehat{r} f_2 g_1 + g_2\right] \\ \dot{\theta} \end{pmatrix} \tag{3.13b}$$

with

$$|A| = \widehat{r}_{\text{g}}^2 + \widehat{r}^2 f_1^2, \tag{3.13c}$$

$$\begin{aligned} g_1 &= -\widehat{r}\ddot{\delta} f_2 - \ddot{\widehat{r}} f_1 - 2(\dot{\delta} - \dot{\theta})\dot{\widehat{r}} f_2 + \widehat{r}(\dot{\delta} - \dot{\theta})^2 f_1 - \cos\Gamma \\ &\quad + \mu \widehat{\mathbf{v}}_{\text{a}}^2 C_{\text{N}} \cos(\Gamma + \theta), \end{aligned} \tag{3.13d}$$

$$\begin{aligned} g_2 &= 2\widehat{r}\dot{\widehat{r}}(\dot{\delta} - \dot{\theta}) + \widehat{r}\dot{\Gamma}^2 f_1 + \widehat{r}^2\ddot{\delta} + \widehat{r}\cos(\delta - \theta) \\ &\quad + \mu \widehat{\mathbf{v}}_{\text{a}}^2 C_{\text{N}} \left(\widehat{x}_{\text{CP}} - \widehat{r}\cos\delta\right). \end{aligned} \tag{3.13e}$$

We can determine the equilibrium states and their stability and introduce a control system to keep the kite's flight altitude a constant. Interested readers may refer to the Ref. [10] for details of these analysis.

3.4 Physics of Kite Strings

In Ref. [12] certain interesting questions about the properties and behaviour of kite strings were addressed. We briefly summarize them here.

The string at the top near the kite must withstand the weight of all the remaining part of the string. *How much stronger does it need to be?* The mass of the string required to fly a kite at, say, 5000 m must have an appreciable weight. Roughly, one may require at least twice this much length of string, that is, 10000 m. Strings come with different diameter,

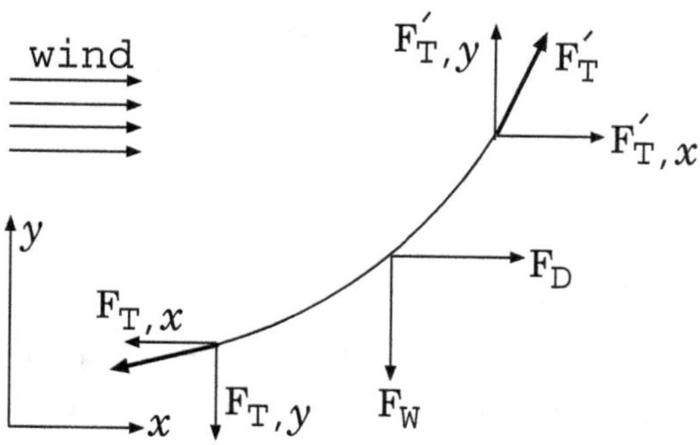

Fig. 3.5: Forces acting on a string of a kite.

tension and linear weight density. Suppose the diameter of the string to be used is, say, 1 mm and the maximum tension is 1000 N. For a string with any diameter and for 10 km its required weight is about 6.6% of the maximum tension. So, for a kite to be set to fly about 5000 m altitude with maximum tension of 1000 N and 1 mm diameter may require 10000 m string with weight $66 \times (1/9.80665)\,\text{kg} = 6.73\,\text{kg}$.

Why do often kites are found the string broken at the bottom near the spool instead of at the top near the kite for high altitudes? Assume that the kite is in equilibrium, that is no acceleration in the horizontal and vertical directions. Then, from Fig. 3.5 [12] according to Newton's second law of motion we write

$$F'_{\mathrm{T},x} - F_{\mathrm{T},x} + F_{\mathrm{D}} = 0, \tag{3.14a}$$

$$F'_{\mathrm{T},y} - F_{\mathrm{T},y} - F_{\mathrm{W}} = 0. \tag{3.14b}$$

Note that $F'_{\mathrm{T},x}$ and $F'_{\mathrm{T},y}$ are due to the drag and lift forces, respectively. According to Eq. (3.14b) the lift is supported by the weight of the string and any excess lift will lead to an additional vertical tension at the bottom. From Eq. (3.14a), we infer that in the case of constant lift, if the kite moves higher and more string is fed out, then the vertical component of tension at the bottom would drop. The point is that the weight of the string is supported at the top whereas the drag is countered at the bottom. Therefore, when the net drag on the string exceeds its breaking strength, then the string will breakup at the bottom near the reel.

The amount of drag exerted on the string can be determined (see the exercise 3.2 at the end of the present chapter). The drag is directly proportional to the diameter of the string while the maximum tension is directly proportional to the cross-sectional area of the string, that is, to the square of the diameter. Therefore, a largest diameter string is a better choice for a successful flight of a kite.

3.5 Conclusion

A kite is a heavier object and it flys by the lift created air in motion over its wing. So, a kite needs moving air to fly. It is the wind pressure that pushes up the kite. There are four forces acting on the kite. The lift force is generated in the direction perpendicular to the wind direction and the drag force is generated in the direction parallel to the wind. The aerodynamical force (lift force) plays the major role in the motion of the kite. The construction and design of the wing and bridle play important roles in making the kite fly in equilibrium. They also decide the stability of the kite along with the tail of the kite. Though the principle of kite flying is based only on Newton's principle, it is quite difficult to develop theories for explaining its motion since the kite has many complicated motions.

3.6 Bibliography

[1] https://en.wikipedia.org/wiki/kite.

[2] http://en.gocelebes.com/kaghati-worlds-first-kite/.

[3] L. Bairstow, E.F. Relf and R. Jones, *The stability of kite ballons: Mathematical investigation*, RM, December 1915, **208**, Aeronautical Research Council, UK.

[4] H. Glauert, *The stability of a body towed by a light wire*, RM, 1930, **1312**, Aeronautical Research Council, UK.

[5] L.W. Bryant, W.S. Brown and N.E. Sweeting, *Collected research on the stability of kites and towed glides*, RM, 1942, **2303**, Aeronautical Research Council, UK.

[6] K. Alexander and J. Stevenson, The Aeronautical J. **105**, 535 (2001).

[7] S. Veldman, H. Bersee, C. Vermeeran and O. Berosma, *Conceptual design of a high altitude kite*, 43rd AIAA/ASME/ASCE/AHS/ASC Structures, Structural Dynamics and Materials Conference, Denver, Colorada, 22-25 (2002).

[8] M. Diehl, L. Magni and G. De Nicolao, Annual Review in Control **28**, 37 (2004).

[9] J.D. Delaurier, J. Aircroft **17**, 305 (1980).

[10] G. Sanchez, The Aeronautical J. **110**, 615 (2006).

[11] L. Salord Losantos and G. Sanchez-Arriaga, J. Aircraft **52**, 660 (2014).
[12] D. Kagan and M. McGie, The Phys. Teach. **35**, 202 (1997).
[13] R.A. Adomaitis, SIAM Review **31**, 478 (1989).
[14] S. Costello, G. Franagois and D. Bonvin, *Real-time optimization for kites*. In Proceedings of the IFAC Workshop on Periodic Control Systems (2013) pages 64-69.
[15] https://www.sciencefriday.com/educational-resources/kite-engineering/.
[16] https://www.gombergkites.com/nkm/why.html.
[17] https://www.fatherly.com/play/how-to-make-fly-diy-kite-using-science/.
[18] https://www.scienceabc.com/pure-sciences/physics-kite-flying-how-tomake-aerodynamic-structure.html.
[19] http://www.gkites.com/howtofly/sl-htf.html.
[20] https://www.my-best-kite.com.

3.7 Exercises

3.1 List out a few things that a kite flyer can learn about the nature of wind in a place.

3.2 Consider a kite flying at an altitude of 3000 m, the diameter of the string is $D = 1\,\mathrm{mm}$, the maximum tension is 1000 N, the wind speed is $v = 6\,\mathrm{m/s}$, the drag coefficient is $\mathrm{C} = 1.1$ and the density of air is $\rho = 1.2\,\mathrm{kg/m}$. Determine the drag force [12].

3.3 Obtain an expression for the length of a string for a kite at an equilibrium at a high altitude of h m by neglecting the weight of the string [12].

3.4 Obtain the generalized potential Q_Γ given by Eq. (3.10a).

3.5 Obtain the generalized potential Q_θ given by Eq. (3.10b).

Part V: Throwable Toys

Chapter 1

Frisbee

For decades all over the world the *frisbee* (a flying disc, a plastic disc curled down at the edges) is one of the popular sports instrument [1-6]. A typical frisbee is shown in Fig. 1.1. Usually frisbees are thrown from person to person. The ancient Indian chakram is considered as an early version of the flying disc. In 1871, William Russel Frisbie (Bridge Port, Connecticut) opened a small bakery-The Frisbie Pie Company. In those days, at the nearby Yale University Frisbie's pies were popular and students enjoyed tossing the empty pie[1] tins. The tins were called as *Frisbies* by the students and the Frisbie-ing meant the act of throwing the pie tins. In 1958 Fred Morrison received a patent for the first production of plastic flying discs and the toy company Wham-O popularized the flying disc by introducing the trademark frisbee.

There are two popular sports with frisbee. The *ultimate game* is played by two teams with seven players in each team. The disc is passed back and

Fig. 1.1: A frisbee.

[1] Pie–A type of food made with vegetables, meat or fruits covered in pastry and baked.

forth by the players from one end of a field to the other. The team having the disc on hand at an instant is known as offence and the other team is defence. The defence team tries to intercept the disc or to make the disc to touch the ground thereby possesses the disc. Another sport is *disc golf*. In this sport, players using a few throws or strokes make a disc to land on a metal basket placed at a distance. Nowadays, in the USA, millions of people use the frisbee as a recreational toy and several thousands in the sports of ultimate frisbee and disc golf.

Researchers examined the aerodynamics of the frisbee by analysing at the airflow around the disc [7,8] and theoretical calculation of its trajectory and analysis of the biomechanical aspects [4]. Numerical modeling of flight of the frisbee is reported [9]. The influences of various factors on the flight dynamics of the frisbee have been analysed [3,10]. It is to be noted that the study of frisbee flies is useful for learning how to throw discs in different situations accurately and designing new golf discs.

First, we describe the flying of the frisbee and bring out the connection between the stable flight of the frisbee and spin of it. Then, we setup the equations of motion for the instantaneous position and velocity of the frisbee and present a numerical simulation result.

1.1 How Does a Frisbee Fly?

We describe the flying of the frisbee [9-11].

Assume that the frisbee is an axially symmetric rigid body of mass m with an elliptical cross-section. The frisbee is different from the discus. The shape of the discus has a plane and an axis of symmetry. The *aerodynamic lift* (or Bernoulli's principle) and the *gyroscopic inertia* (angular momentum) are the two prime physical concepts behind the flying of the frisbee. Gyroscopic inertia is the resistance encountered while attempting to alter the rotation axis of a rotating body and is simply the magnitude of the angular momentum. When we view the spinning disc as a wing freely flying, then the Bernoulli's principle being the cause of the lift whereas the stability being provided by the angular momentum.

A flying disc, when thrown lifts through the air. This is because, when air hits the top edge, the air got splitted and travel above and below the disc. Due to (i) the *angle (of attack)* α with which the disc is thrown and (ii) the curved topside of the disc, the velocity of the air over the top is higher than that below the bottom. According to the Bernoulli's principle, an increase in fluid's velocity leads to a decrease in pressure. As a result,

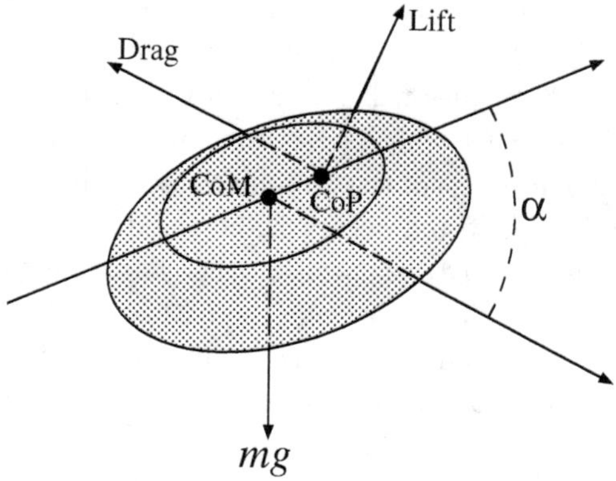

Fig. 1.2: Representation of drag and lift forces of a flying frisbee. They result due to a torque exerted on the frisbee. CoM and CoP refer to the CoM and the CoP, respectively.

there is a pressure difference. The lower pressure above the disc causes it to rise. This lifting occurs close to the centre since the disc is symmetric [11]. When the angle of attack increases, more of the air flow gets deflected downward causing an increase in the lift. A simple example of lift generated by the flow of air is the airplane wing.

The vital element for the stable flight of a frisbee is the angular momentum L, the rotational equivalent of momentum. It is equal to $I\omega$, where I is the moment of inertia and ω is the angular velocity, points in the direction of axis of rotation. It essentially resists changes in the orientation of the disc and generates the stability required for the flight [12]. A disc during its flight would experience torques about any of its major axes causing changing of the course of the disc. *How do these torques arise?* They arise because of the complicated pressure distribution on the disc and the lift and drag forces. The torque due to the lift and drag arises as a result of their action being on the centre of pressure (CoP) instead of on the centre of mass (CoM) (refer Fig. 1.2). In Fig. 1.2 the axes are setup with the x-axis being the direction of traveling of the disc, the positive y-axis is to the left-side and the positive z-axis is upward direction.

There are three moments exerted on the disc. They are *pitch, roll* and *spin down moments*. The pitch moment is with a torque around the y-axis. A positive moment refers to the upward tilting of the leading edge. The roll moment is attached with a torque around the x-axis and the rolling of the disc to the right is referred as a positive moment. The torque along the z-axis is the spin down moment. These three moments can be calculated. The above mentioned torques will give rise a rate of change in the angular momentum $\mathbf{L}$, $\mathrm{d}\mathbf{L}/\mathrm{d}t = \mathbf{M}$ where $\mathbf{M}$ is the total moments exerted on the disc. Because of the decrease of the angular velocity during the flight of the disc, the magnitude of $\mathbf{L}$ would decrease and the direction of $\mathbf{L}$ varies. The frisbee flight is considerably disrupted by the change in the direction of $\mathbf{L}$ (termed as precession $\dot{\psi}$). For a large $\dot{\psi}$ the frisbee would wobble along the x- and y-axes, turn over and fall on the ground. $\dot{\psi}$ can be counteracted by increasing ω. We have $\dot{\psi}\mathbf{L} = \dot{\psi}\mathbf{I}\omega = \mathbf{M}$ which indicates that an increase in ω gives rise to a decrease in $\dot{\psi}$. That is, stability of the disc can be greatly improved by decreasing the rate of precession thereby initial orientation is maintained. The combination of lift and gyroscopic stability provides a simple picture of a flying frisbee.

1.2 Gyroscopic Stability

The frisbee is thrown by giving a rotation along with a linear velocity initially. A frisbee travels a long distance without falling to the ground only due to the initial rotation given to it. A frisbee that is not spinning will not fly very far. The pressure distribution will not be symmetric with respect to the front and back of the frisbee due to its shape. Suppose the point of highest pressure on the bottom side of the disc and the point of lowest pressure on the top surface are near the back of the disc. Then, this difference of pressure will exert a torque which will make the disc to flip over and fall to the ground without having flown very far. When a frisbee is not spinning, the small torque explained above flips the front of the disc up and the frisbee will not have a stable flight.

When the frisbee is given an initial rotation about its axis of symmetry, the disc will have an angular momentum $\mathbf{L}$ about its vertical axis. The moment of inertia is greatest when the axis of rotation is the axis of symmetry. It gives rise to the torque

$$\boldsymbol{\tau} = \frac{\mathrm{d}\mathbf{L}}{\mathrm{d}t} = I\frac{\mathrm{d}\boldsymbol{\omega}}{\mathrm{d}t}, \tag{1.1}$$

where $\boldsymbol{\omega}$ is the angular velocity.

A frisbee rotating about its axis of symmetry tends to remain rotating about its axis of symmetry, giving stability to its flight. As the frisbee is stable to its large moment of inertia and spin, it will not flip over when flying through the air. The higher the rate of spin ω, the frisbee will be more stable.

1.3 Theoretical Treatment

In Sec. 1.1 in the present chapter we pointed out the significance of the force *lift*. This lift on an object is modeled using the lift equation

$$F_{\mathrm{l}} = \frac{1}{2}\rho A C_{\mathrm{l}} v^2 , \tag{1.2}$$

where ρ is the density of the air, v is the velocity of the frisbee relative to the air and A is the surface area of the frisbee. In Eq. (1.2) $v^2/2$ is the dynamic energy. The coefficient of lift C_{l} is given in [4] as

$$C_{\mathrm{l}} = C_{\mathrm{l0}} + C_{\mathrm{l}\alpha}\alpha , \tag{1.3}$$

where α is the angle of attack and C_{l0} and $C_{\mathrm{l}\alpha}$ are constants depend on the physical properties of the frisbee.

Apart from the lift force there is an another aerodynamic force, *drag force*. This acts opposite to the velocity vector of the disc and is modelled through the equation

$$F_{\mathrm{d}} = -\frac{1}{2}\rho A C_{\mathrm{d}} v^2 , \tag{1.4}$$

where C_{d} is the drag coefficient. C_{d} is given by [4]

$$C_{\mathrm{d}} = C_{\mathrm{d0}} + C_{\mathrm{d}\alpha}(\alpha - \alpha_0)^2 , \tag{1.5}$$

where C_{d0}, α_0 and $C_{\mathrm{d}\alpha}$ are constants depending on the physical properties of the object. Refer to Fig. 1.3 for the representation of the lift and drag forces and the angle of attack α.

Next, we proceed to obtain the equation of motion for the instantaneous position and velocity of the frisbee [9]. For this purpose, we separate the horizontal (x) and the vertical (y) components of the forces as

$$F_x = F_{\mathrm{d}} , \tag{1.6a}$$

$$F_y = F_{\mathrm{g}} + F_{\mathrm{l}} . \tag{1.6b}$$

Equation (1.6a) gives

$$m\dot{v}_x = -\rho A C_{\mathrm{d}} v_x^2/2 \tag{1.7}$$

or

$$\ddot{x} + \frac{1}{2m}\rho A C_{\mathrm{d}} v_x^2 = 0 . \tag{1.8a}$$

Equation (1.6b) leads to $m\dot{v}_y = mg + \rho A C_\mathrm{l} v_x^2/2$ or

$$\ddot{y} - g - \frac{1}{2m}\rho A C_\mathrm{l} v_x^2 = 0\,. \tag{1.8b}$$

The theoretical model given by Eqs. (1.8) allows us to analyse the role of various parameters on the dynamics of the frisbee.

Solved Problem 1:

Given that for a frisbee, $C_{\mathrm{l}0} = 0.188$, $C_{\mathrm{l}\alpha} = 2.37$, $C_{\mathrm{d}0} = 0.15$, $C_{\mathrm{d}\alpha} = 1.24$ and $\alpha_0 = 0$ find the angle at which the ratio of the lift force to the drag force is maximum.

We obtain

$$\frac{F_\mathrm{l}}{F_\mathrm{d}} = \frac{C_\mathrm{l}}{C_\mathrm{d}}\,. \tag{1.9}$$

For this ratio to be maximum the condition is

$$\begin{aligned} 0 &= \frac{\mathrm{d}}{\mathrm{d}\alpha}\left(\frac{F_\mathrm{l}}{F_\mathrm{d}}\right) \\ &= \frac{\mathrm{d}}{\mathrm{d}\alpha}\left(\frac{C_\mathrm{l}}{C_\mathrm{d}}\right) \\ &= \frac{1}{C_\mathrm{d}}\frac{\mathrm{d}C_\mathrm{l}}{\mathrm{d}\alpha} + C_\mathrm{l}\frac{\mathrm{d}}{\mathrm{d}\alpha}\left(\frac{1}{C_\mathrm{d}}\right) \\ &= \frac{1}{C_\mathrm{d}}\frac{\mathrm{d}C_\mathrm{l}}{\mathrm{d}\alpha} + C_\mathrm{l}\left(-\frac{1}{C_\mathrm{d}^2}\right)\frac{\mathrm{d}C_\mathrm{d}}{\mathrm{d}\alpha}\,. \end{aligned} \tag{1.10}$$

As $C_\mathrm{l} = C_{\mathrm{l}0} + C_{\mathrm{l}\alpha}\alpha$ and $C_\mathrm{d} = C_{\mathrm{d}0} + C_{\mathrm{d}\alpha}\alpha^2$ we have

$$\frac{\mathrm{d}C_\mathrm{l}}{\mathrm{d}\alpha} = C_{\mathrm{l}\alpha}\,, \qquad \frac{\mathrm{d}C_\mathrm{d}}{\mathrm{d}\alpha} = 2C_{\mathrm{d}\alpha}\alpha\,. \tag{1.11}$$

Using these results in Eq. (1.10), we obtain

$$C_{\mathrm{d}\alpha}C_{\mathrm{l}\alpha}\alpha^2 + 2C_{\mathrm{d}\alpha}C_{\mathrm{l}0}\alpha - C_{\mathrm{l}\alpha}C_{\mathrm{d}0} = 0\,. \tag{1.12}$$

Substituting the given values of the parameters, we get

$$2.9388\alpha^2 + 0.46624\alpha - 0.3555 = 0\,. \tag{1.13}$$

Its positive root is $\alpha_{\max} = 0.27741\,\mathrm{rad} = 15.88^\circ$.

1.4 Numerical Simulation

For numerical simulation purposes, we rewrite Eqs. (1.8) as

$$\dot{x} = v_x\,, \quad \dot{v}_x = -\frac{1}{2m}\rho A C_\mathrm{d} v_x^2\,, \tag{1.14a}$$

$$\dot{y} = v_y\,, \quad \dot{v}_y = g + \frac{1}{2m}\rho A C_\mathrm{l} v_x^2\,. \tag{1.14b}$$

Equations (1.14) can be solved numerically, for example, using the Euler method. The effect of the angle of attack has been reported [9].

We choose $C_{\mathrm{l}0} = 0.188$, $C_{\mathrm{l}\alpha} = 2.37$, $C_{\mathrm{d}0} = 0.15$ and $C_{\mathrm{d}\alpha} = 1.24$. To convert the angle of attack from degrees to radians we use $\alpha \to \alpha\pi/180$ in Eqs. (1.3) and (1.5). Usually the mass lies between 90 g and 175 g. We set $m = 0.175\,\mathrm{kg}$, $A = 0.05\,\mathrm{m}^2$, $\alpha_0 = -4^\circ$, $\rho = 1.23\,\mathrm{kg/m}^3$ and $g = 9.8\,\mathrm{m/s}^2$. Further, in solving the set of Eqs. (1.14), we choose the integration time step $\Delta t = 0.002\mathrm{s}$ and the initial conditions as $x(0) = 0\,\mathrm{m}$, $y(0) = 1\,\mathrm{m}$, $v_x(0) = 10\,\mathrm{m/s}$ and $v_y(0) = 0\,\mathrm{m/s}$. In the simulation, we vary only the angle of attack α from 0° to 90°. The C++ program that is used to solve Eqs. (1.14) employing the Euler method is listed in Sec. 1.8 in the present chapter.

Figure. 1.3 shows x versus y for four values of α. For $\alpha = 5^\circ$ the frisbee without moving up from the initial height, it quickly drops to the ground after flying a short distance. For the angles 20°, 30° and 40° the frisbee gets lifted above and reaching a height $> 1\,\mathrm{m}$ and then travelled towards the ground. We notice that the distance travelled by it along the horizontal direction does not increase monotonically with α, but increases for a certain range of α and then decreases. *How do we account this?*

In order to understand the role of α on the flying dynamics of the frisbee, we compute x_{max} (the maximum horizontal distance travelled by the disc before landing on the ground), y_{max} (the maximum height travelled by the disc), C_l (lift coefficient) and C_d (drag coefficient) as a function of α. The result is presented in Fig. 1.4 for α over the interval $[0^\circ, 90^\circ]$. Observe that $x_{\mathrm{max}} = 13.95\,\mathrm{m}$ at $\alpha = 21.7^\circ$ while $y_{\mathrm{max}} = 5.19\,\mathrm{m}$ at $\alpha = 44.4^\circ$. Both C_l and C_d increase monotonically with α. For $\alpha < 9^\circ$, y_{max} remains as 1 m (the initial value of y, $y(0)$), that is, the frisbee does not fly up but moves downwards and land on the ground due to the very smallness of the lift force. This is the case shown in Fig. 1.3a. When $9^\circ < \alpha < 21.7^\circ$ both y_{max} and x_{max} increase with α as a result of a larger lift force. An example is shown in Fig. 1.3b for $\alpha = 20^\circ$. In the case of $\alpha \in [21.7^\circ, 44.3^\circ]$ the maximum height increases further but x_{max} decreases with α. This is

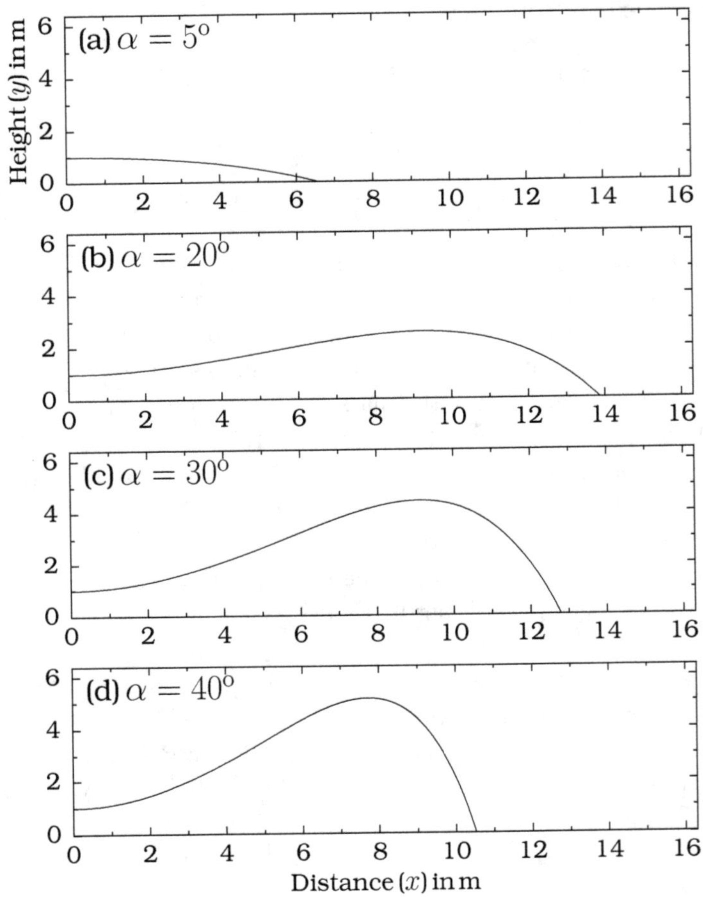

Fig. 1.3: x (horizontal distance) versus y (vertical height) of a frisbee for four selected values of the angle of attack α.

evident from Figs. 1.3b and c. For α in the range 21.7° to 44.3° the frisbee goes considerable higher height, however, because of the much larger drag force it travels a shorter distance. Both x_{max} and y_{max} decrease with α for $\alpha \in [44.3°, 90°]$. However, y_{max} is above the initial height 1 m.

It is to be noted that for a small angle of attack, both drag and lift are small. In this case the flight path becomes flat and fast. For a larger angle of attack the lift as well as the drag are higher making the frisbee to stay longer in air by loosing speed faster. Such throws are easy to catch, but as evident from Figs. 1.3c, they could not reach long distances.

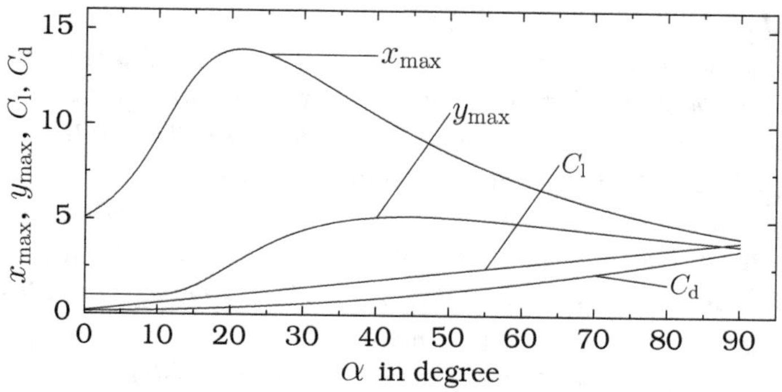

Fig. 1.4: Variation of x_{max}, y_{max}, C_l and C_d versus α.

Estimation of aerodynamic coefficients associated with the flight dynamics of frisbee has been described in [14].

1.5 Conclusion

Though the frisbee is a simple toy, the physical equations describing its motion are quite complicated. The various forces acting on its motion cannot be easily demonstrated. Its motion is decided by gravity, angular momentum, Bernoulli's principle, Newton's third law, the aerodynamical forces lift, drag and three moments acting on the disc due to rotation about the three axes of the disc, namely, pitch, roll and spin. The stability of its flight arising out of the spin given to the frisbee initially can be easily demonstrated to the students by throwing the frisbee by giving an initial rotation. The spinning frisbee will fly a long distance. The effect of the angle of attack can also be demonstrated by throwing the frisbee at different angles. The importance of the rim of the frisbee for generating the lift can be demonstrated by throwing frisbees with rims and without rims. A frisbee without a rim cannot fly as good as a frisbee with a rim. Aerodynamic forces, mostly lift, keeps the frisbee aloft for a long period of time. As lift is influenced by the angle of attack, wind speed and design of the disc, the duration of the flight of the frisbee can be studied with different angles of attack, different types of disc and also at different wind speed.

1.6 Bibliography

[1] M. Schuurmans, New Scientist **127**, 37 (1990).

[2] L.A. Bloomfield, Sci. Am. April, 132 (1999).
[3] R. Lorenz, *Spinning Flight: Dynamics of Frisbees, Boomerangs, Samaras and Skipping Stones* (Springer, New York, London, 2006).
[4] S.A. Hummel, *Frisbee Flight Simulation and Throw Biomechanics* (Master Thesis, University of California Davis, 2003).
[5] B. Schultze, *Biomechanical Aspects of Throwing a Frisbee: A Review* (Bachelor Thesis, Westfalische Wilhelms Universitat Munster, 2016).
[6] S.E.D. Johnson, *Frisbee: A Practitioner's Manual and Definitive Treatise* (Workman Publishing Company, 1975).
[7] G.D. Stilley, *Aerodynamic analysis of the self-sustained flair*, RDTR no.199, Naval Ammunition Dept, Crane, Indiana (1972).
[8] J.R. Potts and W.J. Crowther, *The flow over a rotating disc-wing*, Raes Aerodyn. Res. Conf. Proc., London, U.K., April (2000).
[9] V.R. Morrision, *The Physics of Frisbee*, Electronic Journal of Classical Mechanics and Relativity, Mount Allison University, Canada (2005).
[10] E. Mottoyama, *The physics of flying discs* (2002); https://people.csail.mit.edu/jrennie/discgolf/physics.pdf.
[11] Katherine Keller, *The Physics of Frisbee*, Undergraduate Physics News Letter, December 2004.
[12] P. Pozderac, *Gone with the wind: An investigation into the flight dynamics of discs* (preprint, 2016); http://physics.wooster.edu/JrIS/Files/web-Article_Pozderac.pdf.
[13] S. Hummel and M. Hubbard, *Identification of Frisbee Aerodynamic Coefficients Using Flight Data*, The International Conference on the Engineering of Sport (Kyoto, Japan, 2002).
[14] M. Hubbard and S.A. Hummel, *Simulation of frisbee flight*; https://www.researchgate.net/publication/253842372_Simulation_of_Frisbee_Flight.

1.7 Exercises

1.1 In terms of the Euler angle set of rotations, setup a reference frame for a frisbee. Then, obtain expressions for angular velocity of the disc and moment of inertia.

1.2 What are the effects of the rim of a frisbee on its flight?

1.3 Differentiate the effect of throwing a frisbee with and without a spin.

1.4 Throw a frisbee with its outer edge tilted upward. Describe your observation.

1.5 Derive the equation $F_l = (1/2)\rho A C_l v^2$ applying the Bernoulli's principle.

1.8 Program

The following is the C++ program to solve the dynamical equations for the motion of frisbee by the Euler method.

```
#include ⟨iostream⟩
#include ⟨fstream⟩
#include ⟨iomanip⟩
#include ⟨cmath⟩
int main()
{
    ofstream fo1, fo2;
    double tem= 0.1, delt= 0.002, t= 0.0, x= 0.0, vx= 10.0;
    double y= 1.0, vy= 0.0, delx, delvx, vvx, vvy, temcd, temcl;
    double g= −9.8, m= 0.175, area= 0.05, alpha0= −4.0;
    double cl0= 0.188, cla= 2.37, cd0= 0.15, alpha= 40.0;
    double cl, cd, pi= 4.0*atan(1.0), cda= 1.24, rho= 1.23;
    fo1.open("output1.data");
    fo2.open("output2.data");
    fo1<<setiosflags(ios::fixed|ios::showpoint)<<setprecision(5);
    alpha=alpha*pi/180.0;
    alpha0=alpha0*pi/180.0;
    cd=cd0+cda*(alpha−alpha0)*(alpha−alpha0);
    cl=cl0+cla*alpha;
    temcd=(0.5/m)*rho*area*cd;
    temcl=(0.5/m)*rho*area*cl;
    fo1<<setw(10)<<t<<setw(14)<<x<<setw(14)<<y<<setw(14)
      <<vx<<setw(14)<<vy<<setw(14)<<alpha*180.0/pi<<endl;
    while (y> 0.0)
    {
        delx=delt*funx(vx);
        delvx=delt*funvx(temcd,vx,vy);
```

```
            dely=delt*funy(vy);
            delvy=delt*funvy(temcl,g,vx,vy);
            x=x+delx;
            vx=vx+delvx;
            y=y+dely;
            vy=vy+delvy;
            if(y< 0) break;
            t=t+delt;
            fo1<<setw(10)<<t<<setw(14)<<x<<setw(14)<<y<<setw(14)
               <<vx<<setw(14)<<vy<<endl;
            liftf=0.5*rho*(vx*vx+vy*vy)*area*cl;
            dragf=-0.5*rho*(vx*vx+vy*vy)*area*cd;
            r=fabs(cl/cd);
            fo2<<setw(10)<<t<<setw(14)<<x<<setw(14)<<liftf
               <<setw(14)<<dragf<<setw(14)<<r<<endl;
        }
        fo1.close();
        fo2.close();
        return 1;
}
double funx(double vvx)
{
        return vvx;
}
double funvx(double temcd, double vvx, double vvy)
{
        return -temcd*vvx*vvx;
}
double funy(double vvy)
{
        return vvy;
}
double funvy(double temcl, double g, double vvx, double vvy)
{
        return g+temcl*vvx*vvx;
}
```

Chapter 2

Boomerang

Boomerang is an instrument belonging to the group of flying objects that have the fascinating characteristic of returning to their starting position of throw [1,2]. A typical boomerang with a crescent shape shown in Fig. 2.1 is made of wood and has two wings fashioned together into a single piece, but tilted with respect to each other. The wings are shaped similar to an aeroplane wings. The cross-section of each end is in the shape of an airfoil. The feature of a boomerang is that when you throw it properly (a rather difficult skill), it will come back to you. Its most common path is an ellipse. The direction of flight (left or right) depends on the boomerang and not on the thrower.

Boomerangs are considered as the first heavy flying machine invented by human beings. They were invented thousands of years ago. Evidences of use of such flying objects are found in all the five continents. King Tutankhamen of ancient Egypt who lived 2000 years ago had a collection

Fig. 2.1: A boomerang.

of boomerangs [3,4]. Boomerang appears as a part of a marsh hunting scene on an Egyptian tomb available in The British Museum in London [5]. In 1827 the term boomerang entered the Aboriginal language of New South Wales, Australia. The Sydney Gazette and New South Wales Advertiser, dated 23 December 1804, reported the first witness of a boomerang by western people at Farm Cove (Port Jackson), Australia during a tribal skirmish. Wo-mur-rang as one of the eight aboriginal "Names of Clubs" in 1798 [6]. Later in history, people have left-off using boomerangs except the Maori people of Australia.

Modern boomerangs come in different shapes, sizes, colours, materials and with more than two arms also. Computer-aided designs are also used. Historically, boomerangs were used as hunting weapons and recreational play toys. Nowadays, they are used mainly for sport and leisure. Every year, there are many boomerang competitions both at an individual and at a team level, and they consider longest throw, longest time aloft, number of catches within a fixed time duration and a fast catch [7].

The benchmark study of boomerang is the theoretical and experimental works done by Felix Hess at the University of Groningen, The Netherlands [8,9]. Dynamical analysis of a boomerang has been performed by Wolfgang Bürger [10]. Gabriel Barcelo reported his theoretical investigation of the flight dynamics of the boomerang [11]. Alexander Kuleshov developed a mathematical model of a boomerang flying in undistributed air [12]. King reported an experimental project on boomerangs done by undergraduate students of Dartmouth College, Hanover [13]. Henk Vos presented a construction of a straight boomerang, method of throwing it and a physical explanation of its come back [14].

There are several important questions one may ask about the boomerangs. *How do we throw a boomerang properly? What are the forces acting on the boomerangs and what they do? Why do they fly through the air? How do they come back? Do they come back due to wind motion and curved shape? Do they always come back?* We answer these questions in this chapter.

2.1 Throwing a Boomerang

Throwing a boomerang properly in order to trace out a curved path and return to its starting point, requires a lot of practice but it provides a lot of fun while practicing. Let us enumerate the basics of typical vertical throw [2,16-19] like the one shown in Fig. 2.2a.

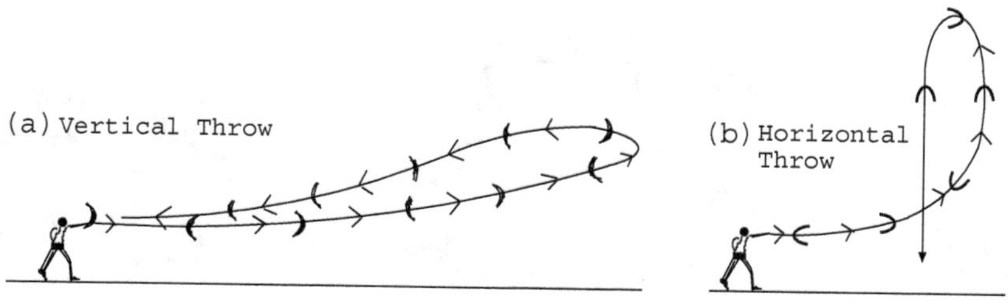

Fig. 2.2: Flight of a boomerang in (a) a typical vertical throw and (b) horizontal throw.

1. Hold the bottom of any one of the wings of a boomerang with the curved side facing you. The boomerang has to be almost perpendicular with the ground (leading to the vertical throw) and tilted 5° to 20° to the right (left) if you are right (left)-handed. The tilt angle is termed as the *angle of attack*.
2. Hold the boomerang in pinch grip, that is, between the thumb and the index finger (or other fingers).
3. If you are a right-handed thrower, then pivot your right-foot outwards, lift your left-leg which makes all your weight on the right. Step forward on to the left-foot to throw it in order to place more weight behind the throw leading to increase in the travel distance of the boomerang. A left-handed person has to do exactly the opposite.
4. In order to make the boomerang to return back to the starting point, it is necessary to spin it while throwing it. You may make it to spin by snapping your wrist while releasing it.
5. To catch the returning boomerang without hurt, wear a padded fingerless gloves in your two hands. The best way to catch it is to wait until it reaches below your shoulder level, then clap it between your two hands. Do not attempt to catch it if it comes at you with very high speed or you lost sight of it. Wear sun glasses to protect your eyes.

A beginner can choose a light weight boomerang which does not need a strong throw. In case of poor quality or an incorrect throw, a boomerang will not return to the starting point. Keeping a boomerang horizontally and throwing it will give a flight similar to the one shown in the Fig. 2.2b, where it rises steeply but falls straight down without returning to the starting point.

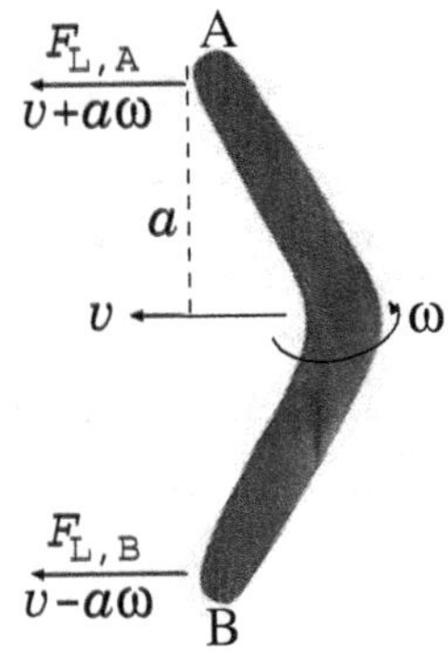

Fig. 2.3: Schematic representation of lift forces and velocities. Here $F_{\mathrm{L,A}} > F_{\mathrm{L,B}}$. For details see the text.

2.2 Forces Acting on a Boomerang

Suppose a boomerang with mass m and radius a is thrown vertically with an angle of attack (tilt) between 5° to 20°. It proceeds away from the starting point at a forward velocity and a rotational velocity. We examine the forces acting on the boomerang [20].

Consider the Fig. 2.3. Let v be the forward speed with which the centre of the boomerang is moving and ω is the angular velocity of its spin. The top edge A moves with the speed $v_{\mathrm{A}} = v + a\omega$ while the bottom edge B moves with the speed $v_{\mathrm{B}} = v - a\omega$.

Note that the wings of the boomerang have an air foil design. That is, in each wing one side is rounded while the other is flat, like a wing of an airplane. As the boomerang moves, air flows faster over the rounded surface and slower over the flat surface. This means according to the Bernoulli effect the air pressure is lower along the rounded surface than the flat surface. The difference in the air pressures gives rise to an aerodynamic lift force.

The aerodynamic lift force $F_{\mathrm{L,A}}$ acting on an air foil surface area S_{A} of the wing with the edge A moving with speed v_{A} in air with density ρ is given by

$$F_{\mathrm{L,A}} = \frac{1}{2} C_{\mathrm{L}} S_{\mathrm{A}}\, \rho\, v_{\mathrm{A}}^2 \,, \tag{2.1}$$

where C_{L} is the lift coefficient. Similarly,

$$F_{\mathrm{L,B}} = \frac{1}{2} C_{\mathrm{L}} S_{\mathrm{B}}\, \rho\, v_{\mathrm{B}}^2 \,. \tag{2.2}$$

Since $v_A > v_B$ we have $F_{L,A} > F_{L,B}$. *What is the consequence of this?*

The other forces to keep in mind are the force of gravity (mg), the force of throw, the force of wind and the torque due to the unequal speed of the wings. These forces have to be balanced in the right way for the boomerang to return to its starting point.

Solved Problem 1:

A typical boomerang with a span of 50 cm is launched at 25 m/s with a spin rate of 10 Hz. Find the ratio of the lift at the upper and lower positions.

We have

$$F_{L,A} = \frac{1}{2} C_L S_A \, \rho \, v_A^2 = \frac{1}{2} C_L S_A \, \rho \, (v + a\omega)^2 \tag{2.3a}$$

$$F_{L,B} = \frac{1}{2} C_L S_B \, \rho \, v_B^2 = \frac{1}{2} C_L S_B \, \rho \, (v - a\omega)^2 \,. \tag{2.3b}$$

As $S_A = S_B$ from Eqs. (2.3) we obtain

$$\frac{F_{L,A}}{F_{L,B}} = \frac{(v + a\omega)^2}{(v - a\omega)^2} \,. \tag{2.4}$$

With $2a = 50\,\text{cm}$ and $f = 10\,\text{Hz}$ we find $\omega a = 2\pi f a = 2\pi \times 10 \times 0.25 = 15.70796\,\text{m/s}$. Then,

$$\frac{F_{L,A}}{F_{L,B}} = \frac{(25 + 15.70796)^2}{(25 - 15.70796)^2} = \frac{(40.70796)^2}{(9.29204)^2} = 19.19272. \tag{2.5}$$

2.3 Why Does a Boomerang Fly?

The flight of the boomerang through the air is due to the air foil design of the wings. As pointed out earlier the air pressure above the wings (above the rounded surface of the wings) is lower than the air pressure below the wings (flat bottom surface). The higher air pressure below the wings is able to push the boomerang upwards, a lift. That is, the aerodynamic lift generated by the wings make it to fly.

As depicted in Fig. 2.3, the top and the bottom edges move with different speeds. *What is the effect of it?* It affects the magnitude of the lift force. The lift force varies quadratically with the speed. Because of the difference in the lift forces, the top edge turns towards the leftward (counter-clockwise) direction. The boomerang rotates and the rotational force is the torque.

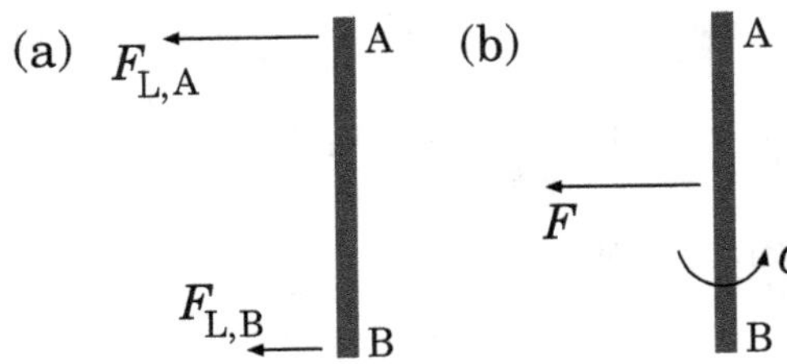

Fig. 2.4: (a) The two unequal lift forces. (b) The equivalent force and a moment (C).

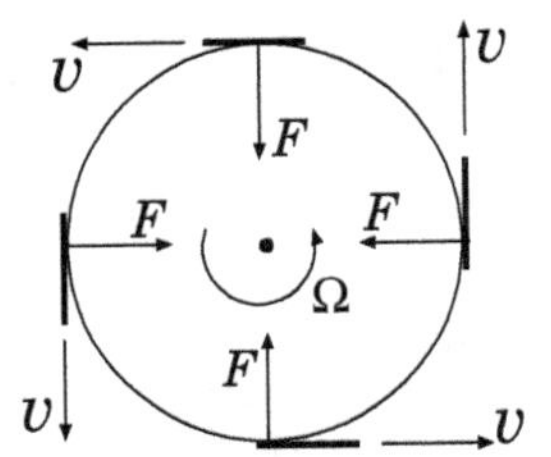

Fig. 2.5: Circular flight path of boomerang.

That is, it is just a rotor, however, not attached to anything. The rotor is moved by the lift force of the wings. Thus, the boomerang flies up into the sky until stopped spinning and pulled down by the gravity.

Referring to Fig. 2.3, we find that the end A moves in the direction of the velocity v and end B moves in the opposite direction to v. As the boomerang is thrown with an angular velocity ω, the velocity at A will be $v + a\omega$. So, as given in Eqs. (2.1) and (2.2), $F_{\mathrm{L,A}} > F_{\mathrm{L,B}}$. The point A is generating more lift than the point B. The two unequal lift forces are shown in a simplified Fig. 2.4a [20]. The two unequal forces can be represented by a single force F and a single couple C as shown in Fig. 2.4b [20].

The centripetal force F makes the boomerang to move on a circular path as shown in Fig. 2.5. If v is the velocity and m is the mass of the boomerang, then the centripetal force is given by $F = mv^2/R$ where R is the radius of the circular path. The constant couple C acting on a gyroscope spinning at an angular velocity ω will cause a steady precession at rate Ω, given by $C = I\omega\Omega$, where I is the moment of inertia of the boomerang. The resulting circular flight path of the boomerang is shown in Fig. 2.5 [20].

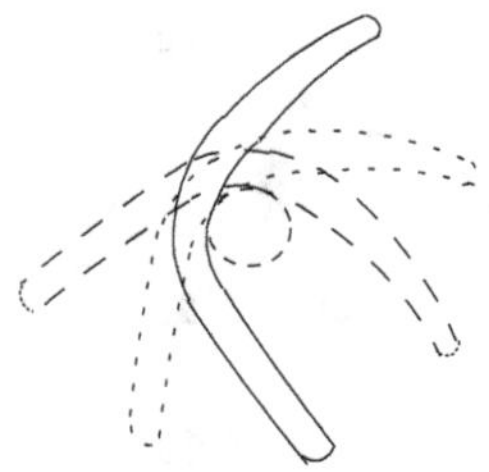

Fig. 2.6: Forming of a cavity.

2.4 Crescent Shape of the Boomerang and its Stability

If a stick is thrown with a force of large magnitude, then it would twirl in the air and fall down on the ground without traveling far away. This is due to the rotation of the stick. The difficulty or ease of rotation is specified by the moment of inertia I.

Consider the three axes of rotation. The x, y and z axes are along the directions of the width, the length and the thickness. For an object in a rotational motion around the z-axis, it is difficult to bring to rest and the moment of inertia about this axis is the largest. When any disturbance tries to dis-enable the rotational motion, then the rotation will be shifted to the y-axis since the moment of inertia is the least around this axis. This can be overcome by making the object in a curved shape so that its width gets increased while the length is reduced [19]. As a result, air resistance increases width-wise so that it would be difficult to rotate about the y-axis. On the other hand, when a crescent shape boomerang is thrown into the air, a doughnut-like cavity is created as shown in Fig. 2.6. The air pressures on the top and bottom surfaces of the wings are balanced by the cavity there by the increase of the stability of the flight [19].

2.5 How Does a Boomerang Come Back?

The key features of a boomerang are its flight, returning back to the starting point and maintaining the same tilt along its path. To make it fly a lift force is necessary. Because the top and bottom edges move with different speeds, the rotational force (torque) works in a counter-clockwise direction making the boomerang to rotate about an axis perpendicular to its plane. Because the lift points along the direction of the rotation axis of the boomerang, the torque works at a right-angle to this axis. This causes the direction of

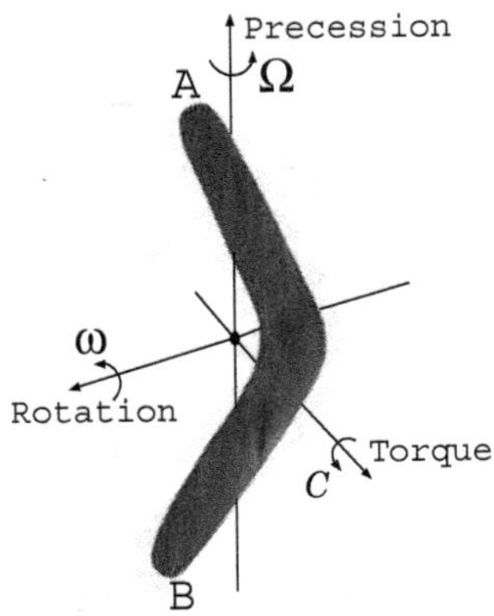

Fig. 2.7: Rotation, torque and precession axes of a boomerang.

the rotation axis without changing the speed of rotation. Now, there is an additional force. The unbalanced lift force and the weight (a force acting at the centre of mass) of the boomerang, causes this additional force that is the force of precession [11]. This force leads to a new rotation which is about a new axis that is different from the torque axis and the rotation axis of the boomerang. The new axis is orthogonal to the other two axes as shown in Fig. 2.7.

The precession is non-coaxial with the rotation of the boomerang. The precession force makes the direction of the movement of the boomerang to turn to the left. The consecutive occurrence of this causes the boomerang to return back along a curved path [19]. As pointed out by Gabriel Barcelo [11] *the boomerang returns to its origin as a result of being a body that is subjected to two simultaneous rotations on different axes.*

The ideal returning path is a circle when the boomerang is not subject to any motive force. During its flight the boomerang will maintain its tilt throughout the path. The rotation due to the precession force does not change the tilt of the plane of the boomerang during the flight. A boomerang has to be *thrown with a tilt in order to have enough lift* [19].

2.6 Conclusion

The boomerang can be used to illustrate certain basics of rigid body dynamics and lift force in aerodynamics. A modified Styrofoam ball can be used as a spherical boomerang [21] (see the exercise 2.5 at the end of the present chapter). As discussed in [19] there are generally three misconceptions as to why a boomerang comes back:

1. It comes back as it is crescent shaped.
2. The wind power turns it back.
3. It follows the same principle like a spinning baseball is curved.

But the real cause for the boomerang to return is the couple C acting on the rotating boomerang causing the gyroscopic precession Ω. The crescent shape is needed for the stability of the motion of the boomerang as discussed in Sec. 2.4.

2.7 Bibliography

[1] M. Hanson, Phys. Edu. **24**, 268 (1989).
[2] M. Hanson, The Phys. Teach. **28**, 142 (1990).
[3] https://web.archieve.org/web/20070630052119.
[4] http://www-gangs.co.uk/boomhistory.htm.
[5] http://www.britishmuseum.org/.
[6] http://www.gutenberg.org/etext/12565.
[7] https://en.wikipedia.org/wiki/Boomerang.
[8] F. Hess, *Boomerangs, Aerodynamics and Motion* (Groningen University, Groningen, 1975).
[9] F. Hess, Sci. Am. **219**, 124 (1968).
[10] W. Burger, Sci. Am. **304** (2002).
[11] G. Barcelo, J. Appl. Math. Phys. **2**, 569 (2014).
[12] A.S. Kuleshov, Procedia Eng. **2**, 3335 (2010).
[13] A.L. King, Am. J. Phys. **43**, 770 (1975).
[14] H. Vos, Am. J. Phys. **53**, 524 (1985).
[15] G. Barcelo, J. Appl. Math. Phys. **3**, 545 (2015).
[16] http://www.gel-boomerang.com/instructions/.
[17] http://www.4physics.com/phydemo/boomerang/boomerang_a.htm.
[18] https://www.wikihow.com/Throw-a-Boomerang.
[19] Y. Nishiyama, Int. J. Pure and Appl. Maths. **78**, 335 (2012).
[20] A.H. Hunt, Boomerang theory (preprint, 2001); http://www.eng.cam.ac.uk/~hemh/boomerangs.htm.
[21] Y. Ogawara, The Phys. Teach. **47**, 555 (2009).

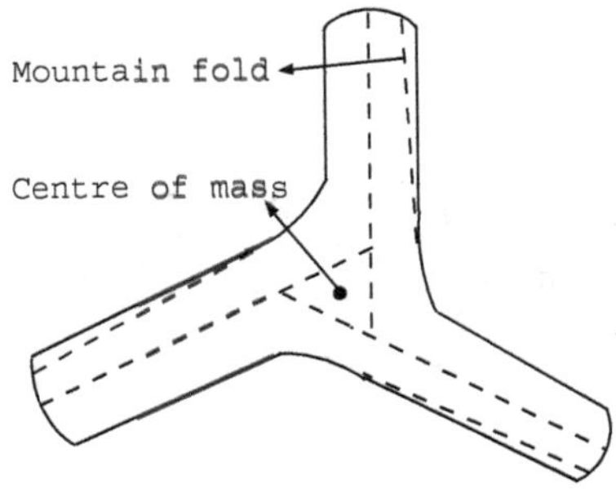

Fig. 2.8: A three arms cardboard boomerang.

2.8 Exercises

2.1 (a) Qualitatively distinguish the flight of a boomerang with aeroplanes or gliders.
(b) Does a boomerang come back due to the presence of wind? Can a boomerang fly and return back in a closed room with no wind motion?

2.2 (a) Why does a horizontal throw of a boomerang following a path similar to the one shown in Fig. 2.2b results in a steep rise and a swift fall?
(b) Pictorially represent the position of a boomerang in a vertical through at various time along with the direction of spin angular velocity and the direction of precession torque.

2.3 Throw a light Styrofoam ball, with many small pieces of its surfaces are removed, to the rear with backspin. Observe that its flight is like a boomerang [21]. What are the forces acting on the ball? Explain its flight.

2.4 Prove that the lift coefficient C_L is dimensionless.

2.5 Make a paper boomerang as shown in Fig. 2.8 using a cardboard of thickness 0.5 to 0.7 mm. Throw it as described in [19] and describe its flight.

Chapter 3

Skipping Stones

Almost every one as a child and adult tried at least once skipping of stones across the surface of the water. Stone skipping is a game of throwing a stone over a water surface so that it bounces-off the surface many times before sinking (see Fig. 3.1). Seeing five or more skips is really exhilarating. It can be played solo. We know that if an object incident on the surface of a fluid, an impulse would be imparted on the object. This impulse will be able to deflect the direction of the flight. When the impulse is sufficiently large, then the object can rebound from the fluid surface. This impulse is the cause for the skipping stone.

Stone skipping is known as *ducks and drakes* in England, *ricochet* in France, *smutting* in Denmark, *little frogs* in Greeks, *frog jumps* in Bengali and in every language it has a unique term.

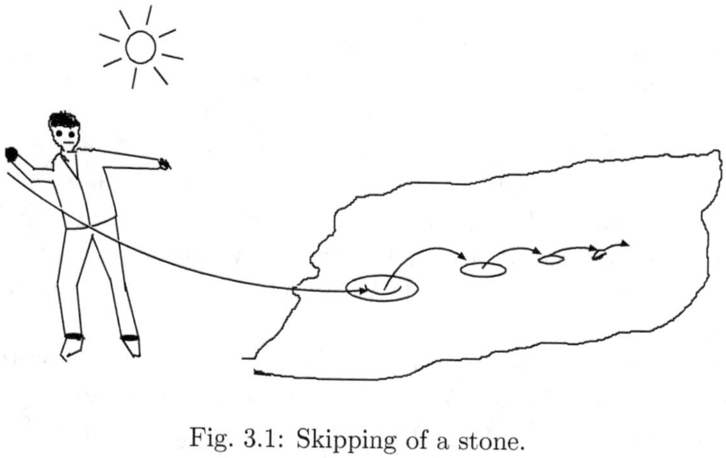

Fig. 3.1: Skipping of a stone.

There are number of stone skipping associations. Skipping stones is indeed a competitive sport. Nowadays, around the world, a variety of skipping stones competitions are held. From 1997 World Stone Skimming Championships is conducted on Easdale Island, Scotland using sea-worn Easdale slate, separately for men and women. The North American Stone Skipping Association is also conducting World Championships on stone skipping from 1989. In Pittsburgh, skipping stones tournament is held annually in the Rock and River festival. In the Mackinac Island, Michigan a Stone Skipping Hall of Fame has been setup for those made outstanding contributions to the sport of stone skipping.

The Guinness World Record for the most consecutive skips of a stone on water was achieved by Kurt Steiner (USA) in Pennsylvania's Allegheny National Forest on 6 September 2013. His throw of a stone produced 88 times touching of the water surface [1]. A video record of the throw of Kurt Steiner can be seen in [2]. Steiner was also the record holder from 2002-07. He had collected more than ten thousand quality rocks to prepare for the best throw.

The sport of skipping stones was known to ancient Greeks. Eskimos used to skip rocks on ice and Bedouins on smooth sand. The Oxford English Dictionary mentioned about skipping stones with regard to *Ducks and Drakes* which appeared first in 1583. The first scientific analysis of stone skipping was performed by Kirston Koths, a chemistry student at Amherst College in Massachusetts in 1968. A mathematical model was developed by Lyderic Bocquet, a French physicist, in 2003 [3]. This model gave a formula for the number of skips (bounces) a stone could be able to do before sinking.

Study of how objects skip over the surface of water has a wide range of applications. Earlier, during the period when seas were ruled by ships, cannonballs have been skipped over the water to bounce onto decks to break the ships and kill the enemies. In 1943 the Royal Air Force used bombs that could bounce across water inspired by the skipping stones. Both theoretical and experimental analysis have been conducted on the understanding of physics of the skipping stones and the role of various factors involved on its motion [3-11]. Several patents have been granted for the methods of skipping stones.

There are several important questions one may ask with respect to stone skipping. *What makes the stone to skip the water surface number of times before sinking? What are the crucial factors involved in the stone skipping? What is the source for stopping of bouncing of the stone? What do we learn through its physics? How many times a stone can bounce-off from the water surface? How do we derive the rules for best throw from the laws of physics?* We answer these questions in this chapter.

3.1 Why and How Do Stones Skip Across the Water?

We can learn much about stone skipping by applying the concepts of physics, particularly, momentum, gravity and hydrodynamics [3-11]. In this section we highlight the physical mechanism involved in the stone skipping.

A typical way of throwing a stone is the following [12].

1. Select a flat water surface. Choose a flat and wide stone roughly the size of your palm. It should sit in your hand comfortably.
2. Keep the index finger along on one side of the stone and the thumb on the other side. Face the water, side on with your feet shoulder width apart and bend your knees slightly.
3. Bend your wrist forward before snapping it forward to throw the stone across the water. But do not bend down and start with your arm close to the ground.
4. When you throw the stone, flick it to give a spin.
5. The stone should hit the water surface with an angle of about 20°.

When a thrown stone hits the water surface, it can either bounce-off its surface or sink. *What are the factors decide the skipping motion of the stone?* The skipping of the stone depends on the following:

1. Its mass.
2. Its initial velocity.
3. The angle between its surface and the water surface.
4. The angle between the surface of the water and its velocity vector.
5. Its horizontal velocity.
6. Its rate of spin.
7. The height from which it is thrown.

Among the above, the angle between the surface of the stone and the water surface is the most important one.

The force of gravity tends to pull the stone down. When the stone reaches the water, it pushes a small quantum of water downwards which in turn produces an upward lift to the stone. This is due to the consideration

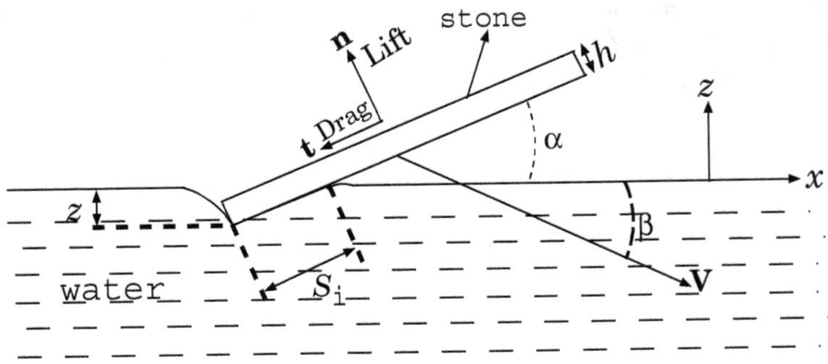

Fig. 3.2: A stone skipping model. For details see the text.

of momentum. According to it as the stone collides with water and pushes some water downwards, the stone is forced upward. This *lifting force* essentially arises from the flow around the incident stone during the course of stone-water collision. The stone will bounce-off the surface if the upward lift is greater than gravity. Its downward velocity is reversed while the horizontal velocity is reduced. Note that if the upward lift force is balanced by the weight of the stone, then it would sink in case its velocity becomes considerably small. That is, there is a minimum velocity of the stone above which the stone would not sink in its first encounter with the water surface. In each collision an amount of energy of the stone is lost due to the friction with water and air. Therefore, in each subsequent skipping the stone is slowed down until it sinks in the water.

3.2 Mathematical Theory

Lyderic Bocquet developed a theoretical model equation for the bouncing process of a skipping stone taking into account of the laws of physics involved [3]. The model is able to provide a formula for the maximum number of bounces of the stone before completely sinking in water. We present the prime features of the theoretical treatment of Bocquet.

3.2.1 *Basic Assumptions*

Consider a flat stone of thickness h and mass m thrown on to the water surface. Figure 3.2 shows the collision of the incident stone with the water [3]. In this figure α is the angle between the water surface and the surface of the stone, $\mathbf{V}$ is the velocity of the stone with the magnitude $V = \sqrt{V_x^2 + V_z^2}$

and z is the depth of the immersed edge of the stone. β is the angle between the water surface and $\mathbf{V}$. Assume that at time $t = 0$ at which the stone is incident on the water surface $V_{z0} \ll V_{x0}$. That is, the largest velocity is in the forward x-direction so that $V^2 = V_x^2 + V_z^2 \approx V_{x0}^2$.

What is the prime force involved in the collision process? The force involved is the lifting force pointed out in Sec. 3.1 in the present chapter. This force is proportional to the area of the immersed surface of the stone (S_i), density of water ρ and quadratic in the velocity $\mathbf{V}$. S_i depends on z. As shown in the Fig. 3.2 this force has components $\mathbf{n}$ and $\mathbf{t}$. One can write

$$\mathbf{F} = \frac{1}{2} C_\mathrm{L} \rho V^2 S_\mathrm{i} \mathbf{n} + \frac{1}{2} C_\mathrm{D} \rho V^2 S_\mathrm{i} \mathbf{t}\,, \tag{3.1}$$

where C_L and C_D are the lift and drag (friction) coefficients, respectively. For small α and β, it is reasonable to assume that C_L and C_D are constants. The incidence angle β is defined as $\tan\beta = V_{z0}/V_{x0}$. Equation (3.1) is valid for higher velocities of the stone and lower values of α and β. *What is the magnitude of the force given by Eq. (3.1) when the stone is fully immersed in water?*

3.2.2 *Equations of Motion*

$\mathbf{F} = m\mathbf{a}$ takes the form (see the exercise 3.2 at the end of this chapter)

$$m\frac{\mathrm{d}^2 x}{\mathrm{d}t^2} = m\frac{\mathrm{d}V_x}{\mathrm{d}t} = -\frac{1}{2}\rho V^2 S_\mathrm{i} \left(C_\mathrm{L}\sin\alpha + C_\mathrm{D}\cos\alpha\right), \tag{3.2a}$$

$$m\frac{\mathrm{d}^2 z}{\mathrm{d}t^2} = m\frac{\mathrm{d}V_z}{\mathrm{d}t} = -mg + \frac{1}{2}\rho V^2 S_\mathrm{i} \left(C_\mathrm{L}\cos\alpha - C_\mathrm{D}\sin\alpha\right), \tag{3.2b}$$

where g is the acceleration due to gravity. Equations (3.2) are coupled equations because of $V^2 = V_x^2 + V_z^2$. Further, they are nonlinear and difficult to find an exact analytical solution in terms of well known functions. However, the assumption $V^2 = V_{x0}^2 + V_{z0}^2 \approx V_{x0}^2$ makes the Eq. (3.2b) decouple from the Eq. (3.2a). Writing $C = C_\mathrm{L}\cos\alpha - C_\mathrm{D}\sin\alpha \approx C_\mathrm{L}$ for small α, Eqs. (3.2) become

$$m\frac{\mathrm{d}^2 x}{\mathrm{d}t^2} = \frac{\mathrm{d}V_x}{\mathrm{d}t} = -\frac{1}{2}\rho V_{x0}^2 S_\mathrm{i} \left(C_\mathrm{L}\sin\alpha + C_\mathrm{D}\cos\alpha\right), \tag{3.3a}$$

$$m\frac{\mathrm{d}^2 z}{\mathrm{d}t^2} = m\frac{\mathrm{d}V_z}{\mathrm{d}t} = -mg + \frac{1}{2}\rho V_{x0}^2 S_\mathrm{i} C\,. \tag{3.3b}$$

For regular shapes we can find simple formulas for S_i. Equations (3.3) describe the motion of the stone during the collision process with water. Motion of the stone in air is simply the ordinary projectile motion.

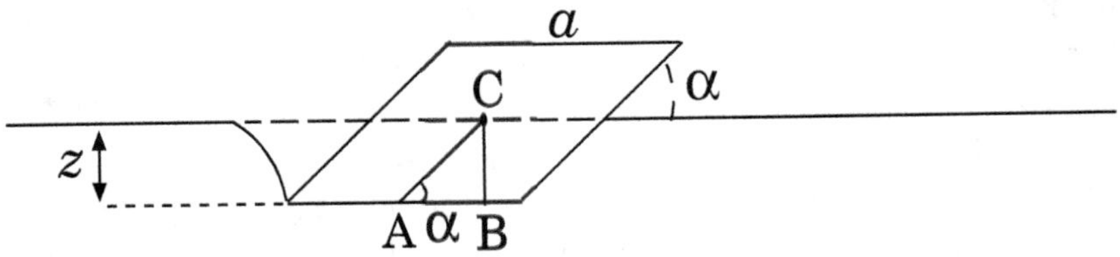

Fig. 3.3: Collision process of a square stone.

3.2.3 *Square Stone*

Suppose, the stone is a square one and consider the Fig. 3.3. The area of the stone immersed in the water is $S_\mathrm{i} = a \times AC$. From the right-angled triangle ABC, $\sin\alpha = BC/AC$, that is, $AC = BC/\sin\alpha = |z|/\sin\alpha$. As $z < 0$ in the Fig. 3.3 we write $S_\mathrm{i} = -az/\sin\alpha$. Then, Eq. (3.3b) is written as

$$\frac{\mathrm{d}^2 z}{\mathrm{d}t^2} = -g - \frac{1}{2m}\rho V_{x0}^2 C \frac{az}{\sin\alpha} \,. \tag{3.4}$$

This equation can be written in a standard form as

$$\frac{\mathrm{d}^2 z}{\mathrm{d}t^2} + \omega_0^2 z = -g\,, \quad \omega_0^2 = \frac{\rho V_{x0}^2 C a}{2m\sin\alpha}\,, \tag{3.5}$$

where ω_0 is the natural frequency.

With the change of variable $\omega_0^2 z + g = y$ Eq. (3.5) becomes

$$\frac{\mathrm{d}^2 y}{\mathrm{d}t^2} + \omega_0^2 y = 0\,. \tag{3.6}$$

Its general solution is

$$y(t) = A\mathrm{e}^{\mathrm{i}\omega_0 t} + B\mathrm{e}^{-\mathrm{i}\omega_0 t}\,. \tag{3.7}$$

Then, the solution $z(t)$ is

$$z(t) = \frac{A}{\omega_0^2}\mathrm{e}^{\mathrm{i}\omega_0 t} + \frac{B}{\omega_0^2}\mathrm{e}^{-\mathrm{i}\omega_0 t} - \frac{g}{\omega_0^2}\,. \tag{3.8}$$

The initial conditions are $z(t=0) = 0$ and $\dot{z}(t=0) = V_{z0}$. The condition $z(0) = 0$ gives $A + B = g$. The other condition $\dot{z}(t=0) = V_{z0}$ gives $A - B = -\mathrm{i}\omega_0 V_{z0}$. Solving these two equations for A and B gives

$$A = \frac{1}{2}\left(g - \mathrm{i}\omega_0 V_{z0}\right)\,, \quad B = \frac{1}{2}\left(g + \mathrm{i}\omega_0 V_{z0}\right)\,. \tag{3.9}$$

$$\begin{aligned} z(t) &= \left(\frac{g - i\omega_0 V_{z0}}{2\omega_0^2}\right) e^{i\omega_0 t} + \left(\frac{g + i\omega_0 V_{z0}}{2\omega_0^2}\right) e^{-i\omega_0 t} - \frac{g}{\omega_0^2} \\ &= \frac{g}{2\omega_0^2}\left(e^{i\omega_0 t} + e^{-i\omega_0 t}\right) - \frac{i\omega_0 V_{z0}}{2\omega_0^2}\left(e^{i\omega_0 t} + e^{-i\omega_0 t}\right) - \frac{g}{\omega_0^2} \\ &= \frac{g}{\omega_0^2}\cos\omega_0 t + \frac{V_{z0}}{\omega_0}\sin\omega_0 t - \frac{g}{\omega_0^2} . \end{aligned} \tag{3.10}$$

The collision process is described by the solution (3.10). The collision process starts, say, at $t = 0$ with $z = 0$. As t increases z increases and reaches a maximum $|z_{\max}|$ at $t = t_{\max}$. Then, from $t_{\max}$, $|z|$ decreases and becomes zero at $t = t_c = 2\pi/\omega_0$. At this time the stone completely leaves the water surface.

It is easy to find an expression for the maximum depth z, $z_{\max}$. It is found as (see the exercise 3.4 at the end of this chapter)

$$|z_{\max}| = \frac{g}{\omega_0^2}\left(1 + \sqrt{1 + \frac{\omega_0^2 V_{z0}^2}{g^2}}\right) . \tag{3.11}$$

3.3 Maximum Number of Skips and the Bounce-Off Condition

Consider a stone skipping with N skips. That is, there are N collisions with water. Note that the initial x-component of $\mathbf{V}$ at the ith collision ($i > 1$) is the final x-component of $\mathbf{V}$ at the end of $(i-1)$th collision with the water. An expression for N can be obtained by calculating the energy loss in collisions.

The loss in kinetic energy $\mathrm{KE_{Loss}}$ in one collision is given by

$$\mathrm{KE_{Loss}} = \frac{1}{2} m V_{x\mathrm{f}}^2 - \frac{1}{2} m V_{x0}^2 , \tag{3.12}$$

where V_{x0} and $V_{x\mathrm{f}}$ are the x-component of velocities before and after collision, respectively. $\mathrm{KE_{Loss}}$ can be expressed in terms of reactive force. For this purpose, multiply the Eq. (3.3a) by V_x and integrate from $t = 0$ to $t = t_c = 2\pi/\omega_0$, collision time:

$$\int_0^{t_c} m V_x \frac{\mathrm{d}V_x}{\mathrm{d}t} \mathrm{d}t = -\int_0^{t_c} V_x(t) F_x(t) \mathrm{d}t , \tag{3.13}$$

where

$$F_x = \frac{1}{2}\rho V_{x0}^2 S_{\mathrm{i}} \bar{C} , \quad \bar{C} = C_{\mathrm{L}} \sin\alpha + C_{\mathrm{D}} \cos\alpha . \tag{3.14}$$

The result is

$$\frac{1}{2}mV_{x\mathrm{f}}^2 - \frac{1}{2}mV_{x0}^2 = -\int_0^{t_c} V_x F_x \mathrm{d}t\,. \tag{3.15}$$

An approximate estimation of the integral in the right-side of the above equation is

$$\int_0^{tc} V_x(t)F_x(t)\mathrm{d}t \approx V_{x0}\int_0^{tc} F_x(t)\mathrm{d}t \approx V_{x0}\int_0^{tc} \mu F_z \mathrm{d}t\,, \tag{3.16}$$

where $\mu = \bar{C}/C$.

It is reasonable to assume that the average value of F_z during the collision process is mg. Then, Eq. (3.15) becomes

$$\begin{aligned} \frac{1}{2}mV_{x\mathrm{f}}^2 - \frac{1}{2}mV_{x0}^2 &= -V_{x0}\int_0^{t_c} \mu mg\mathrm{d}t \\ &= -\mu mgV_{x0}\frac{2\pi}{\omega_0} \\ &= -2\pi\mu mg\sqrt{\frac{2m\sin\alpha}{C\rho a}}\,. \end{aligned} \tag{3.17}$$

Define

$$l = 2\pi\sqrt{\frac{2m\sin\alpha}{C\rho a}}\,. \tag{3.18}$$

Then,

$$\frac{1}{2}mV_{x\mathrm{f}}^2 - \frac{1}{2}mV_{x0}^2 = -\mu mgl\,. \tag{3.19}$$

The energy losses in first, second, ..., Nth collisions are given by

$$\frac{1}{2}mV_{x1}^2 - \frac{1}{2}mV_{x0}^2 = -\mu mgl\,, \tag{3.20a}$$

$$\frac{1}{2}mV_{x2}^2 - \frac{1}{2}mV_{x1}^2 = -\mu mgl\,, \tag{3.20b}$$

$$\vdots$$

$$\frac{1}{2}mV_{xN}^2 - \frac{1}{2}mV_{x(N-1)}^2 = -\mu mgl\,. \tag{3.20n}$$

In each of the subequations μmgl is the dissipated energy (energy loss). This energy is independent of the velocity of the stone. For the stone to bounce-off in the ith collision $\frac{1}{2}mV_{xi}^2 > \mu mgl$. This also indicates that there must be a threshold minimum velocity for V_{xi} for the stone to bounce-off from the water surface.

Adding of all the subequations in (3.20) gives the total loss in the kinetic energy as

$$\frac{1}{2}mV_{xN}^2 - \frac{1}{2}mV_{x0}^2 = -N\mu mgl\,. \tag{3.21}$$

At the final collision, $N = N_{\rm f}$, we have $V_{xN_{\rm f}} = 0$. Now, Eq. (3.21) gives

$$N_{\rm f} = V_{x0}^2/(2\mu gl)\,. \tag{3.22}$$

Actually, $N_{\rm f}$ is an integer part of the quantity in the right-side of the above equation. From Eq. (3.19) we note that if the initial kinetic energy $\frac{1}{2}mV_{x0}^2$ is less than the loss of energy μmgl, then the stone will not bounce-off. Therefore, the *bounce-off condition* is

$$V_{x0} > V_{\rm c} = \sqrt{2\mu gl}\,. \tag{3.23}$$

For $m = 0.1\,\text{kg}$, $a = 0.1\,\text{m}$, $C_{\rm L} \approx C_{\rm D} \approx 1$, $\rho = 1000\,\text{kgm}^{-3}$ and $\alpha \approx \beta \approx 15°$ we find $\mu = 1.732$, $l = 0.17\,\text{m}$ and $V_{\rm c} = 2.4\,\text{ms}^{-1}$. Then, for $V_{x0} = 5\,\text{ms}^{-1}$ we have $N_{\rm f} = 4$ while for $V_{x0} = 10\,\text{ms}^{-1}$ we find $N_{\rm f} = 17$.

Steve Humble [8] conducted experiments to test the theoretical predictions. The experimental results very closely agreed with the theoretically predicted $N_{\rm f}$ and the distance between the successive collisions.

Solved Problem 1:

Calculate the distance between two successive collisions.

Set $(x, z) = (0, 0)$ at the Nth collision and $(x, z) = (\Delta X_N, 0)$ at the $(N+1)$th collision. V_{xN} and V_{zN} are the x- and z-components of velocities, respectively, of the stone, just after the bounce-off from the water after Nth collision. Between two successive collisions when the stone is in air its motion is simply projectile motion. Therefore, we write

$$x = V_{xN}t, \quad z = -\frac{1}{2}gt^2 + V_{zN}t = t\left(-\frac{1}{2}gt + V_{zN}\right). \tag{3.24}$$

At the $(N+1)$th collision $z = 0$. This happens at $t = t' = 2V_{zN}/g$. At this value of t

$$x(t') = \Delta X_N = 2V_{xN}V_{zN}/g\,. \tag{3.25}$$

We have $\tan\beta = V_{z0}/V_{x0} = V_{zN}/V_{xN}$. Then,

$$\Delta X_N = \frac{2V_{xN}^2}{g}\frac{V_{z0}}{V_{x0}}\,. \tag{3.26}$$

From Eqs. (3.21) and (3.22)

$$V_{xN}^2 = V_{x0}^2 - 2N\mu gl = 2N_{\rm f}\mu gl - 2N\mu gl = 2\mu gl(N_{\rm f} - N)\,. \tag{3.27}$$

Now,

$$\Delta X_N = 4\mu l(N_{\rm f} - N)\frac{V_{z0}}{V_{x0}}\,. \tag{3.28}$$

When $N = 0$ the above equation gives

$$\Delta X_0 = 4\mu l N_{\rm f}\frac{V_{z0}}{V_{x0}}\,. \tag{3.29}$$

Using this in Eq. (3.28) for V_{z0}/V_{x0} gives

$$\Delta X_N = \Delta X_0\left(1 - \frac{N}{N_{\rm f}}\right). \tag{3.30}$$

Note that the distance between successive bounces is calculated using standard projectile motion and energy loss equations.

Suppose assume that the stone rebounds elastically in the z-direction. Then, V_z does not depend on the number of collisions. In this case

$$\Delta X_N = \frac{2V_{xN}V_{z0}}{g} \tag{3.31}$$

and

$$\Delta X_0 = \frac{2}{g}V_{x0}V_{z0}\,. \tag{3.32}$$

As $V_{x0}^2 = 2\mu g l N_{\rm f}$ (from Eq. (3.22)) we have

$$\frac{V_{xN}}{V_{x0}} = \sqrt{1 - \frac{N}{N_{\rm f}}}\,. \tag{3.33}$$

Finally,

$$\Delta X_N = \Delta X_0\sqrt{1 - \frac{N}{N_{\rm f}}}\,. \tag{3.34}$$

3.4 Trajectory of the Skipping Stone

In this section we compute the trajectory of the skipping stone. When the stone is in the air, its motion is governed by the Eqs. (3.24). Its motion when it is in contact with water, is described by the Eqs. (3.3). The stone will be in contact with water over the time period $t_{\rm c} = 2\pi/\omega_0$ as determined in Sec.3.2.3 in the present chapter. This time interval and the horizontal and vertical distances during this time interval are very small compared to the time duration and the horizontal and vertical distances the stone is in the air between successive collisions. Therefore, for simplicity we focus on the trajectory of the stone while it is in the air.

Let us choose the values of the parameters as $m = 0.1\,\mathrm{kg}$, $a = 0.1\,\mathrm{m}$, $C_{\mathrm{L}} \approx C_{\mathrm{D}} \approx 1$, $\rho = 1000\,\mathrm{kgm^{-3}}$ and $\alpha \approx \beta \approx 15^{\circ}$. The value of V_{c} is $2.4\,\mathrm{ms^{-1}}$. We choose V_{x0} as the control parameter. Suppose a stone is thrown at $t = 0$ from the position $(x, z) = (0, 0)$ with $V_x(0) = V_{x0}$ and $V_z(0) = V_{z0}$. x and z increase with time and at time $t' = V_{z0}/g$ the vertical height becomes maximum. We have $z(t') = -\frac{1}{2}gt'^2 + V_{z0}t'$, $V_z(t') = 0$, $x(t') = V_{x0}t'$ and $V_x(t') = V_{x0}$. We choose this set of data as the initial condition and redefine this time as 0 and $x = 0$.

The vertical and horizontal velocities at the Nth bounce-off are, from Eq. (3.21), $V_{xN} = \sqrt{V_{x0}^2 - N\mu g l}$ and $V_{zN} = V_{xN}\tan\beta$. At the start of Nth bounce-off the time is set at $t = 0$, $x = 0$ and $z = 0$. The $(N+1)$th collision occurs when $z = 0$ and the corresponding time from Eq. (3.24) is $t' = 2V_{zN}/g$. The values of x and z during the time interval $t = 0$ to $t = t'$ can be determined from the Eq. (3.24). For plotting the trajectory, we can suitably redefine the horizontal and vertical distances and the time.

Figure 3.4 shows the trajectory of a skipping stone for three fixed values of V_{x0}. For $V_{x0} = 2\,\mathrm{m/s} < V_{\mathrm{c}}$ the stone sinks without a single bounce (Fig. 3.4a). In Fig. 3.4b where $V_{x0} = 5\,\mathrm{ms^{-1}} > V_{\mathrm{c}} = 2.4\,\mathrm{ms^{-1}}$, we notice 4 bounces of the stone. The horizontal distance ΔX_1 between first and second collision is also marked. For $V_{x0} = 10\,\mathrm{m/s}$ there are 17 bounces of the stone. The maximum vertical height and the horizontal distance between the bounces decrease with increase in the number of collisions. Figure 3.5 shows N versus ΔX_N for $V_{x0} = 10\,\mathrm{ms^{-1}}$ and V_{x0} versus N.

3.5 The Need to Spin the Stone and the Magic Value of α

In the theoretical treatment we assumed that the angle α is a constant. From $C = C_{\mathrm{L}}\cos\alpha - C_{\mathrm{D}}\sin\alpha$ we note that the force constant C decreases with increase in the value of α. After the first collision, the stone may rotate around the z-axis and hence the rate of change of α is not zero. Consequently, in the second collision there will be only a little chance for α to be in a favorable range for coming out of the water surface. After two or three bounces, the stone will sink completely. Therefore, angular motion has to be stabilized to realize almost a constant α. This is achieved through the spinning of the stone while throwing. The aim is to spin the stone with a sufficient initial rotational angular velocity $\dot{\phi}_0$, so that after the collision the change in α denoted by $\delta\alpha$ is negligible, $\delta\alpha \ll 1$.

A detailed analysis predicted that for $\delta\alpha \ll 1$ the requirement is [3]

$$\dot{\phi}_0 \gg \sqrt{g/R}\,, \tag{3.35}$$

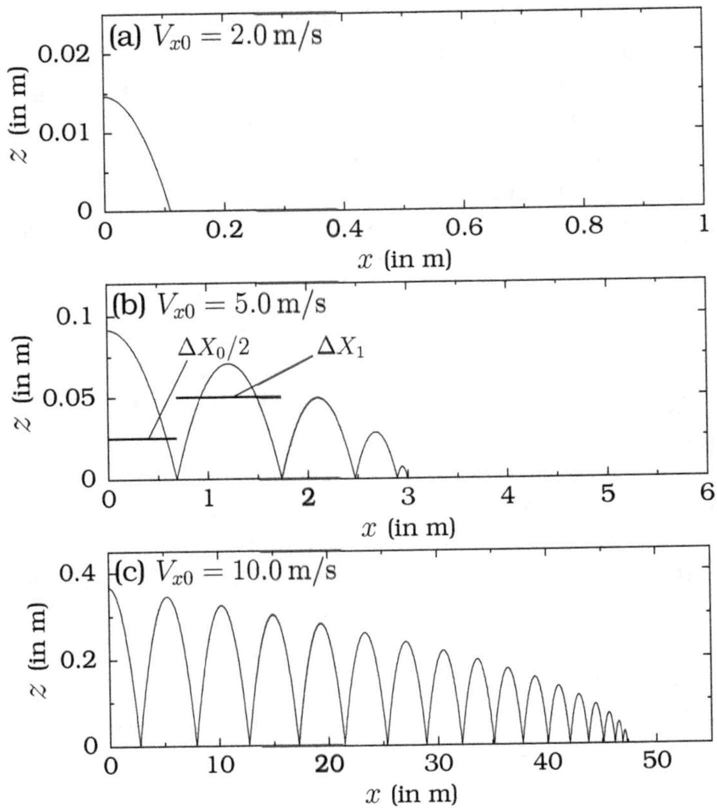

Fig. 3.4: Trajectory of a skipping stone for three values of V_{x0}.

where R is the radius of the stone. The above condition can be easily fulfilled in practice. In view of the above, due to the destabilization arising from the perturbed rotational motion in each collision, the stone will stop bouncing after $N_{\rm f}^{\rm spin}$. It has been found theoretically that [3]

$$N_{\rm f}^{\rm spin} \approx \frac{R\dot{\phi}_0^2}{g}\,. \tag{3.36}$$

That is, even if $N_{\rm f}$ given by the Eq. (3.22) is $\gg 1$ as a result of very high V_{x0} the stone will be stopped roughly after $N_{\rm f}^{\rm spin}$ bounces due to rotational destabilization.

Bocquet and his collaborators [5,6] performed a series of experiments on stone skipping by varying the parameters α, β and the translational velocity for a wide range. The analysis in these parameters space has indicated that

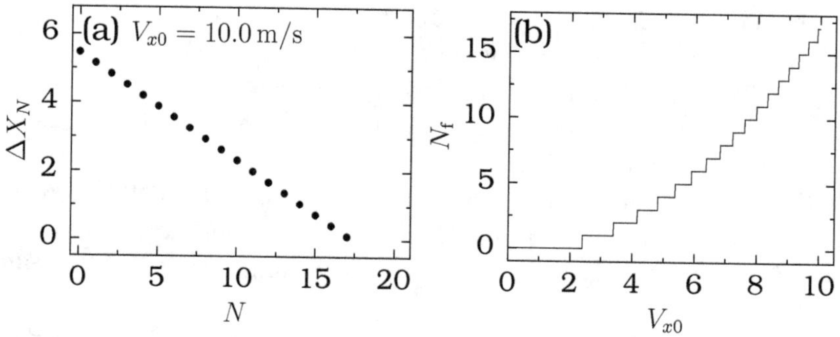

Fig. 3.5: (a) N versus ΔX_N with $V_{x0} = 10\,\mathrm{m/s}$. (b) V_{x0} versus N_f.

the lowest velocity for rebounding is minimum for $\alpha \approx 20^\circ$ and this value of α is termed as the *magic attacking angle*. This value of α is the optimal value for achieving possible maximum number of bounce.

3.6 Conclusion

The analysis of stone skipping does not require advanced mathematics and details of fluid flow around the stone during collision. Thus, the prediction of parameters for championship stone throws in competition is possible. The important outcomes of the analysis of the stone skipping are:

- The crucial factors influencing the bouncing and sinking of a stone are the initial velocity V_{x0} and the angle of attack and initial rotational angular velocity $\dot{\phi}_0$.
- The stone would bounce-off from the water surface only if its initial velocity is above a critical velocity given by Eq. (3.23).
- In each collision, the stone bounces-off after the time interval $2\pi/\omega_0$ where $\omega_0 = \sqrt{\rho V_{x0}^2 Ca/(2m \sin\alpha)}$.
- The attacking angle α is crucial for stone skipping. The angle α about 20° is optimal value for achieving possible maximum number of bounces.

3.7 Bibliography

[1] https://www.guinessworldrecords.com/world-records/most-skips-of-a-skimming-stone.

[2] https://thekidsshouldseethis.com/post/the-2014-stone-skipping-world-record-kurt-steiner-with-88-skips.

[3] L. Bocquet, Am. J. Phys. **71**, 150 (2003).
[4] H. Richard Crane, The Phys. Teach. **26**, 300 (1988).
[5] C. Clanet, F. Hersen and L. Bocquet, Nature **427**, 29 (2004).
[6] L.R. Rosellini, F. Hersen, C. Clanet and L. Bocquet, J. Fluid Mech. **543**, 137 (2005).
[7] S. Nagahiro and Y. Hayakawa, Phys. Rev. Lett. **94**, 174501 (2005).
[8] S. Humble, Teaching Mathematics and Applications **26**, 95 (2007).
[9] I.J. Hewitt, N.J. Balmforth and J.N. McElwaine, J. Fluid Mech. **669**, 328 (2011).
[10] T. Truscott, J. Belden and R. Hurd, Phys. Today **70** (December, 2014).
[11] C.F. Babbs, The Phys. Teach. **57**, 278 (2019).
[12] http://www.skippingstonesdesign.com/lifestyle/what-are-skipping-stones-a-quick-guide-to-skipping-stones/.

3.8 Exercises

3.1 Given the force **F**, Eq. (3.1), with **n** and **t** being its components as shown in Fig. 3.2 find the x- and z-components of **n** and **t**.

3.2 From Eq. (3.1) obtain the equations of motion (3.2).

3.3 Taking into account of energy dissipation Eq. (3.3b) can be modified into

$$m\frac{\mathrm{d}V_z}{\mathrm{d}t} = -mg + F^{(0)}(z) - \zeta(z)V_z\,,$$

where ζ plays the role of an effective friction coefficient. Analyse the effect of the collision process on the z-component of the velocity [6].

3.4 Show that $|z_{\mathrm{max}}|$ is given by Eq. (3.11).

3.5 In a stone skipping experiment, it was found that the skip distances for first and second collisions were found to be 36.9 cm and 30.4 cm, respectively. Find the maximum number of collisions the stone would have made before sinking.

Chapter 4

Bouncing Popper

Drop a rubber ball from different heights with zero initial speed. The ball will never cross its initial height. *Why is it so?* The law of conservation of energy is that energy is neither created nor destroyed. However, it can be transferred from one form to another and from one object to another. The energy of the ball is stored as its position is above the ground. Due to the force of gravity the ball falls down. Part of the energy initially went into the system is converted as sound energy and heat energy. Since the energy comes out of a system is less than the energy that goes in, then the ball cannot bounce higher than the initial height. If there is no friction with the air and the ground surface, then a perfectly elastic ball would bounce back exactly to its initial drop point.

In a high school class, a physics teacher cut a racquetball (a hollow rubber ball) into two equal halves as shown in Fig. 4.1a and shaved a little. The two pieces became little less than the half of the ball. First, one piece of it was dropped normally. It bounced lower than the initial drop point. Nothing exciting happened. The dropped half-ball collided with the floor.

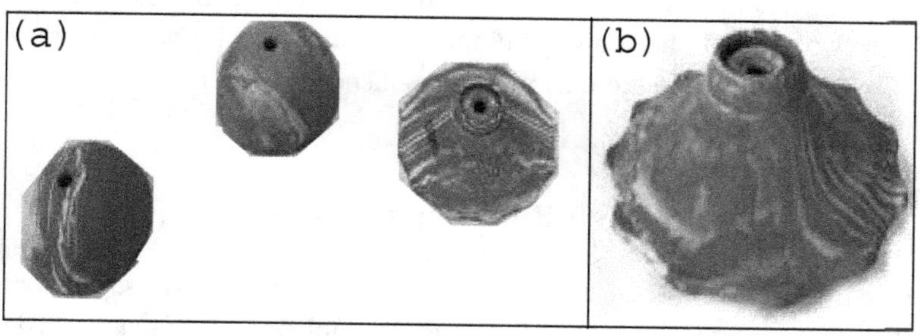

Fig. 4.1: (a) Two half-balls and a half-ball with its centre turned inside out. (b) The right-most one is an enlarged version of the third one.

Fig. 4.2: Bouncing of a popper and a ping-pong ball kept in the bowl of it.

Most of its energy was transferred to the floor, lost to heat and lost to vibrations. Only a fraction of its energy was converted into kinetic energy. Next, the teacher did something in the second half-ball and dropped it gently. Surprisingly, it bounced almost twice the initial drop height. *What was the trick the teacher used?* As the ball is very flexible, before dropping the second half-ball, the teacher by firmly pressing its centre using two thumbs, flipped inside out which created a bulge in its centre. The ball is said to be *loaded*. It was then dropped by keeping the bulged face upward and the flat side facing downward (see Fig. 4.1b). This fascinating half-ball toy is called a *bouncing popper* or a *hopper popper*.

Bouncing poppers are commercially available and are inexpensive. A bouncing popper provides interesting physics insights in the field of mechanics and can be used in a classroom teaching. Students and children enjoy the action of the poppers. There is another interesting fun activity with a popper [2]. Turn the popper inside out. Place a ping-pong ball in the bowl of the popper and keep the bulged part facing downward. Drop it onto a hard surface and ensure that the ping-pong ball is inside the bowl and is above the popper. When the popper hits the surface, the ping-pong ball will fly to an impressive height. This is illustrated in Fig. 4.2.

Here, we present the physics behind the bouncing of a popper. Furthermore, we describe the calculation of time needed for a popper to rise to a maximum height.

4.1 Explanation of High Bouncing of a Popper

When a half-ball is turned inside out, the energy used to do so is stored in it. This kind of stored energy due to the deformation of an elastic object is called *elastic potential energy*. Instead of dropping a popper in its inverted inside out, we can place it on a table or on a floor and observe its jumping to a height. When a popper is in its inverted inside out position on a floor, it is in its unstable position. It will stay for a few seconds in the unstable position if left unperturbed. The duration of time the popper stays in its inverted unstable position can be increased by introducing a hole in the centre [3]. The rubber around the hole radially stretches allowing the rubber to relax. This increases the time the popper stays in the inverted state. It then restores itself. In the case of the dropped or thrown towards the ground when it collides with the ground, it is popped right-side out (that is, returned to its original form) and the elastic potential energy is converted into kinetic energy [1]. In both cases during the restoration time, it exerts a force on to the surface of the floor [3]. This induces a very high acceleration, a reaction force, in the upward direction. This propels the ball to a height higher than the initial drop point and then falls back to the surface.

The initial height from which the popper to be dropped should not be too high. The popper should reach the floor before the restoration begins. *When does the force applied to the surface by the popper begin?*

4.2 Calculation of Initial Speed and the Time to Rise to a Maximum Height

Interestingly, the measurement of average maximum height of the popper allows us to derive many useful information [4].

We start to calculate the initial speed of the popper when it begins to jump from the surface. When the popper is falling towards the floor or rising up from the floor, that is when it is in air, it is a free falling. The only force acting on it is the gravitational force. Denote m, g, h, v_{i} and v_{f} as the mass of the popper, acceleration due to gravity ($9.8\,\mathrm{m/s^2}$), the maximum height of the popper, its initial speed when it begins to leave the surface and the final speed at the maximum height (which is obviously zero), respectively.

If we neglect the energy loss due to the air resistance, the total energy of the popper at the time of leaving the surface is equal to its total energy at the maximum height. This gives

$$\frac{1}{2}mv_{\mathrm{i}}^2 + 0 = \frac{1}{2}mv_{\mathrm{f}}^2 + mgh\,. \tag{4.1}$$

Suppose $m = 20\,\mathrm{g}$ and $h = 2.5\,\mathrm{m}$. Then, with $v_{\mathrm{f}} = 0$ the above equation gives

$$v_{\mathrm{i}}^2 = 2gh = 2 \times 9.8\,\mathrm{ms^{-2}} \times 2.5\,\mathrm{m} = 49\,\mathrm{m^2s^{-2}} \tag{4.2}$$

or

$$v_i = 7\,\mathrm{m/s}\,. \tag{4.3}$$

The initial kinetic energy of the popper is

$$E_{\mathrm{kin,i}} = \frac{1}{2}mv_i^2 = 0.49\,\mathrm{J}\,. \tag{4.4}$$

To determine the time t_h for the popper to rise to its maximum height h, we approximate $v(0+t_h)$ as $v(0)+t_h v'(0)$ (the truncated Taylor series approximation). Here $t = 0$ is the time at which the popper leaves the surface. Since $v(0) = v_i$, $v(0+t_h) = v_f = 0$ and $v'(0) = -g$ we write [4]

$$t_h = -\frac{v_f - v_i}{g} = -\frac{(0 - 7\,\mathrm{ms}^{-1})}{9.8\,\mathrm{ms}^{-2}} = 0.71\,\mathrm{s}\,. \tag{4.5}$$

Lapp has reported a surprising result on the magnitude of acceleration of the popper during the pop, that is during the restoration of the initial shape [4]. For a popper with radius 1.25 cm and the speed at the beginning of jumping as $7\,\mathrm{ms}^{-1}$, the acceleration of the popper is found as $1960\,\mathrm{ms}^{-2}$ (see the exercise 4.5 at the end of this chapter).

Solved Problem 1:

A popper of mass 20 g is turned inside out by giving an energy of 0.5 J. A ping-pong ball of mass 2 g is placed in the bowl of the popper. By keeping the bulged part facing downward, the popper is dropped on a hard surface from a height of 2 m, by keeping the ping-pong ball inside the bowl. If 10% of the energy is transferred to the ping-pong ball when the popper falls on the hard surface, calculate what will be the height to which the ping-pong ball will fly. Assume that the frictional losses are negligible.

The elastic potential energy is 0.5 J. As the popper with the ping-pong ball is dropped from a height 2 m,

$$\begin{aligned}\text{the gravitational potential energy} &= mgh \\ &= (20+2)\times 10^{-3}\times 9.8\times 2 \\ &= 0.4312\,\mathrm{J}\,.\end{aligned} \tag{4.6}$$

Then,

$$\text{total energy} = (0.5 + 0.4312)\,\mathrm{J} = 0.9312\,\mathrm{J}\,. \tag{4.7}$$

10% of this energy is given to the ping-pong ball. If h is the height to which the ball fly, then from $mgh = 0.1 \times 0.9312$ J we find

$$h = \frac{0.09312}{2\times 10^{-3}\times 9.8} = 4.7510\,\mathrm{m}. \tag{4.8}$$

4.3 Conclusion

Simple experiments with a bouncing popper can be done to explain many physical concepts to school students. The law of conservation of energy is a fundamental law of physics and it can be explained using the bouncing popper. The students can be taught by deforming an elastic object, energy can be given to that elastic body and that energy is called *elastic potential energy*. They can also be taught that this elastic potential energy is converted into kinetic energy when the popper hits the ground. By observing different rebound heights of the popper dropped from the same height, the losses due to friction can be explained. As discussed in [4], the students can be trained in using various kinematic equations to analyse motion using just the maximum height to which the popper jumps. Many advanced experiments with bouncing poppers can be done using a high-speed video recording instrument as described in [3].

4.4 Bibliography

[1] https://phun.physics.virginia.edu/demos/hopper.html.
[2] http://blog.teachersource.com/2009/12/30/teaching-energy-using-dropper-poppers/.
[3] M. Vollmer and K.P. Mollmann, The Phys. Teach. **53**, 489 (2015).
[4] D.R. Lapp, Phys. Edu. **43**, 492 (2008).

4.5 Exercises

4.1 The potential energy of a bouncing popper with mass 20 g is 0.31 J when it is on a floor before jumping. Calculate the maximum height it can jump.

4.2 Assume that a bouncing popper is a compressed linear spring with an average spring constant $k = 2500\,\text{N/m}$ and its centre part moves up around 1.25 cm. Calculate the stored potential energy. Next, assume the mass of the bouncing popper as 20 g. Neglecting frictional losses and drag find
(a) the initial velocity with which the popper will rebound,
(b) the acceleration of the popper if the time taken to flip is 5 msec and
(c) the effective length during which the force acts while bulging out.

4.3 A bouncing popper with mass 20 g jumped to an average height 2.5 m. Estimate the potential energy it gained in the field of gravity.

4.4 Place a popper with the bulge facing above. Note down the approximate height to which it bounced. Next, keep it on a finger tip. Note the approximate height it jumped. Which case produced a higher height? Why?

4.5 A bouncing popper restores its initial shape when it is on a surface. In this process, it exerts a force against the surface which acts over a distance that is approximately equal to its radius r. Assume that $r = 1.25\,\text{cm}$, the speed at the beginning of restoration is $0\,\text{m/s}$ and the speed at the time of jumping away from the surface is $7\,\text{m/s}$. Calculate
(a) the acceleration a of the popper over the distance $h = 1.25\,\text{cm}$,
(b) the time taken to reach the maximum height,
(c) the magnitude of the force exerted and the work done by this force for the distance $h = 1.25\,\text{cm}$.

Part VI: Heat and Thermodynamic Toys

Chapter 1

Drinking Bird

One of the most successful scientific toys of all times is the drinking bird (Fig. 1.1). It is also called as *dunking bird*, *dippy bird* and *dunking duck*. The drinking bird toy was invented by Miles V. Sullivan, an inventor-scientist at Bell Labs., Murray Hill, New Jersey, USA, in 1945 and patented in the next year. It is a thermodynamic device demonstrating liquid-vapour equilibrium and evaporation [1,2] and using this work can be done from a temperature difference. *The toy exhibits oscillation without a direct supply of energy*. Essentially, it combines engineering, classical mechanics and thermodynamics. The working of this toy makes use of the capillary effect, vapour pressure and latent heat of vapourization.

The capillary effect refers to the ability of a liquid to flow in narrow spaces without applying any force or even in opposition to external forces like gravity. Since this effect causes a liquid to rise in a capillary tube, it is referred in general as capillary effect. This effect causes the drawing-up of liquid in porous materials like paper and cloth. Also, the plants draw up water from soil to branches only due to this capillary effect. In some liquids like mercury, the liquid level will go down in a capillary tube. Whether a liquid would ascend or descend on a narrow tube depends on the *cohesion force* and *adhesion force*. Cohesion forces are the attractive forces that bind the molecules of the liquid together. It is this force that gives rise to surface tension which forces the liquid to have a minimum surface. Adhesion forces are the attractive forces between dissimilar atoms. A liquid will rise only of the adhesion forces between its molecules.

The vapour pressure of a liquid is the equilibrium pressure of a vapour above its liquid (or solid) in a closed container. The evaporation of a liquid or solid gives to this pressure. The vapour pressure depends on

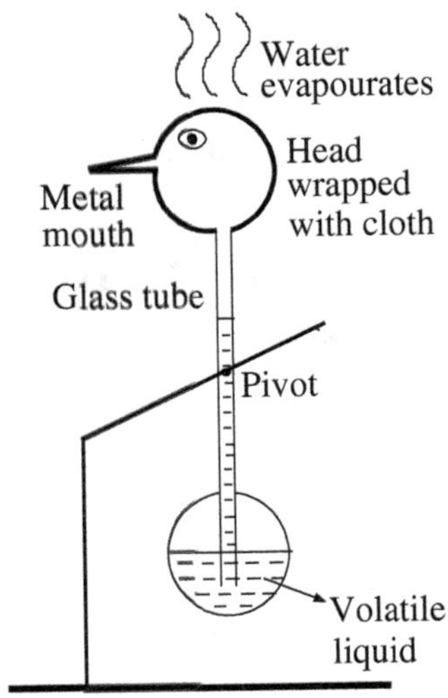

Fig. 1.1: The structure of a typical drinking bird toy.

the temperature. As the increase of temperature leads to increase of evaporation, the vapour pressure increases with temperature.

The input energy required to change the state from liquid to vapour at constant temperature is called the *latent heat of vapourization*. If the latent heat for evaporation is taken from the liquid, then the liquid will cool down as its thermal energy decreases as seen in the cooling of water in mud pot.

Various scientific aspects of the drinking bird have been reported in [1-13]. It has been proposed as a device for pumping water in Egypt [14], as well as to produce work by raising a small mass [9].

In this chapter, first we describe the drinking bird toy. Next, we present the various stages of its working and obtain an expression for the period of oscillation of the bird. Then, we consider the sunbird where heating the bottom bulb by sun light is used in place of evaporation of the external liquid.

1.1 Description of the Drinking Bird Setup

A typical drinking bird shown in Fig. 1.1 consists of the following. The toy has two hollow glass bulbs connected by a long glass tube that extend almost entirely into the lower bulb (tail part). The upper bulb is like a head and it has a narrow metal peak. There is a cotton string inside the peak and is connected with the water absorbing cloth covering the head. In the bottom bulb some easily evapourating liquid is placed so that the connecting glass is filled with vapour. The bottom of the long tube is submerged in the liquid of the bottom bulb. The bird has two plastic legs with a pivot connection which allows the bird to rotate its body freely.

The internal liquid is the critical part of the toy. The boiling point of the internal liquid must be around room temperature. Often methylene chloride is used. Alternate internal liquids are methylic alcohol, chloroform, ethyl acetate, ethylic alcohol, Freon 11(trichlorofluoromethene) and diethyl ether. In the construction of the toy, the internal air should be removed.

1.2 Working and Performance of the Toy

The typical working of the drinking bird toy consists of four steps [15].

1. Drinking process.
2. Head down to head up (upright) position.
3. Drying of external liquid in the upright position.
4. Dunking stroke - head to head down position.

Initially, the bird is upright and the internal liquid is fully in the tail part, that is in the bottom bulb. The internal liquid and the vapour in the body are in an equilibrium state. To initiate the motion of the bird, we have to dip its beak in the glass of water kept infront of it. Water is absorbed by the cotton string in the beak. The water moves to the cloth around the head due to capillarity effect. The cloth on the head becomes wet. As the water in the head evaporates the outside of the head cools. This is because faster molecules quickly escape, leaving the slower and cooler ones on the head as the head supplies the latent heat of vapourization to the escaping molecules. The vapour in the head gets condensed in small drops. The cooling of the outside of the head cools the vapour inside the head. Consequently, due to the ideal gas law, the reduction in temperature and condensation decreases the pressure in the head. The pressure in the tail (bottom bulb) is relatively higher than in the head. The higher pressure in the tail causes the internal liquid to move up in the tube to the cooled head. The centre of gravity (CoG) also rises. As the internal liquid enters the head the bird is *top-heavy*. Now, the CoG is above the hinges. Hence, it tips forward. When it is in a completely forward position, it drinks water kept in the glass in front of it and the bottom of the connecting tube is no longer submerged in liquid. This part of motion of the bird is called *dip*.

For a short time interval, the bird is in the dip (drinking) state. At this stage, the lower end of the tube is above the surface of the internal liquid. Some of the liquid vapour goes to the head from the body. Due to dipping, part of the liquid in the connecting glass tube drains back into the lower bulb. This makes the CoG to come below the pivot point. The bird now comes to its upright position. The internal liquid in the bottom bulb returns to the room temperature. A new cycle is then being started due to the continuous water evaporation from the head and the rise of the internal liquid. The oscillation of the bird continues as long as there is an external liquid in the head.

To force the liquid to move up the tube, the internal liquid should possess the property of yielding a large change in pressure (ΔP) for a small change in temperature. Near the room temperature we find

$$\begin{aligned} &\Delta P(\mathrm{CCl_3F}) > \Delta P(\mathrm{CF_3CHCl_2}) > \Delta P((\mathrm{C_2H_5})_2\mathrm{O}) \\ &\quad > \Delta P(\mathrm{CH_2Cl_2}) > \Delta P(\mathrm{CHCl_3}) > \Delta P(\mathrm{H_2O}) \,. \end{aligned} \tag{1.1}$$

The drinking bird is considered as a heat engine, since mechanical work is produced from heat differences.

1.3 Period of Oscillation

Our purpose now is to find an expression for the time period of oscillation of the bird [12].

With z being the level of the internal liquid in the tube with respect to the body's surface level, ρ_I being the density of the internal liquid and Δz being the change in z, the quantity ΔP in one cycle is given by

$$\Delta P = -\rho_\mathrm{I} g \Delta z \,. \tag{1.2}$$

Denote Δm_E is the mass of the external liquid outside the head that evaporated during one cycle (period). ΔT (a negative quantity) is the decrease in temperature in one cycle inside the head and Δh_E is the specific evaporation enthalpy of the liquid outside the head. Then, one can write

$$C\Delta T = -\Delta m_\mathrm{E} \Delta h_\mathrm{E} \,, \tag{1.3}$$

where C is an effective heat capacity of the head.

According to the Clausius–Clapeyron equation [16]

$$\Delta T = \frac{\Delta P}{B} \,, \quad B = \frac{\Delta h_\mathrm{I} P_\mathrm{I}(T_\mathrm{R})}{R T_\mathrm{R}^2} \,, \tag{1.4}$$

where ΔP is the decrease in pressure inside the head over one cycle, Δh_{I} is the molar vapourization enthalpy of the liquid (CH_2Cl_2) inside the body of the bird, R is the ideal gas constant and $P_{\mathrm{I}}(T_{\mathrm{R}})$ is the vapour pressure at room temperature T_{R}. $P_{\mathrm{I}}(T_{\mathrm{R}})$ is given by the solution of the Clausius–Clapeyron equation as [16]

$$P_{\mathrm{I}} = P_0 \exp\left[-\frac{\Delta h_{\mathrm{I}}}{R}\frac{(T_{\mathrm{I,b}} - T_{\mathrm{R}})}{T_{\mathrm{I,b}}T_{\mathrm{R}}}\right], \tag{1.5}$$

where the normal atmospheric pressure $P_0 = 1.013 \times 10^5\,\mathrm{Pa}$, the normal boiling point of the internal liquid $T_{\mathrm{I,b}} = 313.15\,\mathrm{K}$. Here $P_{\mathrm{I}}(T_{\mathrm{R}}) = 0.535 \times 10^5\,\mathrm{Pa}$. Moreover, τ is sensitive to the air humidity if the external liquid is water. Experimental results [12] predict

$$\tau = \kappa(100 - H)^{-\beta}, \tag{1.6}$$

where H is the relative humidity and κ and β are parameters.

From Eqs. (1.2) and Eq. (1.4) we write

$$\Delta T = -\frac{\rho_{\mathrm{I}} g \Delta z}{B}. \tag{1.7}$$

Assume that the evaporation rate $\dot{m}_{\mathrm{E}}$ of the external liquid (which is a negative quantity) is almost constant over a few cycles, then the period τ is given by

$$\tau = -\frac{\Delta m_{\mathrm{E}}}{\dot{m}_{\mathrm{E}}} = \frac{C\Delta T}{\dot{m}_{\mathrm{E}}\Delta h_{\mathrm{E}}}. \tag{1.8}$$

Solved Problem 1:

Setup the equation of motion for the small angle (θ) oscillation of the drinking bird and then obtain an expression for the time period of oscillation [17].

Consider the Fig. 1.2 depicting the parameters L, H, CoG (centre of gravity) and the reference point O at the beginning of a cycle. Just before the bird dip the quantities are denoted by primed quantities. Let $I_{\mathrm{H,0}}$ be the moment of inertia of the bird about the axis parallel to its axis of drinking and the axis passing through CoG.

The new moment of inertia denoted as $I_{\mathrm{H}}(x)$ about the new axis of rotation that is parallel to the normal axis of drinking defined at a distance x from the point O is

$$I_{\mathrm{H}}(x) = I_{\mathrm{H,0}} + M(L - x)^2, \tag{1.9}$$

where M is the mass of the bird. The bird starts to oscillate when it is released after displacing it through a small angle θ from the vertical.

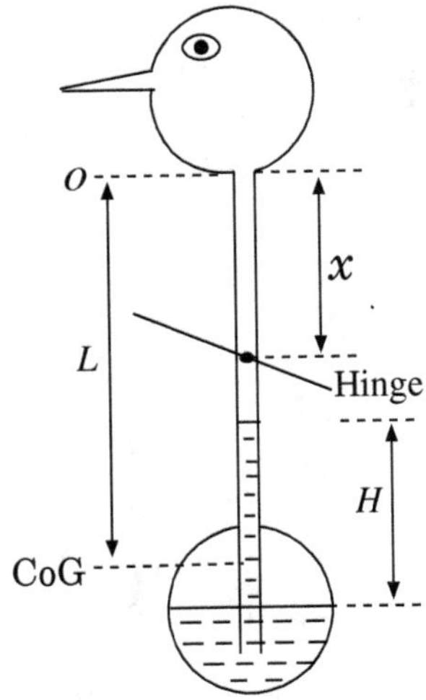

Fig. 1.2: A drinking bird toy.

The resulting force is $Mg(L-x)\sin\theta \approx Mg(L-x)\theta$, so that the equation of motion becomes

$$I_{\mathrm{H}}\ddot{\theta} = -Mg(L-x)\theta$$

or

$$\ddot{\theta} + \omega^2\theta = 0, \quad \omega^2 = \frac{Mg(L-x)}{I_{\mathrm{H}}}. \tag{1.10}$$

It is the equation of motion of the linear harmonic oscillator with period of oscillation τ given by

$$\tau = \frac{2\pi}{\omega} = 2\pi\sqrt{\frac{I_{\mathrm{H}}}{Mg(L-x)}} = 2\pi\sqrt{\frac{(I_{\mathrm{H},0}/M) + (L-x)^2}{g(L-x)}}. \tag{1.11}$$

Defining $I_{\mathrm{H},0} = MK_{\mathrm{H}}^2$ with the radius of gyration for an axis through the CoG of the bird as K_{H}. Then,

$$\tau = 2\pi\sqrt{\frac{K_{\mathrm{H}}^2 + (L-x)^2}{g(L-x)}}. \tag{1.12}$$

1.4 Sunbird

The motion of a usual drinking bird depends on air humidity. Further, the height of the toy is limited by the consideration of a natural cooling mechanism. Robert Mentzer [10] proposed the so-called *sunbird*. Guemez and his collaborators [18,19] analyzed both theoretically and experimentally its working.

In the sunbird toy, the process of evaporation of the external liquid is replaced by heating its bottom bulb and keeping its head at room temperature. The bottom bulb is black painted and is exposed to the sun or light from a bulb to increase the temperature. This increases the vapour pressure of the internal liquid thereby leading to the rise of the liquid level in the tube. The sunbird displays a motion similar to the drinking bird.

For a sunbird, whose black painted tail part is exposed to a light bulb kept at a distance d from it, it is possible to find an expression for the time period of oscillation [19]. Let V be the voltage applied to the bulb, I is the current, A is the area of the bird exposed to the light, ϵ is a factor related to the absorption characteristics of the body (for a dull black surface $\epsilon \approx 1$) and α is the power loss, namely, the fraction of electrical power converted into thermal radiation by the bulb. If U is the power absorbed by the tail part, then we can write

$$\dot{U} = \frac{\epsilon A \alpha I V}{f(d)}, \tag{1.13}$$

where f is a function of d. For a spherical bulb $f = 4\pi d^2$. The temperature variation $\mathrm{d}T$ is given by

$$\mathrm{d}T = \frac{\rho_\mathrm{I} g \mathrm{d}z}{B}. \tag{1.14}$$

In the above equation, z is the height of the liquid from the surface level in the body, ρ_I is the density of the liquid and B is a constant characteristic of the bird. Initially, the bird takes a few minutes for the first time to dip and then begins a periodic oscillation. Denote the period as τ while the initial time (the time elapsed before the 1st dip) as τ_0. We have

$$\frac{\mathrm{d}T}{\mathrm{d}t} = \frac{\dot{U}}{C} = \frac{\epsilon A \alpha I V}{C f(d)}, \tag{1.15}$$

where the effective heat capacity C is given by

$$C = m_\mathrm{I} c_\mathrm{I} + m_\mathrm{g}^\mathrm{B} c_\mathrm{g} + m_\mathrm{t} c_\mathrm{g}, \tag{1.16}$$

where c_g and c_I are the specific heats of the glass and the internal liquid, respectively, m_I, m_t and m_g^B are the masses of the internal liquid, tube and glass, respectively.

From Eqs. (1.14) and (1.15) we obtain

$$\mathrm{d}t = \frac{\rho_I g C f(d)}{\epsilon A B \alpha I V} \mathrm{d}z\,, \tag{1.17}$$

that is,

$$\tau_0 = \frac{\rho_I g C f(d) z_{\max}}{\epsilon A B \alpha I V}\,, \tag{1.18}$$

then,

$$\tau = \beta \tau_0, \quad \beta < 1. \tag{1.19}$$

For a detailed discussion on simulation results see [18].

1.5 Drinking Bird of Second Kind

Nadine Abraham and Peter Palffy-Muhoray [15] designed a drinking bird which displays motion that of the classical drinking bird but without temperature difference. Their device has a triangular wing made of sponge and there is no need for internal liquid. The bird is top heavy when the wing is dry. So, it leans forward and dips into a cup of water. The tip of the sponge got immersed in the water. The sponge absorbs water and the water level rises in the sponge due to capillary action. As a result of the presence of water, the weight of the sponge increases and the bird becomes bottom heavy. The bird then returns to its upright position. Due to the evaporation of water the wing dries, the bird becomes top heavy and learns forward again. The above cycle thus repeats. The period of oscillation of this bird is found to be very large compared to a typical drinking bird with internal liquid.

1.6 Conclusion

The drinking bird can be used as an educational tool as it can be used as a demonstration apparatus on liquid-vapour equilibrium and evaporation. The students can learn that evaporation can cool a system when its latent heat of evaporation is taken from the system. It can teach also that falling temperature reduces the pressure as well as. They can also learn how the reduction of temperature and pressure lead to condensation of vapour into liquid. Further, one can also understand that a drinking bird is a thermal engine because it can be used to produce work from a temperature difference. The students can do a project about the dependence of the period of oscillation of a drinking bird on the humidity using a digital hydrometer and a simple setup as given in [18]. They can also do projects on the analysis of the dependence of the period of oscillation with different external liquids. Furthermore, the sunbird can be used to demonstrate the inverse square-law dependence of thermal radiation.

1.7 Bibliography

[1] J.S. Miller, Am. J. Phys. **26**, 42 (1958).
[2] D.L. Frank, J. Chem. Edu. **50**, 211 (1973).
[3] J.L. Gaines, Am. J. Phys. **27**, 189 (1959).
[4] H.E. Stockman, Am. J. Phys. **29**, 335 (1961).
[5] H.E. Stockman, Am. J. Phys. **29**, 374 (1961).
[6] R. Plumb, J. Chem. Educ. **50**, 212 (1973).
[7] R. Plumb, J. Chem. Educ. **52**, 728 (1975).
[8] H. Bent and H. Teague, J. Coll. Sci. Teach. **8**, 18 (1978).
[9] C. Bauchhuber, Am. J. Phys. **51**, 259 (1983).
[10] R. Mentzer, The Phys. Teach. **31**, 126 (1993).
[11] D. Rathjen, Exploring **18**, 7 (1994).
[12] J. Guemez, R. Valiente, C. Fiolhais and M. Fiolhais, Am. J. Phys. **71**, 1257 (2003).
[13] R. Lorenz, Am. J. Phys. **74**, 677 (2006).
[14] R.B. Murrow, "A simple heat engine of possible utility in primitive environments", August 1966, RAND Corporation.
[15] N. Abraham and P. Palffy-Muhoray, Am. J. Phys. **72**, 782 (2004).
[16] M.W. Zemansky and R.H. Dittman, *Heat and Thermodynamics* (McGraw-Hill, Singapore, 1997).
[17] L.M. Ng and Y.S. Ng, Phys. Edu. **28**, 320 (1993).
[18] J. Guemez, R. Valiente, C. Fiolhais and M. Fiolhais, Phys. Teach. **42**, 307 (2004).
[19] J. Guemez, R. Valiente, C. Fiolhais and M. Fiolhais, Am. J. Phys. **71**, 1264 (2003).

1.8 Exercises

1.1 a) Describe what will happen when you touch the bottom bulb of the bird when it will be in the upright position.

b) List the various physical principles involved in the action of the drinking bird.

c) What is the source of energy of the drinking bird?

1.2 Obtain the expressions for the moment of inertia and torque of the drinking bird when the internal liquid inside the tube reaches the height z [12].

1.3 It is found for a drinking bird using CH_4O as an external liquid, that the evaporation rate initially to be -5.91×10^{-4}g/sec and its period of oscillation to be 1.5 sec. After some time, the evaporation rate reduces to -0.96×10^{-4}g/sec. What will be its period of oscillation at that time?

1.4 A drinking bird operating with CH_3Cl as the external liquid has a period 2 sec when its rate of evaporation is -21.6×10^{-4}g/sec. What will be its period when it operates with C_2H_6O as external liquid when its evaporation rate is -4.4×10^{-4}g/sec. The specific evaporation enthalpies of CH_3Cl and C_2H_6O are, respectively, 247.02 and 841.55.

1.5 In a study of the variation of the oscillating frequency of a drinking bird with water as external liquid, it was found that the oscillating period for a humidity of 94% to be 250 sec and the period for a humidity of 86% to be 90 sec. Find the values of the parameters β and κ.

Chapter 2

Putt-Putt Boat

For more than hundred years, everyone from children to physicists have been fascinated by the putt-putt toy boat (Fig. 2.1). Heat from a candle in the boat fires jets of water from the pipes at its rear side, thereby the boat is pushed forward and then water is drawn from the pipes for the next cycle. The boat is also called *phutt-phutt, pop-pop, pouet-pouet, toc-toc, puf-puf* boat [1] and *pulsating water engine.*

The toy boat was introduced by the French Thomas Piot in 1891. In 1933 J.G. Baker made a discussion of the boat [2]. Later, Stuart Mackay [3] described the operation of the boat, and Finnie and Curl [4] examined the mechanism of propulsion of the boat. Patents have been filed for the designs of the boat [5-9].

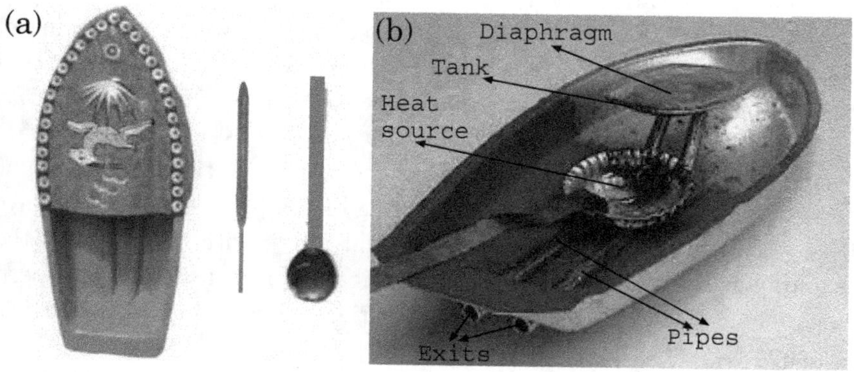

Fig. 2.1: (a) A putt-putt boat with an ink filler and a candle (heat source). (b) A putt-putt boat with the top cover removed showing the arrangement of the tank, diaphragm, a heat source and two pipes and their exits.

Commercially available putt-putt toy boat is made of tinplate. Figure 2.1b shows the inner parts of the boat. It has a tank with a thin deformable *diaphragm* (made of copper) covering a shallow chamber. The tinplate cover hiding the tank can be easily removable and one can clearly see the tank and the diaphragm. The top and the bottom covers of the tank are closely spaced. From the front bottom of the chamber, two pipes extend to the rear of the boat. A single pipe can also be used, however, use of two pipes is a better choice. Because the tank has to be filled with water the use of two pipes allows injection of water into one pipe while the air in the tank escapes through the other pipe. The boiler is usually in the shape of a box or cylinder and the top of it (diaphragm) is slightly concave. To heat the water in the tank a heat source like a candle or a small oil burner is placed below the tank as shown in Fig. 2.1b.

First, the tank is filled with water by injecting it through one pipe using an ink filler or a syringe. Water is to be injected until it comes out from the other tube. A commercially sold putt-putt boat set contains an ink filler and a candle also. Place the boat in a tub of water without allowing any water to escape from the pipes. Light the candle and place it below the tank. In a few seconds the tank will be heated up enough to create steam from the water inside. The steam is pushed into the pipes and the water is ejected in the form of jets through the exits of the pipes. The boat will move forward with the putt-putt sound. As the ejection of water takes place the steam in the pipes starts to condensate, the outside water enters into the pipes, goes to the tank and the next cycle starts. The boat continues its forward motion until the candle goes out.

There are several important questions concerned with the working of the toy boat. *What is the cause of the putt-putt sound? Does the putt-putt sound responsible for the propulsion of the boat? Why does the boat move even though water is sucked in? Are two pipes necessary for the boat to function?* In order to find the answers to these questions and to know what is actually happening in the working of the boat, experiments have been conducted with transparent tank and pipes, with and without the diaphragm, using only one pipe, visualizing the flow using a high-speed camera and with hydrogen bubbles measuring velocities in flow by particle velocimetry [4,10-14].

In this chapter, through a qualitative description, we explain the working of the toy boat. We bring out the answers to the various questions raised above about the working of the boat.

2.1 Principle of Operation of the Putt-Putt Boat – Cyclic Ejection and Suction of Water

In this section we explain the working of the putt-putt boat [4,10-15].

When the bottom of the tank is heated, the water in it evaporates (but not the water in the pipes) and the steam is produced. Note that the tank is not completely filled with water and only part of it has water and the rest is filled with air. This is clearly evident from the design and arrangement of the tank and the pipes as shown in Fig. 2.1b. The steam in the tank expands, pushed out of it and moves into the pipes over a short length. The pressure at the open ends of the pipes is at atmospheric pressure, while the pressure in the boiler is higher than the atmospheric pressure. The higher pressure accelerates the water columns in the pipes to go down. This forces the cooler water in the pipes out of the opening at the rear side of the boat. The water comes out in the form of *jets* like a smoke laden exhaust from an automobile. These jets *propels the boat forward.*

The tank and pipes, after the expulsion of most of water from them are filled with steam. The steam in the tank is hot since it is heated by the heat source. But the steam that moved into the pipes quickly cools due to the conduction of heat from steam to the pipes and then to the surrounding water on which the boat is floating. Consequently, condensation of steam and sudden decrease in pressure in the pipes take place. The atmospheric pressure now pushes the outside surrounding water into the pipes. In this way pipes are refilled and a new cycle of firing of jets of water starts.

In each cycle, the water in the tank evaporates, steam is produced, water is expelled out of the pipes in the form of jets, the boat is pushed forward and then water is sucked in from the pipes (due to the condensation of steam in the pipes) for the next cycle. The candle flame powers this delicate cycle. An *ejection and suction of water* essentially constitute one cycle of operation of the boat. Initially, sufficient water is injected into the pipes in order for the tank to have steam. If there is no steam in the pipes to condense, then it cann't draw the water on its own. The observation is that the engine in the putt-putt boat is like a *piston engine* with cyclic strokes of expansion and compression.

Solved Problem 1:

A putt-putt boat is moving at the speed $v = 0.10\,\text{m/s}$ and the frequency of the cycle is $10\,\text{Hz}$. Assume that the flow of water in a pipe is sinusoidal.

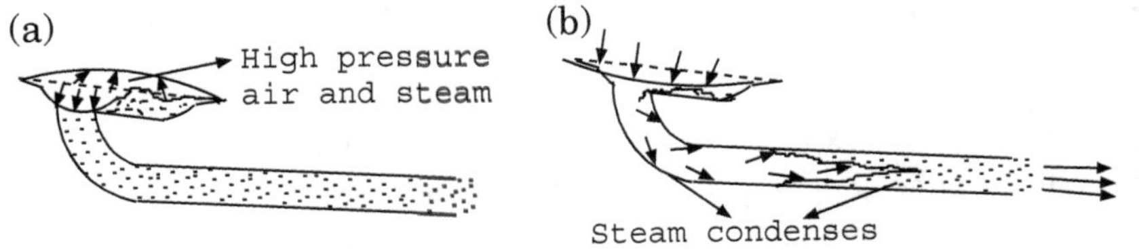

Fig. 2.2: Change in the shape of the diaphragm due to the (a) expansion of the steam and (b) ejection of the water through the pipes.

Obtain an expression for maximum speed and calculate the half-amplitude of the displacement.

For a putt-putt boat moving with the speed v the water jet in the output has to be at least one per cycle faster than v. That is, $V_{\max} > v$. As the flow of water in the pipe is sinusoidal each point of the water profile moves by $x = A\sin(2\pi ft)$ where f is the frequency of the cycle and A is the half-amplitude of the displacement. Then, the speed of the water in the pipe is

$$V = \mathrm{d}x/\mathrm{d}t = 2\pi fA\cos(2\pi ft)\,. \tag{2.1}$$

$V_{\max}$ is $2\pi fA$ and it should be $> v$. For the given $f = 10\,\mathrm{Hz}$ and $v = 0.10\,\mathrm{m/s}$ we obtain $A > v/(2\pi f) = 1.6\,\mathrm{mm}$.

2.2 Origin of the Putt-Putt Sound and the Need of Two Pipes

The putt-putt or pop-pop sound is produced by the top and hard thin copper cover of the tank [11-15]. This cover acts as a diaphragm. It is initially in concave shape. During the ejection and suction of water, the volume and the pressure of the steam in the tank vary. When the steam expands, the diaphragm pulges *outward* and while the water in the pipes are pushed out, it pulges *inward*. That is, the diaphragm changes from *concavity* to *convexity* and back (similar to an oil can snap action) thereby undergoing vibratory motion as shown in Fig.2.2 [13].

A plot of the amplitude of the sound recorded versus time had shown two bursts of sound pulses in each cycle of operation of the boat [12]. The average frequency of the putt-putt sound recorded was 16 Hz (*What does it mean?*). Each putt is like an impulse loading on the diaphragm and produced high-frequency oscillations in the sound which is what we hear [12]. We all love the putt-putt sound but this sound and diaphragm are

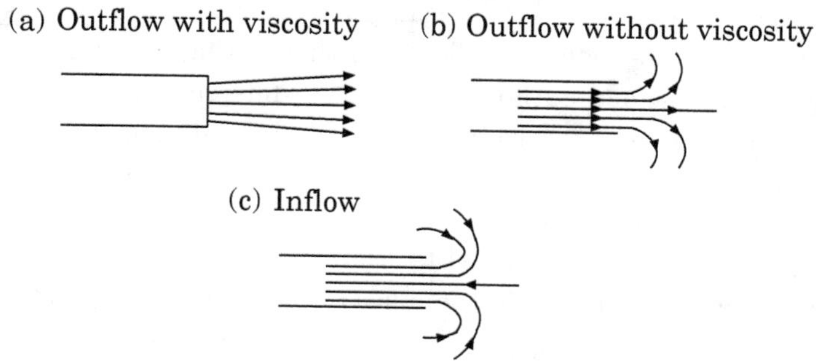

Fig. 2.3: Water leaves a pipe (a) as jets when viscosity is nonzero and (b) in all directions in the case of absence of viscosity. (c) Water flows from all around into a pipe.

not necessary for good working of the boat. Thickening of the diaphragm results in a putt-putt soundless good working boat. That is, the presence of a thin flexible diaphragm is to produce putt-putt sound and has nothing to do with the ejection and suction of water.

Why are two pipes? As mentioned earlier, first we need to fill the pipe(s) and the tank with water before applying heat. If only one pipe is used, then water cannot get into the tank since air in it has to come out. The second pipe allows the air to come out when water is injected through the first pipe. After filling the tank, if one of the pipes is closed, then also jets of water come out from the other pipe, however, the speed of the boat is reduced. Remember that expelling and sucking of water takes place in both pipes.

2.3 Mechanism of Propulsion

In each cycle of operation of the boat, as water is alternately expelled and sucked one may expect that according to Newton's third law of motion for every action there is an equal and opposite reaction and hence zero net force. But there is a nonzero net force causing the boat to move forward. *Does it happen due to the sharp front part and blunt square shaped rear part?*

Let us analyse the outflow and inflow of the water [11-15]. During the outflow water comes out of the pipes in the form of narrow jets as shown in Fig. 2.3a [12]. In the case of ideal fluid with zero viscosity, the outflow will be as in Fig. 2.3b. Due to the viscosity the water jets are separated from the surrounding water. During the inflow as shown in Fig. 2.3c water enters the pipe in all directions and there is no jet-like flow. As a result, the thrust during inflow becomes much smaller than that during the outflow. Due to viscosity, the inflow and outflow patterns are different. However, the boat would be able to move forward in the case of an ideal fluid [16]. Therefore, in addition to the above mentioned asymmetry in inflow and outflow, other asymmetries also contribute to the forward motion of the boat. The asymmetry in the hull of the boat has to be taken into account. The outflow and inflow water movements in the pipes also display asymmetry. Because the pipes go down from the tank to the rear side of the boat, the water can move faster towards the outlet than forward to the tank. Further, the sucked water transfers its momentum to the boat when hits on the tank. The reaction force pulling the boat backwards is compensated by the pushing of the water as it strikes the inner of the tank [15]. In summary, the total force exerted on the boat during expulsion of water in the form of jets is higher than the force exerted while sucking of water. Hence, the boat moves forward.

2.4 Conclusion

Many of the physical principles on which the putt-putt boat works can be demonstrated to the students by doing simple experiments with this toy. Many interesting experiments can be done with this simple toy. The dependence of speed of the toy with the heating power can be studied using different sizes of candles. The dependence of the water cycle in the pipes with heating power can also be studied. It can be demonstrated that the frequency of water cycle is almost independent of the heating power above a certain minimum power. It can be verified that the toy speed has practically no influence on the frequency of the water cycle. One of the pipes can be plugged and the decrease of the frequency of the water cycle can be demonstrated. Students can be asked to touch the pipes to find them cooler even after a long period of operation of the toy.

We have seen that the boat moves in the forward direction due to the jet of water going out and the reaction force it exerts on the boat, just like in a rocket. When the water is sucked into the pipes, the reaction force does not have sufficient value to make the boat move backwards. This is due to the fact that water enters the pipes from all directions, not as a jet in one direction. This difference of reaction forces in both cases can be demonstrated to the students by using a straw and a burning candle. They can easily blow out a burning candle by blowing air through the straw

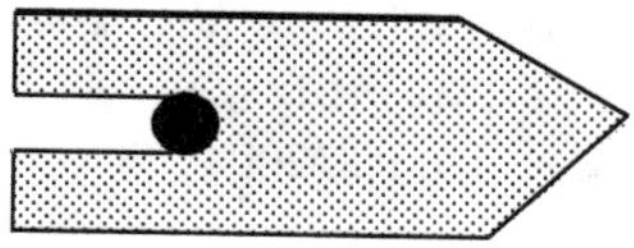

Fig. 2.4: A simple paper boat.

even if the candle is one foot away. But they cannot blow out a candle by sucking air through the straw unless the candle is very near.

A toothpaste powered boat can be easily made. Cut a small paper boat as shown in Fig. 2.4. Cut out a small slit at the bottom. Place a small piece of toothpaste at the bottom of the slit and place it in water. It will move like a boat as the toothpaste start dissolving and expanding on contact with water.

2.5 Bibliography

[1] B. Harley, *Toy Steam* (Argus Books, 1978).

[2] J.G. Baker, Trans. ASME **55**, AMP-55-2, 5 (1933).

[3] R. Stuart Mackay, Am. J. Phys. **26**, 583 (1958).

[4] I. Finnie and R.L. Curl, Am. J. Phys. **31**, 289 (1963).

[5] T. Piot, *Improvement in steam generators*, UK Patent 20,081 dt Oct. 15, 1892.

[6] C.J. McHugh, *Power-proppelled boat*, U.S. Patent 1,200,960 dt Oct. 10, 1916.

[7] W.F. Purcell, *Propelling device*, U.S. Patent 1,480,836 dt Jan. 15, 1924.

[8] C.J. McHugh, *Power-proppelled boat*, U.S. Patent 1,596,934 dt Aug. 24, 1926.

[9] P. Jones, *Tot boat*, U.S. Patent, 1,993,670 dt Mar. 5, 1935.

[10] P.R. Payne, S.W. Greenwood and R.C. Brown, *Recent developments with the water pulse jet*, Society of Automative Engineers, paper number 789077 (1978).

[11] H.R. Crane, The Phys. Teach. **35**, 176 (1997).

[12] V. Sharadha and J.H. Arakeri, Resonance **19**, 66 (2004).

[13] J. Bindon, *The secret working of a transparent pop-pop engine*; https://www.sciencetoymaker.org/wp-content/uploads/2017/03/bindon-article-9_04.pdf.

[14] J.Y. Renaud, *Propulsion of a boat by means of a pop-pop engine*; http://www.eclecticspace.net/poppop/jy/pop-pop_engine_V2.pdf.

[15] https://en.wikipedia.org/wiki/Pop_pop_boat.
[16] A. Jenkins, Eur. J. Phys. **32**, 1213 (2011).
[17] J.H. Arakeri, Resonance **19**, 92 (2004).

2.6 Exercises

2.1 State the Newton's laws that are part of the process of the putt-putt boat.

2.2 Enumerate the step by step process taking place in the putt-putt boat.

2.3 List out various facts influencing the performances of the toy boat [14].

2.4 Applying control volume analysis find the thrusts due to outflow and inflow of water [17].

2.5 Write the momentum conservation equation for the putt-putt boat when it reached uniform speed, say, along x-direction.

Chapter 3

Crookes Radiometer

In 1873 Sir William Crookes, a chemist, devised a special type of radiometer during his investigation of infrared radiation and the element thallium [1]. He introduced it in the study of attraction or repulsion occurring from radiation. A typical radiometer (Fig. 3.1) also called a *light mill*, is a glass bulb with a partial vacuum, and containing a rotor with four light weighted metal wing mills mounted on a vertical pivot. Each vane (wing) is black coated on one side and silver coated on the other side. That is, the black side is light absorbing, while the other side is light reflective. When the device is exposed to direct light from the sun or a flashlight (that is, heat), or infrared radiation or heat of a hand, then the rotor rotates, and the wing mill spins. The black coated sides retreat from the light source and

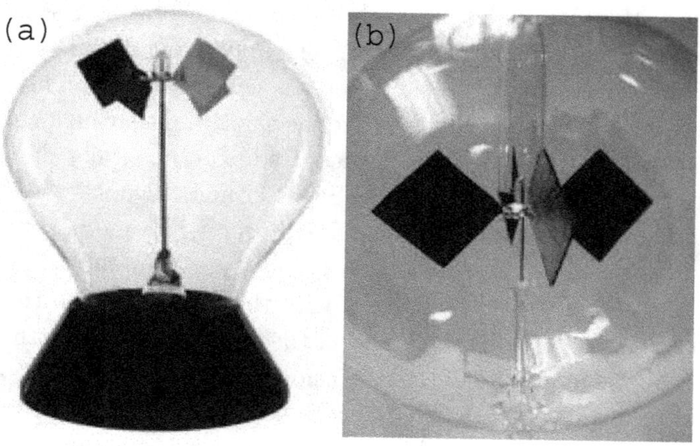

Fig. 3.1: (a) A crookes radiometer and (b) magnification of wings of a radiometer.

the other sides advance. Note that the rotation of the rotor is completely powered by light! Further, the light energy is transformed into the kinetic energy of the wings.

The radiometer can be used to measure the intensity of electromagnetic radiation since more intense light gives rise to a faster rotation of the wings. The prefix radio in the name radiometer refers to the Latin word radius for a ray (electromagnetic radiation) and the suffix meter means that the device can be used for measurement of electromagnetic radiation intensity.

On 7 April 1875 Crookes demonstrated the working of a radiometer to the Royal Society (London). Crookes in 1911 donated several original radiometers to the Royal Society. The Crookes radiometer had received a great interest by Osborne Reynolds and Albert Einstein [2-5]. Nowadays, the Crookes radiometer is used for demonstration of a heat engine functioning with light energy and without batteries. The radiometric force is the key force in the radiometer. In particular, the forces originating from the area and the edge effect corresponding to thermal transpiration are shown to be important [6,7].

Now, we briefly mention the physical mechanism of the radiometer. When the device is exposed to an intense light, the black sides of the vane absorb relatively more amount of light than the silvered sides due to a larger absorption coefficient. The black sides thus become hotter than the other sides. The hotter air molecules near the black sides of the vane move around the edges of these sides to the cooler (silvered) sides. A force is generated consequently. This resulting force acts towards the cooler side. As a result, the black sides are pushed thereby generating rotation of the rotor on which the vanes are mounted. The underlying mechanism is called *thermal transpiration*. If the radiometer is completely or highly vacuum, then there are no air molecules to strike the vanes and they no longer rotate.

Marsh and his collaborators [8] measured experimentally the temperature between the hot and cold sides. Schuster and Reynolds measured the forces acting on the vanes [2,9]. Scandurra proposed a method to enhance the radiometric effect [10]. Woodruff pointed out that radiation pressure and the increase in overall pressure are not the sources behind the working of a radiometer [11]. He highlighted that the force exerted along the edges of the vanes by the molecules of the gas inside the radiometer is crucial. Hladkouski and Pinchuk have shown that a radiometer can be treated as a heat engine [12]. Lu analysed, both experimentally and through Monte Carlo simulation, the optimal operation pressure inside the Crookes radiometer [13]. Wolfe and his collaborators developed a horizontal vane radiometer and analyzed its working principle [14]. Crookes radiometer has modern applications in micromachines [15]. It can be employed as an electric generator in an optical induction generator of renewable energy applications [16]. Molecular diameters can be determined using a Crookes radiometer-based systems [17].

We first discuss whether the radiometer works due to the pressure increase of the gas against warmer surface due to molecular impacts. Next, we describe the thermal forces that are responsible for the rotation of the vanes in the radiometer. We bring out the thermodynamic explanation of the function of the Crookes radiometer. Finally, we point out the possibility of backward motion of the vanes of the radiometer.

3.1 Working Principle of the Radiometer

An explanation for the working of the radiometer based on the momentum transferred by the individual molecular impact was given by Reynolds [2]. The molecules of the residual gas bombarding the warmer surface of the plate rebound with increased vigor and so import a greater recoil to the warmer surface than the cold surface. Thus, these impacts increase the pressure on the warm surface. As the warm surface is the blackened, absorption of the light is more by this side and the radiometer must rotate with the white side advancing.

Woodruff [11] has found the increase of pressure due to molecular impact on the warmer surface with a temperature difference with the colder surface. If a molecule of mass m collides with velocity v and rebounds with velocity $-v$, then the momentum transferred to the surface will be $2mv$ as the momentum has to be conserved. If the rebounding molecule gains in speed, due to higher temperature, from v to $v + \mathrm{d}v$, the momentum transferred will be $m(v + v + \mathrm{d}v) = 2mv + m\mathrm{d}v$.

As the average kinetic energy is proportional to the temperature T, we have

$$\frac{1}{2}mv^2 = aT, \tag{3.1}$$

where a is a proportionality constant. If the hot surface has a temperature $T + \mathrm{d}T$, then the average kinetic energy for temperature $T + \mathrm{d}T$ is given by the relation

$$\frac{1}{2}m(v + \mathrm{d}v)^2 = a(T + \mathrm{d}T). \tag{3.2}$$

Subtracting Eq. (3.1) from Eq. (3.2) and as dT and dv are very small, we obtain

$$mv\mathrm{d}v = a\mathrm{d}T. \tag{3.3}$$

Since the pressure is proportional to the momentum transferred, we get the ratio of the pressure on the hot plate to cold plate as

$$\frac{P+\mathrm{d}P}{P} = \frac{2mv+m\mathrm{d}v}{2mv} = 1 + \frac{\mathrm{d}v}{2v}. \tag{3.4}$$

Dividing Eq. (3.3) by Eq. (3.1) gives

$$\frac{\mathrm{d}v}{v} = \frac{\mathrm{d}T}{2T}, \tag{3.5}$$

which when substituted in Eq. (3.4) gives

$$\frac{\mathrm{d}P}{P} = \frac{\mathrm{d}T}{4T}. \tag{3.6}$$

This pressure acts on the hot (black) surface of the plate. So, at a first sight, this pressure would be considered as the reason for the rotation of the radiometer. But it is not correct. This argument did not take into account the fact that the rate at which molecules are impinging on the surface. It has been pointed out [18] that the rapidly rebounding molecules from the warmer surface are more effective in stopping other molecules hitting the warmer surface.

3.2 Radiometric Forces

Consider a thin vane kept in a rarefied gas. If there exists a nonequilibrium temperature gradient across or along the vane, then a force acts on it. This process is known as *thermal transpiration* or *thermal diffusion*. This thermal force acting on the surface of the vane is termed as *radiometric force*. This force is exerted by the thermally driven flow of the air molecules. Radiometric forces are crucial in micro- and nano-scale systems and low density environments. In 1825 Augustin Fresnel made a first observation of a radiometric force by changing the pressure in a chamber. He reported that this force was not due to the evaporation of vapor from the hot surfaces or by the convection currents in the air.

In a Crookes radiometer, the radiometric forces are induced by the incident energy. This happens between the black side and the silvered side of the vanes. In the region where the molecules are free, the prime force is due to the change in the momentum flux of the molecules those leaving the hot surface (with a larger thermal velocity) and the cold surface. As these two surfaces are not at the same temperature, the momentum fluxes associated with these two surfaces are also not the same. There exists a pressure on both sides of each of the vane. This gives rise a force called *area force* since it depends on the area of the vane [15]. This force acts from the hot surface to the cold surface, that is, from the black side to the silvered side. This force becomes dominant if the ratio of the molecular mean free path length to the size of the vane called *Knudsen number* (K_{n}) is high.

When the Knudsen number is low, then the force arising due to the difference in the momentum fluxes exists only in the neighbourhood of the vane's edge. This *edge force* [15] is dominant for low values of K_{n}. The prime sources for this force are the thermal tangential stresses occurring in the gas surrounding the edge. Such stresses are due to the temperature gradient along the vane's edge through the process known as *thermal creep* [15]. This process is a transitional flow regime with the flow taking place from the silvered surface to the black surface. This thermal creep effect makes air to *creep* along a surface having a temperature gradient.

The thermal creep gives rise to another kind of force called *shear force*. This force is realizable for low values of K_{n} and acts on the vane's lateral edge in the direction opposite to the other two forces – area force and edge force. Therefore, one can consider that the net force exerted on the vane in a radiometer is essentially the balance between the area and the edge forces and the shear force. The direction of rotation of the vanes is caused by the area and the edge effects with the the black surfaces trailing the silvered surfaces.

Figure 3.2 depicts the regimes where the three above mentioned forces act [10]. The ratio h/d, where h and d are the thickness and the size, respectively, of the vane, and the value of K_{n} determines the relative contribution of the three forces [15]. For $h/d > 1$, the shear force is dominant and is negligible for $h/d \ll 1$ when $K_{\mathrm{n}} \ll 1$. The area force becomes dominant for $K_{\mathrm{n}} > 0.1$ and small h/d.

The radiometric effect has been observed in a microelectromechanical system, atomic force microscopy [19-21], light-driven micromachines [22] and microactuator [23]. A radiometric based propulsion system has been also developed [24].

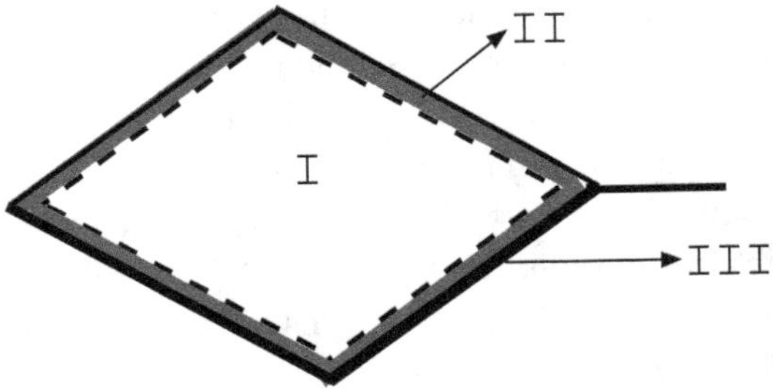

Fig. 3.2: I, II and III are the regimes where the area force, edge force and shear force, respectively, acts on a vane of a radiometer. The region-I includes the black and silvered sides of the vane.

3.3 Thermodynamic Explanation

We present here the physics behind the working of a Crookes radiometer [25,26].

Expose the radiometer to light energy. The black surfaces of the vanes absorb more light energy than the silvered surface. The black surfaces of the vane become hotter than the silvered surfaces. The black sides retreat from the light source and the silvered sides advance.

Crookes first thought that light radiation pressure on the black sides was causing the rotor to rotate similar to water in a mill. But this explanation has a difficulty. As the light incident on the black side would be absorbed and light incident on the silvered side would be reflected, the result should be that twice as much radiation on the silvered side than the black side should appear. If this would be the case, then the rotor should spin with the black side leading the silvered side, but actually the opposite is what is happening. So, the radiation pressure is too weak to make the rotor to spin in the opposite direction.

What would be the cause of the spinning of the rotor? We turn our attention from the radiation pressure to the air pressure. The air molecules are heated up when they are incident on the black surface. Reynolds considered thermal transpiration [3]. According to this, if the vanes have porous, then air in the radiometer flows through the porous and creates a temperature difference on the sides of the vanes. If the two sides of the vanes have the same air pressure, then the flow of the air occurs from the silvered side (colder side) to the black side (hotter side). In case that the vanes remain stationary, then the pressure in the black side is higher.

Because the faces of the vanes are not porous, we focus on the edges of the sides. The air molecules moving faster (and almost freely due to low pressure) from the hotter side hit the edges obliquely (at an angle) and exert a relatively stronger force than the colder molecules. This effect is the thermal creep defined in the previous section. Because of the tangential forces arise around the edges as a result of the air molecules moving around the edges, the vane moves from the hotter air towards the cooler air.

In the Crookes radiometer, the molecules that are heated up after some time touch the glass surface of the bulb. Then, they loose heat through the glass and are cooled again. The heated glass bulb looses heat to the external air. This process makes the internal temperature steady and the two sides of the vanes posses a temperature gradient. Due to the conduction of heat from the black side, the silvered side is hotter than the internal air, however, it is cooler than the black side.

Solved Problem 1:

Show that the rotor should spin with the black side leading the silvered side due to light radiation pressure. Assume that reflectivity $R = 1$ for the silver side and $R = 0$ for the black side.

Now, assume that an equal amount of light falls on both sides. If p is the radiation pressure, then in the blackened side the momentum transferred to the wing is just p, as there is no reflection. For the silvered side, light is reflected back completely with the same momentum p in the opposite direction. So, the pressure change is $2p$. If A is the area of the plate, then the net force acting on the plate will be $Ap - 2Ap = -Ap$. This force will act from the silvered side to the darkened side. So, the rotor will spin with the black side leading to the silvered side.

3.4 Backward Motion of Vanes

It is possible to witness backward motion of vanes of a radiometer. Woodruff [11] and Crawford [27] pointed out how to do this. Place the radiometer in the freezer of a refrigerator. The black surface cools more quickly than the silvered surface. In a few seconds the vanes begin rotating backwards. That is, the silvered sides are now receding and the black sides are advancing. After a few minutes the rotation stops when both sides acquire the same temperature, that is, equilibrium is reached. Next, place the radiometer in a warm room. *What will happen?* Both black and silvered sides start to warm up. However, the black sides warm up quickly. Therefore, the vanes start rotating in the forward direction. This rotation happens for a few minutes. When both sides have attained the same temperature the rotation ceases.

To start the motion of the vanes in the forward direction without exposing the radiometer to light we can place it in a warm oven. The vanes begin to rotate in the forward direction and stop after a few minutes. Now, all the sides are at the same temperature but higher than the room temperature. Next, remove the radiometer from the oven and place it in the room. The vanes rotate backwards and halts after some time.

3.5 Conclusion

In the Crookes radiometer, rotation is usually induced by exposing it to a light, either artificial or natural or infrared radiation. The heat from a nearby hand itself is even enough to make the rotor to turn. The underlying mechanism is not a radiation pressure [28]. Also, it does not work due to the pressure increase of the gas against the warmer surface. The pressure is compensated by the reduction of collision of molecules in the warmer surface. The radiometer works due to forces exerted along the edges of the vanes. Since a temperature gradient is present at the edge, gases creep along the surface of the edge. This movement of the gases on the edges of the vanes causes a tangential force around the edges. This force acts away from the warmer gas and towards the cooler gas. Consequently, a greater force acts on the blackened side of the vane and makes the silvered side to advance forward. The Crookes radiometer can be used to demonstrate both black body absorption and black body radiation.

Experiments with the Crookes radiometer can be designed for students to do the following:

1. Identify the possible forms of energy including magnetic energy that can give rise to the motion of the rotor.
2. Identify the types of energy conversions involved.
3. Compare the underlying energy conversion with the various energy conversions that we notice in our daily life.
4. Develop an understanding of the interactions of matter and energy and transfer of energy.

In the so-called acoustic radiometer [28], radiation pressure is the dominant source for rotation of the two planes. In places of black side and silvered side a thick acoustic absorbing foam and a thick aluminum plate, respectively, are used. That is, one side is acoustically absorptive and the other is acoustically reflective. High-intensity noise plays the role of light energy. Rotation in the acoustic radiometer takes place in the direction of the total radiation pressure.

3.6 Bibliography

[1] W. Crookes, Philos. Trans. R. Soc. London **164**, 501 (1874).
[2] O. Reynolds, Philos. Trans. R. Soc. London **166**, 725 (1876).
[3] O. Reynolds, Philos. Trans. R. Soc. London **170**, 727 (1879).
[4] J.C. Maxwell, Philos. Trans. R. Soc. London **170**, 231 (1879).
[5] A. Einstein, Z. Phys. **27**, 1 (1924).
[6] W.H. Westphal, Z. Phys. **1**, 92 (1920).
[7] N. Selden, C. Ngalande, S. Gimelshein, E.P. Muntz, A. Alexeenko and A. Ketsdever, Phys. Rev. E **79**, 041201 (2009).
[8] H.E. Marsh, E. Condon and L.B. Loeb, J. Opt. Soc. Am. **11**, 257 (1925).
[9] A. Schuster, Philos. Trans. R. Soc. London **166**, 715 (1876).
[10] M. Scandurra, Enhanced radiometric forces, arXiv: physics/0402011 v1 [physics.class-ph] 3 Feb 2004.
[11] A.E. Woodruff, The Phys. Teach. **6**, 358 (1968).
[12] V.I. Hladkouski and A.I. Pinchuk, The Phys. Teach. **53**, 19 (2015).
[13] Z. Lu, J. Vac. Sci. Technol. **23**, 1531 (2005).
[14] D. Wolfe, A. Larraza and A. Garcia, Phys. Fluids **28**, 037103 (2016).
[15] A. Ketsdever, N. Gimelshein, S. Gimelshein and N. Selden, Vacuum **86**, 1644 (2012).
[16] D. Delaine, S. Herbert and A. Fontecchio, Proc. SPIE 7787, Novel Optical Systems Design and Optimization XIII, 77870P (25 August 2010).
[17] J.H. Smith, J. Chem. Edu. **47**, 580 (1970).
[18] P.G. Tait and J. Dewar, Nature **12**, 217 (1875).
[19] A. Passian, A. Wig, F. Meriaudeau, T.L. Ferrell, T. Thundat, J. Appl. Phys. **92**, 6326 (2002).
[20] A. Passian, R.J. Warmack, A. Wig, R.H. Farahi, F. Meriaudeau, T.L. Ferrell, and T. Thundat, Ultramicroscopy **97**, 401 (2003).

[21] A. Passian, R.J. Warmack, T.L. Ferrell and T. Thundat, Phys. Rev. Lett. **90**, 124503 (2003).

[22] A. Lereu, A. Passian, R.J. Warmack, T.L. Ferrell and T. Thundat, Appl. Phys. Lett. **84**, 1013 (2004).

[23] D.C. Wadsworth, E.P. Muntz, G. Pham-Van-Diep and P. Keely, Crookes radiometer and micromechanical actuators. In: 19th Int. Symp. on Rarefied Gas Dynamics (Oxford University Press, Oxford, 1995) pp708.

[24] M. Scandurra, Radiometric propulsion system. U.S. Patent Application Publication, 2006/0001569, January 2006.

[25] P. Gibbs, How does a light mill work?; http://www.math.ucr.edu/home/baez/physics/Genera/Light-Mill/light-mill.html.

[26] https://en.wikikedia.org/wiki/Crookes_radiometer.

[27] F.S. Crawford, Am. J. Phys. **53**, 1105 (1985).

[28] B. Denardo and T.G. Simmons, Am. J. Phys. **72**, 843 (2004).

3.7 Exercises

3.1 Expose a radiometer to [12]
(a) the light from an incandescent lamp with electric power of, say, 60 W and a minimum luminous power of 800 lm,
(b) the light from a fluorescent lamp with 13 W and 800 lm and
(c) a thermal heater with 60 W power.
Explain your observations.

3.2 What will happen if there is a strong vacuum or a high pressure inside the radiometer?

3.3 Explain your observation
(a) when you place your hands around the glass without touching it and
(b) when you touch the glass bulb by your hands
in the absence of a light source.

3.4 What will be effect of increase of the surface of the warmer side?

3.5 Due to the conduction of heat from the black side, the silvered side is hotter than the internal air. What can be done to avoid this?

Part VII: Electric and Magnetic Toys

Chapter 1

The Gauss Rifle

The Gauss rifle also known as the magnetic cannon or the Gaussian gun or the Gauss accelerator, makes use of a chain of magnetic reaction to hit a target by a steel ball. The Gauss rifle is not only easy to construct but also simple to understand and explain.

A Gauss rifle can be easily constructed [1-5] using a wooden or brass or plastic ruler with a slot down the centre, a few number of 8-12 mm square or spherical neodymium magnets (for example, NdFeB), cellophane or electrical tape and seven, say, $\sim$ 8 mm radius nickel-coated steel identical ball bearings. Place the magnets on the centre of the ruler at equal distance, for example, at the markings 3, 6 and 9 inches. Keep the north pole faces of the magnets to point in the same direction. To make the position of the magnets fixed and to avoid the jumping of them to nearby, one electrical tape can be wrapped around the ruler and the magnets. Place two balls touching north pole faces of each of the magnet as shown in Fig. 1.1. Use the remaining ball bearing as the incoming projectile.

Activities with the Gauss rifle can be used to explore the concepts of force versus energy, escape velocity, breaking bonds, conservation of momentum and potential energy of a magnetic field. A fun competition for the last ball moving the furthest and fastest can be held [3]. Methods for

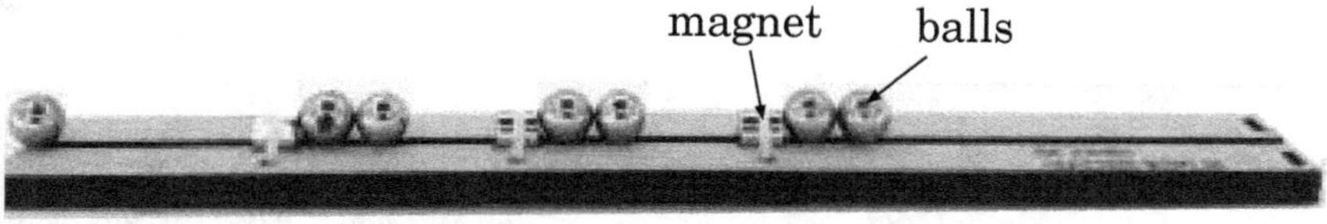

Fig. 1.1: A Gauss rifle.

measuring change in the potential energy and change in the kinetic energy are described in the Refs. [1,2]. The Gauss rifle is utilized in energy education [1,6-8]. Construction of a gravitational analog of the Gauss rifle is described in [9]. Gauss guns are much heavier than other options and more over, better performance can be achieved in an air rifle which is much lighter.

1.1 Working of a Gauss Rifle

Release the first ball, the projectile, towards the south pole of the first magnet with a low velocity. The incoming ball is accelerated by the magnetic force acting on it. The ball strikes the magnet with a force and a kinetic energy, say, 1 unit. This kinetic energy is transferred to the magnet and then to the ball touching that one. That is, there is a propagation of momentum into the chain of balls. As there is no ball to the right of the third ball, it starts with a kinetic energy and leaves from the magnetic attraction of the first magnet. This is the first stage. An amount of energy is lost in this process. *What are the sources of energy loss?* The loss of energy is due to the impacts, rolling friction and air resistance and a magnetic force on the ball at the stage of leaving.

Note that it is easy for the third ball to move away than the second ball which is in touch with the magnet. The third ball starts its motion with the kinetic energy 1 unit minus the energy lost in the above first stage of process. As the third ball moves towards the second magnet due to the pull of the magnet the speed of the third ball increases and hence its kinetic energy. It hits the second magnet, with the speed much higher than that of its initial speed. This in turn makes the fifth ball to detach from the fourth ball, accelerated and then hits the third magnet. This process continues and the last ball comes out very fast and can be used to destroy a target. After the last ball came out there is one ball on each side of the three magnets. *What will happen if the track of the Gauss rifle is circular? What will happen if plastic balls are used instead of steel balls?*

1.1.1 *Number of Balls to be Used*

How many balls to be placed after each magnet? How do we increase the speed of the rolling balls? How many stages can we use? First, if few balls are placed then, all the balls can be within the magnetic force of the magnet. When the number of balls are too many, then some of them may be out of the action of the magnetic force. This would reduce the speed of the end ball. Secondly, the energy of the last ball depends on the mass and the speed of the balls hitting the magnets. If these balls are smaller in size, then the leaving balls have much greater speeds. Thirdly, theoretically, adding more stages can make the ball to roll faster. However, in practice, due to the loss of energy arising from the various sources mentioned earlier the speed is increased by a less amount. Therefore, a few stages is appropriate for a better performance.

1.1.2 *Field, Force and Magnetic Moment*

In the Gauss rifle, the presence of magnets is significant. It provides inhomogeneity. Also, it creates a magenetic field which in turn magnetizes the steel ball. As a result, there is a strong attraction between steel balls. This attraction is strong enough to prevent the rebounding of the impacting ball [6]. We wish to know the magnetic field and the magnetic force associated with the magnet. For a uniformly magnetized sphere, the magnetic induction B at a distance r from its centre is given by (see the exercise 1.1 at the end of this chapter)

$$B(r) = \frac{\mu_0 M_0}{2\pi r^3}, \tag{1.1}$$

where M_0 is the magnetic dipole moment and μ_0 is the vacuum magnetic permeability. This relation can be used to determine M_0 experimentally. The magnetic field produced by the magnet and the magnetization of the steel ball interacts and leads to the magnetic force

$$\mathbf{F}_{\mathrm{M}}(r) = -\nabla(\mathbf{m}_{\mathrm{ball}}(r) \cdot \mathbf{B}(r)), \tag{1.2}$$

where $\mathbf{m}_{\mathrm{ball}}(r)$ is the induced magnetic moment of the steel ball. According to classical magnetostatics, the magnetic moment of a sphere of relative magnetic permeability μ_{r} and radius R placed in a constant magnetic induction field $\mathbf{B}$ is given by

$$\mathbf{m}_{\mathrm{ball}} = \frac{4\pi R^3(\mu_{\mathrm{r}} - 1)}{\mu_0(\mu_{\mathrm{r}} + 2)}\mathbf{B}_0. \tag{1.3}$$

For ferromagnetic steel $\mu_{\mathrm{r}} \gg 1$. In the case of the steel ball is placed in a dipole field given by Eq. (1.1), the force on the steel ball can be worked out as

$$F_{\mathrm{M}}(r) = \frac{6\mu_0 R^3 M_0^2}{\pi r^7}. \tag{1.4}$$

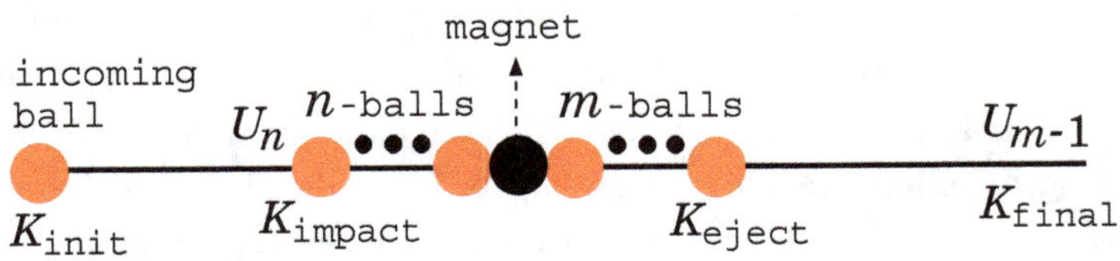

Fig. 1.2: Schematic of a single stage Gauss rifle.

Solved Problem 1:

Prove the Eq. (1.4).

Consider the $\mathbf{m}_{\text{ball}}$ given by Eq. (1.3). For $\mu_{\text{r}} \gg 1$

$$\mathbf{m}_{\text{ball}} \approx \frac{4\pi R^3}{\mu_0}\mathbf{B} \tag{1.5}$$

Then,

$$F_{\text{M}}(r) = -\nabla(\mathbf{m}_{\text{ball}}.\mathbf{B}) = \frac{4\pi R^3}{\mu_0}B^2. \tag{1.6}$$

As $B(r) = \mu_0 M_0/(2\pi r^3)$ we obtain

$$F_{\text{M}}(r) = -\frac{\mu_0 R^3 M_0^2}{\pi}\nabla\left(\frac{1}{r^6}\right) = \frac{6\mu_0 R^3 M_0^2}{\pi r^7}. \tag{1.7}$$

1.2 Optimization of a Single Stage Gauss Rifle

Consider a single stage Gauss rifle with n-balls on the left-hand side of the magnet and m-balls on the righ-hand side as shown in Fig. 1.2. The incoming ball is accelerated due to the magnetic attraction and hence its kinetic energy increases from its initial value K_{init}. Its magnetic (potential) energy decreases to U_n when it impacts with the first of the n-balls array. So, using conservation of energy we can write the kinetic energy of impact as

$$K_{\text{impact}} = K_{\text{init}} + U_n. \tag{1.8}$$

This energy is transformed to the last ball of the m-balls array on the right-hand side of the magnet which gets an energy K_{eject}. The efficiency of transfer of energy K_{eject} to the K_{impact} energy is defined as

$$\eta = \frac{K_{\text{eject}}}{K_{\text{impact}}}. \tag{1.9}$$

The ejected ball is attracted by the magnet. Therefore, its kinetic energy decreases as its potential energy increases. If the final kinetic energy is K_{final} and the potential energy at its escape point is U_{m-1}, then we can write, using the conservation of energy

$$K_{\text{eject}} = K_{\text{final}} + U_{m-1}. \tag{1.10}$$

From Eqs. (1.9) and (1.10) we get

$$K_{\text{final}} = \eta K_{\text{impact}} - U_{m-1}. \tag{1.11}$$

Using Eq. (1.8) in Eq. (1.11) we obtain

$$K_{\text{final}} = \eta\left(K_{\text{init}} + U_n\right) - U_{m-1}. \tag{1.12}$$

For optimization of a single stage Gauss rifle the number of balls n and m must be determined. The number of balls not only determine the potentials U_n and U_{m-1} but also η. Less number of balls will increase U_n and more number of m balls will decrease U_{m-1} and the total nnumber of the balls will affect η. The optimal energy gain is obtained with no ball on the left-side of the magnet ($n = 0$) and also the strength of the magnet used. In that case K_{impact} will be more. The optimal configuration also requires minimizing U_{m-1}. Hence, m should be large. But increase in m will also decrease the efficiency of energy transfer from the impact to ejection. Based on experimental studies on the energy yield η for various number of balls m and n the kinetic energy of the ejected ball as a function of the initial kinetic energy and properties of the system is found as [6]

$$K_{\text{final}} = \eta(m + n + 1)\left(K_{\text{init}} + U_n\right) - U_{m-1}. \tag{1.13}$$

The maximal value of K_{final} is obtained by minimizing the total loss $[1 - \eta(m + 1n + 1)U_n + U_{m-1}$. U_{m-1} can be neglected in the losses when $m > n + 2$. The optimal configuration is obtained when $n = 0$ and $m = 3$. For $n = 0$ and $m = 3$ we find from Eq. (1.10)

$$\left(K_{\text{final}}\right)_{\text{optimum}} = 4\eta\left(K_{\text{init}} + U_0\right) - U_2 \approx 4\eta\left(K_{\text{init}} + U_0\right). \tag{1.14}$$

Usually the initial kinetic energy K_{init} of the ball is zero. Hence, finding $U - 0$ will give the final optimum velocity if η can be determined by some experimental measurements.

1.3 Maximum Kinetic Energy for N Successive Rifles

The method of optimization of the single stage discussed in the previous section has been extended to N-successive rifles [6]. Use of many stages

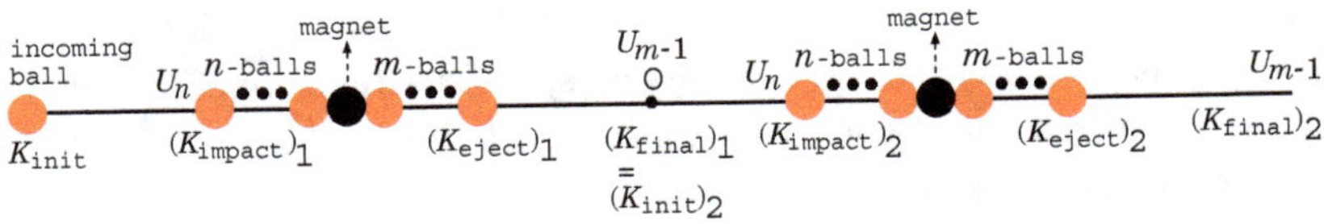

Fig. 1.3: Schematic of two successive Gauss rifles.

will increase the ejected kinetic energy of the ball in the last stage. Let us consider a two stage rifles as shown in Fig. 1.3. The ball has initially energy $(K_{\text{init}})_1$ and it is accelerated by the first magnet. We have

$$(K_{\text{impact}})_1 = ((K_{\text{init}})_1 + U_n)\,, \tag{1.15a}$$

$$(K_{\text{eject}})_1 = \eta\,((K_{\text{init}})_1 + U_n)\,, \tag{1.15b}$$

$$(K_{\text{final}})_1 = \eta\,((K_{\text{init}})_1 + U_n) - U_{m-1}. \tag{1.15c}$$

At point O the potential energy becomes zero as the attraction due to the south pole of $(\text{magnet})_1$ is cancelled by the attraction of the north pole of $(\text{magnet})_2$. Therefore, $(K_{\text{init}})_2 = (K_{\text{final}})_1$. Further,

$$(K_{\text{impact}})_2 = ((K_{\text{final}})_1 + U_n)\,, \tag{1.16a}$$

$$(K_{\text{eject}})_2 = \eta\,(K_{\text{impact}}) = \eta\,((K_{\text{final}})_1 + U_n)\,, \tag{1.16b}$$

$$(K_{\text{final}})_2 = \eta\,((K_{\text{final}})_1 + U_n) - U_{m-1}. \tag{1.16c}$$

Substituting for $(K_{\text{final}})_1$ from Eq. (1.15c) we obtain

$$\begin{aligned}(K_{\text{final}})_2 &= \eta\,[\eta\,((K_{\text{init}})_1 + U_n) - U_{m-1} + U_n] - U_{m-1}\\ &= \eta^2\,(K_{\text{init}})_1 + \eta U_n + \eta^2 U_n - (\eta U_{m-1} + U_{m-1})\\ &= \eta^2\,(K_{\text{init}})_1 + U_n \sum_{i=1}^{2} \eta^i - U_{m-1} \sum_{i=0}^{2-1} \eta^i. \end{aligned} \tag{1.17}$$

Continuing this calculation for N-stage $(K_{\text{final}})_N$ is obtained as

$$(K_{\text{final}})_N = \eta^N\,(K_{\text{init}})_1 + U_n \sum_{i=1}^{N} \eta^i - U_{m-1} \sum_{i=0}^{N-1} \eta^i. \tag{1.18}$$

The kinetic energy of the ejected ball increases with the number of stages. The energy gain in each stage is a constant and independent of the initial energy $(K_{\text{init}})_1$. The dominant losses of energy is proportional to the impact energy. So, as N increases losses also increases. The maximum achievable kinetic energy is found to be

$$K_{\text{max}} = \frac{\eta U_n - U_{m-1}}{1-\eta}. \tag{1.19}$$

This is the *saturation energy*.

1.4 Conclusion

Gauss rifle is relatively easy to assemble. It can be used as a teaching tool to even school students. It can be used to demonstrate the concepts of conservation of energy, conversion of magnetic energy to kinetic energy and magnetic force and potential. It is quite easy to measure the final kinetic energy of the ejected ball by making it to fall from an edge of a table and measuring the distance of fall on the ground. It is quite easy to study the change of kinetic energy of the ejected ball by changing n and m balls in the rifle. It can be easily demonstrated that for $n = 0$, maximum velocity is obtained. Also, students can learn that increasing m will increase the final velocity up to some value. The students can be given a project to make more efficient gun.

1.5 Bibliography

[1] J.A. Rabchuk, The Phys. Teach. **41**, 158 (2003).
[2] D. Kagan, The Phys. Teach. **42**, 24 (2004).
[3] O. Chittasirinuwat, T. Kruatong and B. Paosawatyanyong, Phys. Edu. **46**, 318 (2011).
[4] https://sci-toys.com/scitoys/magnets/gauss.html.
[5] https://www.kjmagnetics.com/blog.asp?p=gauss-guns.
[6] A. Chemin, P. Besserve, A. Caussarieu, N. Taberlet and N. Plihon, Am. J. Phys. **85**, 495 (2017).
[7] L.J. Atkins, C. Erstad, P. Gudeman, J. McGowan, K. Mulhern, K. Prader, G. Rodriguez, A. Showaker and A. Timmons, The Phys. Teach. **52**, 152 (2014).
[8] B.W. Dreyfus, J. Gouvea, B.D. Geller, V. Sawtelle, C. Turpen and E.F. Redish, Am. J. Phys. **82**, 403 (2014).
[9] L. Atkins, A. Bolliou, H. Irving and D. Jackson, The Phys. Teach. **57**, 520 (2019).

1.6 Exercises

1.1 Show that the magnetic induction $\mathbf{B}$ at a distance r from the centre of a short magnetic dipole moment $\mathbf{M}_0$ is $\mu_0\mathbf{M}_0/(2\pi r^3)$.

1.2 A ferromagnetic steel ball will have an induced magnetic dipole moment $\mathbf{m}$ due to the induction field $\mathbf{B}(r)$ of a permanent magnet placed at a distance r from it. Show that (i) if $\mathbf{m}$ is independent of r, then the force between the ball and the magnet varies as $1/r^4$ and (ii) if $\mathbf{m}$ is proportional to $\mathbf{B}(r)$, then the force between them varies as r^7.

1.3 Consider n balls placed in the left-hand side of a magnet and a steel ball is released from a long distance from the left-side on a frictionless rails to hit these balls due to the attraction of the magnet. Find the potential energy of the ball U_n when it kits the n balls away. Find the ratio of U_0/U_1 and U_0/U_2.

1.4 What is the work done if the ball-1 is released with the initial velocity (a) $v_{\rm i} = 0$ and (b) $v_{\rm i} \neq 0$ and hits the first magnet with the velocity $v_{\rm f}$?

1.5 The last ball of a Gauss rifle is allowed to fly horizontally off a table top of height h. The horizontal range of the ball is R. What is the velocity with which the ball is ejected?

Chapter 2

The Levitron

The Levitron, a levitating spinner system, is a remarkable toy which mesmerizes people of all ages by levitating a small handspun magnetized top in air for several minutes without being in touch with any external object [1-8]. It works without any battery or power source. The toy set consists of (i) a square plastic or wooden base, (ii) a small top and (iii) a thin plastic plate (a lifter). The sectional view of the toy is shown in Fig. 2.1. Inside the base, there is a square or a ring shaped magnet and is magnetized perpendicular to the plane surface. Its centre has a large circular hole forming an unmagnetized region. The top is a rotationally symmetric rigid body and is typically a nonmagnetic spindle inserted in a permanent magnetic material of flat and toroidal shape. The lifter is for raising the spinning for levitation and is placed on the base. The north pole of the base is oriented upward while the north pole of the top is facing downward.

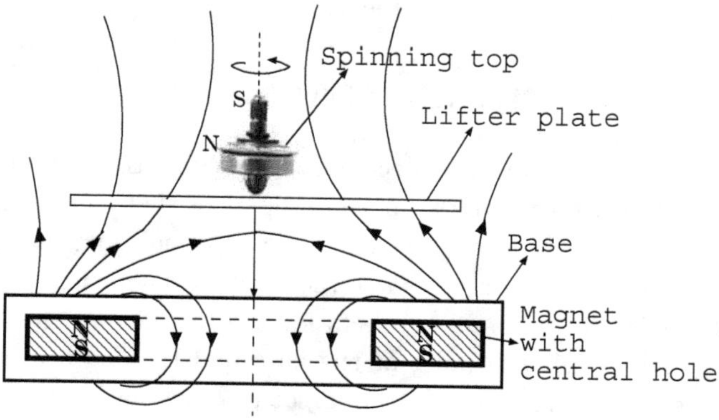

Fig. 2.1: A sectional view of a Levitron toy.

The top is operated by spinning it on the plastic plate with a rotational speed of 25-50 rotations per second. The plate is raised slowly by hand until the top raises above, leave the plate and spin in mid air about a few centimeters above the base. Then, the plastic plate is removed. The top spins in air for a few minutes. The air friction slows the speed of the top and finally it falls straight down. The Levitron appeared to violate the so-called Earnshaw's theorem. According to this theorem no static arrangements of magnetic charges can be stable. If we try to float the magnetic top above the base magnet, then the top (which is a loose magnet) would flip over and the opposite poles got attracted. But in the Levitron, the top is not stationary and is spinning. That is, the levitation in a Levitron is not a violation of Earnshaw's theorem. It is possible to suspend a superconducting disc below or above a permanent magnet. This phenomenon is called *quantum levitation* or *quantum trapping*.

Roy Harrigan invented a device capable of levitating a spinning permanent magnetic top in 1976 and patented it in 1983 [1]. In 1993 Hones and Harrigan jointly attemped to commercialize a levitating top toy but did not succeed [7]. After this, Hones and his father published a patent in 1995 on a levitating top with a base made of a square shaped permanent magnet [2]. Simple theories for the operation of the Levitron are reported [5-7]. Construction of a Levitron is described in [3,4]. Acoustic Levitrons using resonant ultrasonic standing waves capable of levitating small objects are constructed and demonstrated [9,10]. Michael V. Berry posted answers to frequently asked questions about the Levitron [11]. Numerical simulation reproducing various salient features of dynamics of a Levitron is presented [12]. The non-quantum mechanical basics of NMR has been illustrated using inclined Levitron experiments [13].

2.1 Qualitative Analysis of Levitation

Let us briefly point out the mechanism of holding the spinning top [3-8,11].

Essentially, there are two prime properties that causes the Levitron to levitate stably. They are the magnetic repulsion and gyroscopic effects. The magnetic repulsion gives rise the force for levitation. The spinning of the top provides gyroscopic effects for the stability of the levitation. As shown in Fig. 2.1, the poles of the base and the top are pointing in the opposite direction. Consider the forces on the top due to the presence of the base. On the north pole of the top, there is a repulsion force from the north pole of the base and the attractive force from the south pole of the top. Similarly, there are two forces on the south pole of the top, an attractive force from the north pole of the base and a repulsion force from the south pole of the base. Since the north poles of the top and the base are closer, the north-north repulsion is dominant and the top is repelled (magnetically). There is another force acting on the top. This force is the force of gravity which pulls the top downward. At a point the north-north repulsion and the downward pull of gravity balances, cancels each other and the net force is zero. At this equilibrium point the top floats. This is the mechanism of holding the spinning top up. Due to the friction with air the levitated spinning top slowly loses its rotational energy and its rotational speed slows down and finally falls down.

Note that, the levitation in the toy in the vertical direction is due to the balance between the weight of the top (the force due to gravity) and the magnetic force in opposite directions. This points out that weight and the level of the base are crucial parameters for stable levitation. Further, the magnetic field is sensitive to the temperature. Even a change in the temperature less than 1^oC can change the field and make the weight of the top too heavy or too light. The base is to be highly horizontal and a small tilt in the base can cause the top to fall to one side. In view of these difficulties, in a commercially available Levitron the manufacturer provides a set of O-ring shaped plastic or rubber or copper washers to the top in order to adjust its weight. The level of the base can be adjusted through the adjustable legs.

Apart from the adjustment of the weight of the top and the level of the base, gyroscopic effects also play a role on the stability. When the top got tilted a torque results on the top because of the force of attraction between the opposite poles and gravitational force. This torque pull it down to the base. The angular momentum tries to resist the top to this torque and keeps it mostly upright. When the top moves out of the centre, then it makes itself to reorient to the local magnetic field and precesses around the centre.

The initial angular speed of the top is also crucial. If spun too slowly, then the top would fell down and fly-off sideways. On the other hand, too fast spin does not make the orientation of the top to follow the magnetic field lines as it moves resulting in sliding-off it.

What will happen if a top is brought to the equilibrium point without spinning it? The magnetic torque due to the magnetic field of the base turns the top to point the south pole downwards. The top would fall down due to the south-north attraction and the gravity. In the case of the spinning top, the above mentioned magnetic torque acts gyroscopically and rotates the top almost in the vertical direction of the magnetic field. Hence, the top is prevented from overturn [11].

The top is made of ceramic and it should not be a metal. This is because [11] the electrons in the top would be pushed side ways by the magnetic field of the base. These electrons would flow forming eddy currents in a metal top. These currents would be dissipated due to the resistivity of the metal leading to dissipation of rotational energy of the top. Consequently, the top would slow down and fall. To avoid this, the top is made of ceramic which is an insulator [11].

2.2 Dynamical Origin of Stable Float

We follow the treatment in [6]. Assume that the top is a small, heavy, symmetrical rigid body with mass m and is a magnetic dipole with the dipole moment $\boldsymbol{\mu}$ with magnitude μ. The position of the centre of mass (CoM) and the centre of the dipole are the same and denoted by $\mathbf{r} = (x, y, z)$. $\boldsymbol{\mu}$ is directed along the symmetry axis. The magnetic field provided by the base is $\mathbf{B}(\mathbf{r})$. The top is fast, that is, the angular momentum is along the axis of the spin of the top and its magnitude is larger than the precession. Here $\mathbf{L} = I\omega(\boldsymbol{\mu}/\mu)$ where ω is the constant angular spin frequency and I is the rotational moment of inertia about the symmetry axis which is parallel to $\boldsymbol{\mu}$.

2.2.1 *Total Force and Total Potential*

With z-axis as vertically upwards, the gravitational potential energy $V_{\mathrm{g}} = mgz$ and the gravitational force is

$$\mathbf{F}_{\mathrm{g}} = -mg\hat{z}\,, \tag{2.1}$$

where $\hat{z}$ is the unit vector along the z-axis. In the field $\mathbf{B}$, the magnetic force exerted by the field on a dipole of moment $\boldsymbol{\mu}$ and the magnetic potential energy are

$$\mathbf{F}_B = \boldsymbol{\mu} \cdot \nabla \mathbf{B}, \quad V_B(\mathbf{r}) = -\boldsymbol{\mu} \cdot \mathbf{B}(\mathbf{r})\,. \tag{2.2}$$

Hence, the total force and the total potential energy of the top are

$$\mathbf{F} = \boldsymbol{\mu} \cdot \nabla \mathbf{B} - mg\hat{z} \tag{2.3a}$$

and

$$V(\mathbf{r}) = -\boldsymbol{\mu} \cdot \mathbf{B}(\mathbf{r}) + mgz\,. \tag{2.3b}$$

2.2.2 *Equilibrium Point*

As noted earlier, at the point of stable floating of the spinning top, the net force becomes zero and that point (equilibrium point) corresponds to the minimum of the potential. The equilibrium point is given by $\nabla V(\mathbf{r}) = 0$. The condition for the equilibrium to be stable is the second derivative to be > 0. We require $\partial^2 V/\partial z^2 > 0$ for vertical stability and $\partial^2 V/\partial x^2 > 0$ and $\partial^2 V/\partial y^2 > 0$ for horizontal stability.

We set the equilibrium position as $z = z_0 = 0$. Denote z and r, respectively, small axial and radial displacements from the equilibrium. B_0 is the field strength at $z = 0$, $r = 0$. The magnetic field in a current-free region is rotationally symmetric about the vertical (z) axis. Using the cylindrical coordinates (r, ϕ, z) we write

$$\mathbf{B} = B_z(r,z)\hat{z} + B_r(r,z)\hat{r}, \quad r^2 = x^2 + y^2. \tag{2.4}$$

The power series expansion of the z-component and radial component of the field around the equilibrium point up to second-order are [14] (for details see the solved problem 1 in this chapter)

$$B_z = B_0\left[1 + \alpha_1 z + \alpha_2\left(z^2 - \frac{r^2}{2}\right)\right], \tag{2.5a}$$

$$B_r = B_0\left[-\alpha_1\frac{r}{2} - \alpha_2 z r\right], \tag{2.5b}$$

where

$$\alpha_n = \frac{1}{n!}\left[\frac{1}{B_z}\frac{\partial^n B_z}{\partial z^n}\right]_{z=z_0, r=r_0} \tag{2.5c}$$

and assume that $B_0 > 0$. In Eqs. (2.5) z and r are small axial and radial excursions, respectively, of the top from the equilibrium point of floating.

At the equilibrium point the requirement is $\mathbf{F} = 0$. With $\boldsymbol{\mu} = \mu\hat{z}$ and $\nabla\mathbf{B} = \nabla B_z = \alpha_1 B_0$ this requirement gives

$$\mu\alpha_1 B_0 - mg = 0. \tag{2.6}$$

Substituting for α_1 we have

$$\mu\frac{\partial B_z}{\partial z}\bigg|_{r=0,z=0} - mg = 0 \text{ or } \beta_1 = \frac{\partial B_z}{\partial z}\bigg|_{r=0,z=0} = \frac{mg}{\mu}. \tag{2.7}$$

Because $mg > 0$ and $B_0 > 0$ the requirement for equilibrium (from Eq. (2.6)) is $\mu\alpha_1 > 0$. Further, from Eq. (2.7) we note that mg/μ has to match with β_1, that is, *the force of gravity is to be balanced by the magnetic field gradient*. As g is a constant, the ratio m/μ has to be adjusted suitably. This can be achieved by adding small weights to the top. For this purpose, a few light-weight O-ring washers are provided with commercially available Levitrons. We proceed to determine the conditions for stability of the equilibrium.

Solved Problem 1:

For the **B** given by Eq. (2.4) obtain the power series expansion upto second-order given by Eqs. (2.5).

Outside the source, the field acting on the top satisfies the Maxwell's equations for magnetostatics $\nabla \cdot \mathbf{B} = 0$ and $\nabla \times \mathbf{B} = 0$. For the field given by Eq. (2.4) these equations take the form

$$\nabla \cdot \mathbf{B} = \frac{1}{r}\frac{\partial}{\partial r}(rB_r) + \frac{\partial B_z}{\partial z} = 0\,, \tag{2.8a}$$

$$(\nabla \times \mathbf{B})_\phi = \frac{\partial B_r}{\partial z} - \frac{\partial B_z}{\partial r} = 0\,. \tag{2.8b}$$

With $z = 0$ as the equilibrium point, the point of levitation, let us expand the axial component B_z and the radial component B_r of the field **B** about $z = 0$.

We write the series expansion of B_z and B_r upto second-order in z and r as [15]

$$B_z = B_0 + B_1 z + B_2 z^2 + B_3 r + B_4 r^2 + B_5 rz\,, \tag{2.9a}$$

$$B_r = C_0 + C_1 z + C_2 z^2 + C_3 r + C_4 r^2 + C_5 rz\,. \tag{2.9b}$$

The above expansions of B_z and B_r should satisfy the conditions given by Eqs. (2.8) [15]. Equation (2.8a) gives

$$\frac{1}{r}\left(C_0 + C_1 z + C_2 z^2\right) + 2C_3 + 3C_4 r + 2C_5 z + B_1 + 2B_2 z + B_5 r = 0\,. \tag{2.10}$$

This equation is true only if

$$C_0 = C_1 = C_2 = 0, \;\; C_3 = -\frac{B_1}{2}, \;\; C_4 = -\frac{B_5}{3}, \;\; C_5 = -B_2\,. \tag{2.11}$$

Then, Eq. (2.9b) becomes

$$B_r = -\frac{1}{2}B_1 r - \frac{1}{3}B_5 r^2 - B_2 rz\,. \tag{2.12}$$

Next, Eq. (2.8b) gives

$$(-B_2 - 2B_4)r - B_5 z - B_3 = 0\,. \tag{2.13}$$

That is,

$$B_3 = 0, \;\; B_4 = -\frac{1}{2}B_2, \;\; B_5 = 0. \tag{2.14}$$

Now,

$$B_z = B_0 + B_1 z + B_2 z^2 - \frac{1}{2} B_2 r^2, \quad B_r = -\frac{1}{2} B_1 r - B_2 z r\,. \tag{2.15}$$

In the above

$$B_0 = B_z|_{z=0,r=0}, \quad B_1 = B_z'|_{z=0,r=0}, \quad B_2 = \frac{1}{2} B_z''|_{z=0,r=0}\,, \tag{2.16}$$

where prime denotes differentiation with respect to z. For convenience, we rewrite B_z and B_r as in Eq. (2.5).

2.2.3 *Stability Analysis*

To determine the stability of the equilibrium, we consider the perturbation analysis on $\mathbf{F}$ given by Eq. (2.3a).

If the top is so fast enough, then irrespective of radial and axial deviations of the top from equilibrium, the gyroscopic action keeps $\boldsymbol{\mu}$ in the vertical alignment ($\boldsymbol{\mu} \to \mu\hat{z}$). Now, $\mathbf{F}_B = \boldsymbol{\mu} \cdot \nabla \mathbf{B}$ becomes

$$\mathbf{F}_B = \mu \frac{\partial \mathbf{B}}{\partial z}\,. \tag{2.17}$$

Considering the action of the series expansion of $\mathbf{B}$ given by Eq. (2.5) and subtracting the field at equilibrium the perturbation force from (2.17) is (see the solved problem 2 in the present chapter)

$$\mathbf{F}_B^{(\mathrm{p})} = \mu B_0 \alpha_2 (2z\hat{z} - r\hat{r})\,. \tag{2.18}$$

If $\mu\alpha_2 < 0$, then the z-component of the perturbation force is negative and the top will be stable to any axial displacement. But in this case, the radial component of the perturbation force is positive (repulsive force) and the top is unstable to a radial displacement. When $\mu\alpha_2 > 0$ the top will be unstable to axial displacement but stable to a radial displacement. Therefore, $\boldsymbol{\mu} \to \mu\hat{z}$ does not give rise to stability.

Solved Problem 2:

Obtain the perturbation force $\mathbf{F}_B^{(\mathrm{p})} = \mu B_0 \alpha_2 (2z\hat{z} - r\hat{r})$ from $\mathbf{F}_B = \mu \partial \mathbf{B}/\partial z$.

We have

$$\mathbf{F}_B^{(\mathrm{p})} = \mu \frac{\partial \mathbf{B}}{\partial z} = \mu \frac{\partial}{\partial z} (B_z \mathbf{z} + B_r \mathbf{r})\,, \tag{2.19}$$

where B_z and B_r are given by Eqs. (2.5a) and (2.5b), respectively. Substitution of the expressions for B_z and B_r in Eq. (2.19) gives

$$\begin{aligned} \mathbf{F}_B^{(\mathrm{p})} &= \mu B_0 [\alpha_1 + 2\alpha_2 z]\,\hat{z} + \mu B_0 (-\alpha_2 r)\,\mathbf{r} \\ &= \mu B_0 [(\alpha_1 + 2\alpha_2 z)\,\hat{z} - \alpha_2 r \mathbf{r}]\,. \end{aligned} \tag{2.20}$$

$(\mathbf{F}_B)_0$ at the equilibrium point $z = r = 0$ is obtained from Eq. (2.20) as

$$(\mathbf{F}_B)_0 = \mu B \alpha_1 \hat{z}. \tag{2.21}$$

So, the perturbation force is

$$\begin{aligned} \mathbf{F}_B^{(\mathrm{p})} &= \mathbf{F}_B - (\mathbf{F}_B)_0 \\ &= \mu B_0 \alpha_2 (2z\mathbf{z} - r\mathbf{r}) \, . \end{aligned} \tag{2.22}$$

The levitated spinning top stays parallel to the field $\mathbf{B}$ for all excursions from the equilibrium. Then,

$$\boldsymbol{\mu} = \mu \frac{\mathbf{B}}{B}, \quad B = |\mathbf{B}| \, . \tag{2.23}$$

Then, $\mathbf{F}_B = \boldsymbol{\mu} \cdot \nabla \mathbf{B}$ becomes

$$\mathbf{F}_B = \mu \frac{\mathbf{B}}{B} \cdot \nabla \mathbf{B} = \frac{1}{2} \frac{\mu}{B} \nabla B^2 \tag{2.24}$$

To proceed further, we consider B^2 and the expansions of $\partial B^2/\partial z$ and $\partial B^2/\partial r$ as

$$B^2 = B_0^2 \left[1 + 2\alpha_1 z + 2\alpha_2 z^2 + \alpha_1^2 z^2 + \cdots \right], \tag{2.25a}$$

$$\frac{\partial B^2}{\partial z} = B_0^2 \left[2\alpha_1 + 2\left(\alpha_1^2 + 2\alpha_2\right) z + \cdots \right], \tag{2.25b}$$

$$\frac{\partial B^2}{\partial r} = B_0^2 \left[2\left(\frac{1}{4}\alpha_1^2 + \alpha_2\right) r + \cdots \right]. \tag{2.25c}$$

Substitution of Eqs. (2.25) in Eq. (2.24) gives the z-component of the force as

$$\begin{aligned} F_{B,z} &= \frac{1}{2} \frac{\mu}{B} \frac{\partial B^2}{\partial z} \\ &= \mu B_0 \left[\alpha_1 + \left(\alpha_1^2 + 2\alpha_2\right) z + \cdots \right] \left[1 - \alpha_1 z + \cdots \right] \\ &= \mu B_0 \left[\alpha_1 + 2\alpha_2 z + \cdots \right]. \end{aligned} \tag{2.26}$$

The first term in the right-side of the above equation is the force at the equilibrium. The remaining part is the perturbation force. The linear part of it is

$$F_{B,z}^{(\mathrm{p})} = 2\mu B_0 \alpha_2 z \, . \tag{2.27}$$

The radial component of the perturbation force is

$$F_{B,r}^{(\mathrm{p})} = \frac{\mu}{2B} \frac{\partial B^2}{\partial r} \approx \frac{1}{4} \mu B_0 \left(\alpha_1^2 - 4\alpha_2\right) r \, . \tag{2.28}$$

For stability the forces $F^{\mathrm{p}}_{B,z}$ and $F^{\mathrm{p}}_{B,r}$ should be attractive. The conditions for stable equilibrium and these forces to be attractive are

$$\mu\alpha_1 > 0\,, \tag{2.29a}$$

$$\mu\alpha_2 < 0\,, \tag{2.29b}$$

$$\mu\left(\alpha_1^2 - 4\alpha_2\right) < 0\,. \tag{2.29c}$$

What will happen if $\mu > 0$*?* In this case (2.29a) and (2.29b) are satisfied for $\alpha_1 > 0$ and $\alpha_2 < 0$, respectively. But the condition (2.29c) is not satisfied. Therefore, the requirements for the existence of stable equilibrium are

$$\mu < 0, \quad \alpha_1 < 0, \quad \alpha_2 > 0, \quad \alpha_1^2 > 4\alpha_2\,. \tag{2.30}$$

These conditions can be met by a properly designed base magnet.

2.2.4 *Stability region for a Ring Dipole*

Let us apply the stability analysis to the simple case of the base being a ring dipole, however, the base is often a square plate. The ring dipole is a line dipole which is bent into a circle. Its magnetic axis is, say, oriented parallel to $\hat{z}$. B_z is given by [16]

$$B_z = \frac{\mu}{4\pi R^3}\left[\frac{2(z/R)^2 - 1}{((z/R)^2+1)^{5/2}}\right]. \tag{2.31}$$

In the above, μ is the dipole moment of the ring and R is the radius of the ring and z is the vertical distance from the plane of the ring (base). Figure 2.2 reports the variation of B_z, α_1, α_2 and $\alpha_1^2 - 4\alpha_2$ (in arbitrary units) as a function of normalized distance z/R. The stable region can be identified with $B_z > 0$ and the conditions given by Eq. (2.30). The field B_z is > 0 for $z > R/\sqrt{2}$ and it becomes a positive maximum at $z = \sqrt{3/2}R$. For $z \in [1.694R, 1.825R]$ (the solid thick interval of z in Fig. 2.2 the stability conditions are satisfied. The stabilities of vertical axis and horizontal axis Levitrons have been analysed [17].

The vertical and horixontal stabilities of a beachball levitated by a vertical jet of air are discussed in Ref. [18].

2.3 Conclusion

The Levitron is a remarkable spinning permanent magnet toy which levitates in air over a permanent magnetic base. Earnshaw's theorem rules out one magnetic dipole levitating another magnetic dipole if both the dipoles

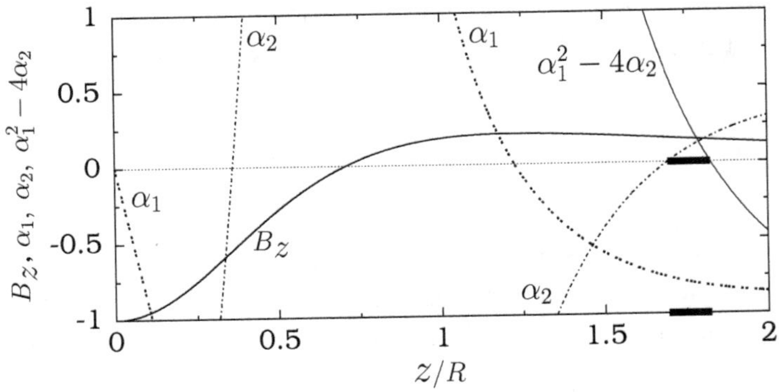

Fig. 2.2: Plot of B_z, α_1, α_2 and $\alpha_1^2 - 4\alpha_2$ versus z/R. The interval of z/R where the equilibrium point becomes stable is marked by a thick line segment on z/R-axis.

are stationary. But in the Levitron, the floating magnet is spinning. So, it is quite possible for levitation provided that there are magnetic field gradients.

The toy is not affected by the magnetic field of the earth since this field is very weak. But the equilibrium of the levitated spinning top could be well affected by even any small magnets or magnetizable objects nearer to the system. The tops such as spinning eggs, PhiTop, tippe top and an ordinary top with a sharp edge need a smooth surface to display a stable motion. In contrast, the magnetized top considered in this chapter is able to float in the air above a magnetic base. The magnetic spinning top cannot be stably floated in the air if it is held rigidly with the vertical axis. It has to tilt as it moves off-axis.

A simple theory for the Levitron discussed in [6] is presented. This theory is based on a magnetic dipole interaction and Taylor series expansion of the magnetic field to predict certain features of the equilibrium and stability of the Levitron. The magnetic field gradient is the main contributing factor for the stability of the spinning top. Both theory and experiments find that the stability region is very small and the mass of the top has to be adjusted to a narrow range of the order of 1% for the levitation of the top. So, bringing the top to levitate will not be quite easy. It is to be remembered that this theory does not explain gyroscopic precession and nutation of the top. Moreover, this model has not taken into account the strongly coupled axial and radial motions. For advanced theories on Levitron one may refer to the Refs. [5,7]. In [18] geometries of certain new Levitron are described.

2.4 Bibliography

[1] R.M. Harrigan, US Patent 4.382,245 (1983).
[2] E.W. Hones and W.G. Hones, US Patent 4,404,062 (1995).
[3] D. Kagan, The Phys. Teach. **31**, 432 (1993).
[4] R. Edge, The Phys. Teach. **33**, 252 (1995).
[5] M.V. Berry, Proc. Roy. Soc. Lond. A **452**, 1207 (1996).
[6] T.B. Jones, M. Washizu and R. Gans, J. Appl. Phys. **82**, 883 (1997).
[7] M.D. Simon, L.O. Heflinger and S.L. Ridgway, Am. J. Phys. **65**, 286 (1997).
[8] http://www.4physics.com/phy_demo/levitron/LevitronScience.htm #How%20does%20the%20LEVITRON.
[9] R. Cordaro and C.F. Cordaro, The Phys. Teach. **24**, 416 (1986).
[10] R.S. Schappe and C. Barbosa, The Phys. Teach. **55**, 6 (2017).
[11] https://web.archieve.org/web/19961109065520/http://www/lauralee. com/physics.htm.
[12] A.T. Perez and P.G. Sanchez, Am. J. Phys. **83**, 133 (2015).
[13] M.M. Michaelis, Am. J. Phys. **80**, 949 (2012).
[14] L.M. Holmes, J. Appl. Phys. **49**, 3102 (1978).
[15] K.T. McDonald, The Levitron (preprint, 1997); http://physics.princeton.edu/~mcdonald/examples/levitron.pdf
[16] E. Weber, *Electromagnetic Theory* (Dover, New York, 1965) p.125-127.
[17] M.M. Michaelis and D.B. Taylor, Eur. J. Phys. **36**, 065003 (2015).
[18] K.T. McDonald, Am. J. Phys. **68**, 388 (2000).
[19] M. Michaelis, B. Bingham, M. Chariton and C. Aled Isaac, Eur. J. Phys. **42**, 015001 (2021).

2.5 Exercises

2.1 Setup the equations of motion for the components of the magnetic dipole moment $\boldsymbol{\mu}$ and the CoM $\mathbf{r}$ of the spinning top of the Levitron. Assume that the top is spinning with the constant angular frequency ω and the mass of the top is m.

2.2 Consider the motion of the spinning top in the generalized coordinate system $(x, y, z, \phi, \theta, \psi)$. Setup the associated Lagrangian and Lagrange's equations.

2.3 Consider the potential $V(r, z) = mgz - \boldsymbol{\mu}\cdot\mathbf{B}(r, z)$. Obtain the condition for the equilibrium point $(r, z) = (0, 0)$ to be stable.

2.4 Let the base of a Levitron be a current loop with radius a and the field is

$$B_z = \frac{Aa^3}{(a^2 + z^2)^{3/2}}, \tag{2.32}$$

where A is the maximum magnetic field. Determine the stability interval of z for stable levitation [15].

2.5 Assume that the magnetic moment $\boldsymbol{\mu}$ of the top is always oriented in the vertically downward $(-z)$ direction and the repulsive magnetic field $\mathbf{B}$ from the base magnet is primarily in the vertical upward $(+z)$ direction. Show that the stability condition for levitation is not satisfied for this arrangement.

Chapter 3

Electric Train

An electric toy train locomotion is essentially an electromechanical device in which the electrical and mechanical parts are coupled. Electric trains are used to demonstrate certain concepts like friction, electrical/mechanical energy transfer, electrically induced torque, resistance, inductance, electromotive force of a motor and the influence of applied voltage in electricity and magnetism. A typical electric toy train [1] (Fig. 3.1a) contains wheel assemblies in its front and rear sides. These assemblies pivot independently to negotiate turns. In an assembly, usually there are four conducting wheels and they move on a conducting track. These wheels are coupled through a flywheels set and small size gears to a permanent magnet dc driving motor. Slipping between the wheels and the track does not occur.

We follow the treatment of an electric toy train described in [1]. We consider a toy train with electric power supplied by a constant voltage source and assume the case of the train motion on a horizontal track. In developing a theoretical model, we take into account of the influence of resistance, inductance, applied voltage, electromotive force of the motor, friction and the electrically induced torque. The equations of motion for the speed of the train and the current in the system are decoupled second-order linear ordinary differential equations. We obtain the steady state speed of the train and the current and analyse their stability. We describe the computation of certain parameters in the model equations.

3.1 Mechanical Description

We start with presenting the underlying mechanical aspects of the toy train system including the various parameters of the system, the force and the torques arising in the train [1]. Figure 3.1b depicts a motor-flywheel setup

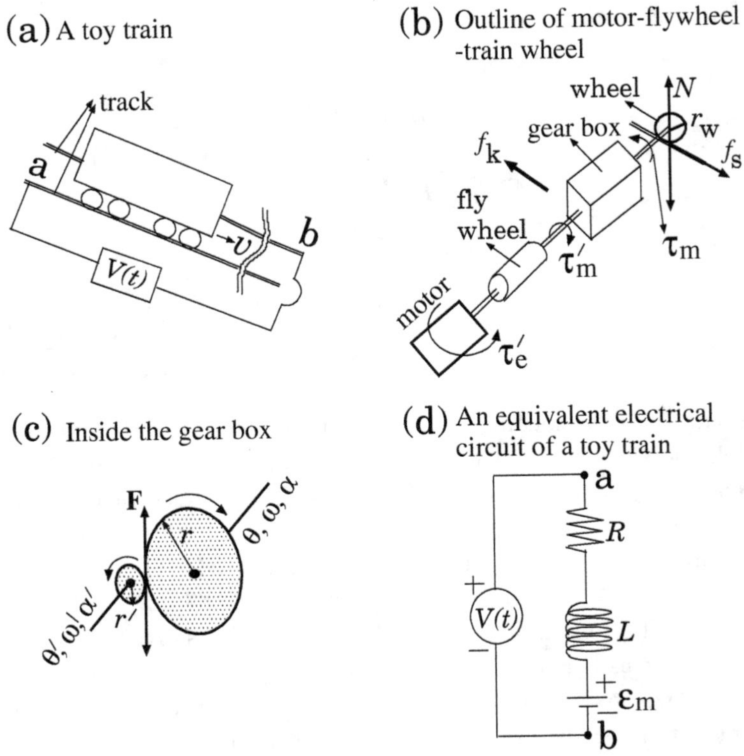

Fig. 3.1: (a) An outline of a typical electric toy train. (b) A motor-flywheel arrangement coupled to a wheel of an electric train [1]. (c) Two gears in a gear box. (d) An equivalent electic circuit of the toy train shown in (a). The resistance of the train and the track is R and L is the total inductance. (Figures 3.1b and c are reproduced from Am. J. Phys. **72**, 863 (2004), with permission of American Association of Physics Teachers. Copyright 2004 D.P. Wick and M.W. Ramsdell, licensed under a Creative Commons Attribution License.)

for a wheel and the forces and the torques involved. The electric power is supplied to the motor via the track by means of sliding contacts adjacent to the wheels. The operating voltage of the train is usually 3 V−12 V. Denote M, N and r_w as the total mass of the train, the net normal force and the radius of the wheel, respectively. r and r' are the radii of two gears as indicated in Fig. 3.1c. The angular displacement θ, the angular speed ω and the angular acceleration α of the flywheel (the big wheel in Fig. 3.1c) are related to those (denoted as θ', ω' and α') of the small wheel as

$$\theta' = G\theta, \quad \omega' = G\omega, \quad \alpha' = G\alpha, \tag{3.1}$$

where G is the motor-to-load gear ratio. The difference between the angular speeds of the motor shaft and wheel are described by G and its value can be determined by counting the number of turns of the flywheel (small wheel) required for one complete rotation of the (big)wheel.

The motor-flywheel system is coupled to the train wheel by the contact force $\mathbf{F}$ (see Fig. 3.1c) arising from action-reaction. This coupling leads to $r'\theta' = r\theta$. Between the wheels and the track a static friction force f_s exists. When the circuit is closed, f_s acts in the forward direction. This is opposed by the retarding friction force. The magnitude of this force increases and reaches a miximum value. Denote the time at which this happens as t_c. At this time, the current i in the circuit attains a critical value i_c and acceleration of the train starts. Now, the retarding friction force is essentially an effective sliding friction force (f_k). This force accounts for the minimum voltage V_{min} needed for the train to start its motion.

A current flowing through the coil in the motor induces a torque τ'_e. This torque is transmitted through the gear box to the wheel. τ'_e is given by

$$\tau'_e = |\boldsymbol{\mu} \times \mathbf{B}| = NBA|\sin\theta'|i = ci\,, \tag{3.2}$$

where $\boldsymbol{\mu}$ is the magnetic dipole moment. $|\boldsymbol{\mu}| = NAi$ where N is the number of turns in the coil, A is the area enclosed by the coil and $c = NBA|\sin\theta'|$ is an electromechanical coupling constant. τ_m and τ'_m associated with the subsystems as marked in Fig. 3.1b are given by

$$\tau_m = k\omega = kv/r_w\,, \quad \tau'_m = k'\omega' = k'G\omega = k'Gv/r_w\,, \tag{3.3}$$

where k and k' are the damping coefficients of the corresponding subsystems and v is the speed of the train.

3.2 Equations of Motion

Let us setup the equations of motion for the speed v of the train and the current i in the circuit for time $t > t_c$ [1]. Consider the acceleration, that is, rate of change of the speed of the train associated with the translation motion of the train and the rotational motion of the two subsystems, namely, motor-flywheel and train wheel.

For the translation motion, as the static friction force f_s acts in the direction of the motion of the train and the sliding friction force f_k acts in the opposite direction, according to Newton's second law of motion we write

$$M\dot{v} = f_{\mathrm{s}} - f_{\mathrm{k}}\,. \tag{3.4a}$$

For the rotational motion of the motor-flywheel subsystem, we take into account of the angular momentum $L' = I'\alpha'$ where I' is the moment of inertia. $I'\alpha'$ is equal to the net torque of this subsystem. The torques involved are (refer Fig. 3.1b) τ'_{e}, τ'_{m} and $\tau' = r'F$. Then,

$$\tau'_{\mathrm{e}} - \tau'_{\mathrm{m}} - r'F = I'\alpha'\,. \tag{3.4b}$$

For the rotational motion of the train wheel (from Fig. 3.1b)

$$rF - \tau_{\mathrm{m}} - r_{\mathrm{w}} f_{\mathrm{s}} = I\alpha\,, \tag{3.4c}$$

where I is the moment of inertia of the train wheel subsystem.

Elimination of f_{s} in Eq. (3.4a) using Eq. (3.4c), using Eq. (3.4b) for F and then using Eqs. (3.2) and (3.3) and performing some simple mathematics result in the equation

$$I_{\mathrm{eq}}\dot{v} = cGr_{\mathrm{w}}i - k_{\mathrm{eq}}v - r_{\mathrm{w}}^2 f_{\mathrm{k}}\,, \tag{3.5}$$

where

$$I_{\mathrm{eq}} = I + G^2 I' + Mr_{\mathrm{w}}^2\,, \quad k_{\mathrm{eq}} = k + G^2 k'\,. \tag{3.6}$$

For $t < t_{\mathrm{c}}$ the train is at rest and there is no induced emf because the coil is not in rotation. For $t > t_{\mathrm{c}}$ the train is in motion and the motor coil rotating through the magnetic field generates the emf ε_{m} and is equal to $-\mathrm{d}\Phi_B/\mathrm{d}t$ where Φ_B is the magnetic flux. For the coil, the magnetic flux Φ_B is

$$\Phi_B = N\int_A \mathbf{B}\cdot\mathbf{d}\mathrm{A} = NBA\cos\theta'\,. \tag{3.7}$$

Then, with the use of Eqs. (3.1) and (3.2) $\varepsilon_{\mathrm{m}} = -\mathrm{d}\Phi_B/\mathrm{d}t$ gives

$$\varepsilon_{\mathrm{m}} = NBA|\sin\theta'|\,\dot{\theta}' = (\tau'_{\mathrm{e}}/i)\dot{\theta}' = c\dot{\theta}' = cG\dot{\theta} = cGv/r_{\mathrm{w}}\,. \tag{3.8}$$

v and ε_{m} are related and are not independent quantities. The current i in the train circuit will vary with v (or ε_{m}). We can find the time evolution of i. The train circuit essentially has circuit elements R, L, ε_{m} and $V(t)$. An equivalent electrical circuit for the train system circuit is shown in Fig. 3.1d. For this circuit the Kirchhoff's voltage rule gives

$$L\dot{i} + iR + \varepsilon_{\mathrm{m}} = V(t), \quad \varepsilon_{\mathrm{m}} = cGv/r_{\mathrm{w}}\,. \tag{3.9}$$

The equations for v and i given by the Eqs. (3.5) and (3.9) are coupled equations. They can be decoupled as

$$\alpha\ddot{v} + \beta\dot{v} + v = \gamma\,, \tag{3.10a}$$

$$\alpha\ddot{i} + \beta\dot{i} + i = \delta\,, \tag{3.10b}$$

where

$$\alpha = I_{\mathrm{eq}}L/(R\sigma)\,, \quad \sigma = k_{\mathrm{eq}} + c^2G^2/R\,, \tag{3.10c}$$

$$\beta = (RI_{\mathrm{eq}} + Lk_{\mathrm{eq}})/(R\sigma)\,, \quad \gamma = \left(cGr_{\mathrm{w}}V - Rr_{\mathrm{w}}^2 f_{\mathrm{k}}\right)/(R\sigma)\,, \tag{3.10d}$$

$$\delta = \left(cGr_{\mathrm{w}}f_{\mathrm{k}} + k_{\mathrm{eq}}V + I_{\mathrm{eq}}\dot{V}\right)/(R\sigma)\,. \tag{3.10e}$$

Solved Problem 1:

Determine the steady state values of v and i and analyse their stability for $V(t) = V$, a constant applied voltage.

Consider the equation of motion for v. We follow the stability anaysis discussed in Sec. 1.8. We rewrite the Eq. (3.10a) as

$$\dot{v} = w\,, \quad \dot{w} = -\frac{\beta}{\alpha}w - \frac{1}{\alpha}v + \frac{\gamma}{\alpha}\,. \tag{3.11}$$

The steady state values of v and w are obtained by substituting $\dot{v} = \dot{w} = 0$. This gives $(v_{\mathrm{s}}, w_{\mathrm{s}}) = (\gamma, 0)$ where γ is given by Eq. (3.10d). For the forward motion of the toy train the requirement is $\gamma > 0$. As all the parameters in Eq. (3.10d) are positive, for $\gamma > 0$ the condition is

$$V > V_{\mathrm{min}} = \frac{Rf_{\mathrm{k}}r_{\mathrm{w}}}{cG}\,. \tag{3.12}$$

That is, for $V < V_{\mathrm{min}}$ the train will be at rest. For $V > V_{\mathrm{min}}$ the train will be at rest for $t < t_{\mathrm{c}}$ and will move in the forward direction for $t > t_{\mathrm{c}}$ and its steady state speed is $v_{\mathrm{s}} = \gamma$. This speed increases linearly with $V > V_{\mathrm{min}}$.

The stability determining eigenvalues are the roots of the equation

$$\begin{vmatrix} -\lambda & 1 \\ -1/\alpha & -\beta/\alpha - \lambda \end{vmatrix} = 0. \tag{3.13}$$

Expanding the determinant gives $\lambda^2 + (\beta/\alpha)\lambda + 1/\alpha = 0$. Its roots are

$$\lambda_{\pm} = \frac{1}{2\alpha}\left[-\beta \pm \sqrt{\beta^2 - 4\alpha}\right]\,. \tag{3.14}$$

The condition for stable v_{s} is real part of both the eigenvalues must be negative. As α and β are positive quantities, the stability condition is satisfied without any restriction on the values of β and α.

Next, from Eq. (3.10b) the steady state current i_s is

$$i_s = \delta = (cGr_w f_k + k_{eq}V)/(R\sigma). \tag{3.15}$$

The steady state current is always positive. The stability determining eigenvalues are given by Eq. (3.14). i_s is thus stable.

3.3 Determination of the Parameters

For a given toy train, certain parameters can be measured directly and others can be computed by experiments [1]. We can measure the values of r_w, R, L and G directly. The values of c, k_{eq} and f_k can be determined through experiments. To find these three parameters consider v_s and i_s. Rewrite the expressions for them as

$$v_s = \gamma = S_\gamma V - I_\gamma, \quad i_s = \delta = S_\delta V + I_\delta, \tag{3.16}$$

where

$$S_\gamma = \frac{cGr_w}{R\sigma}, \quad I_\gamma = \frac{r_w^2 f_k}{\sigma}, \quad S_\delta = \frac{k_{eq}}{R\sigma}, \quad I_\delta = \frac{cGr_w f_k}{R\sigma}. \tag{3.17}$$

By varying the voltage V from, say, 0 V to 12 V in steps of 1 V the distance travelled by the train for t seconds and the current flowing in the circuit can be measured. For each V, several trails can be perfomed and the average speed per second and current can be computed. Straight-line fits to V versus v_s and V versus i_s provide the values of S_γ, I_γ, S_δ and I_δ. Experiments performed for a toy train confirmed the linear relations (3.16) [1]. In Eq. (3.17) the quantity σ is given by Eq. (3.10c). The four subequations in Eq. (3.17) are system of nonlinear equations for the parameters cG, R, k_{eq} and f_k. By solving them, algebraic expressions for these four parameters can be obtained in terms of the quantites r_w, S_γ, I_γ, S_δ and I_δ (see the exercise 3 at the end of this chapter).

For $V > V_{min}$ the toy train will move in the forward direction for $t > t_c$. At $V = V_{min}$ the value of v_s is 0. This gives, from the first subequation of (3.16), the value of V_{min} as $I_\gamma/S_\gamma = Rf_k r_w/(cG)$ (refer Eq. (3.12)). The current corresponding to the applied voltage V_{min} is the critical current i_c necessary for the electrically induced torque of the motor to overcome the retarding friction force f_k. Substitution of $V = V_{min} = I_\gamma/S_\gamma$ in the second subequation of (3.16) gives

$$i_c = S_\delta V_{min} + I_\delta = \frac{S_\delta}{S_\gamma} I_\gamma + I_\delta. \tag{3.18}$$

Suppose it is found in an experiment with the electronic toy train of wheel radius $r_w = 0.6\,\text{cm}$, velocity $v_s = 0.18\,\text{m/s}$ for $V = 5\,\text{V}$ and $v_s = 0.41\,\text{m/s}$ for $V = 10\,\text{V}$. Also, the current $i_s = 0.105\,\text{A}$ for $V = 5\,\text{V}$ and $i_s = 0.14\,\text{A}$ for $V = 10\,\text{V}$. Using these data we can determine the parameters f_k, k_{eq}, R and cG. From the first subequation Eq. (3.16) we write $v_{s_1} - v_{s_2} = S_\gamma (V_1 - V_2)$. This gives

$$S_\gamma = \frac{v_{s_1} - v_{s_2}}{V_1 - V_2} = \frac{0.18 - 0.41}{5 - 10} = 0.046\,. \tag{3.19}$$

Further,

$$I_\gamma = S_\gamma V - v_s = 0.046 \times 5 - 0.18 = 0.05\,\text{A}\,. \tag{3.20}$$

Using the second subequation of Eq. (3.16) we find

$$S_\delta = \frac{i_{s_1} - i_{s_2}}{V_1 - V_2} = \frac{0.105 - 0.14}{5 - 10} = 0.007 \tag{3.21}$$

and

$$I_\delta = i_s - S_\delta V = 0.105 - 0.007 \times 5 = 0.07\,\text{A}. \tag{3.22}$$

From the first and the last subequations of Eq. (3.17) we find

$$f_k = \frac{I_\delta}{S_\gamma} = \frac{0.07}{0.046} = 1.52\,\text{N}\,. \tag{3.23}$$

For the determination of the expressions for R, k_{eq} and cG see the exercise 3 at the end of this chapter. We use the obtained expressions. We obtain

$$R = \frac{1}{S_\gamma\left(\dfrac{S_\delta}{S_\gamma} + \dfrac{I_\delta}{I_\gamma}\right)} = \frac{1}{0.046\left(\dfrac{0.007}{0.046} + \dfrac{0.07}{0.05}\right)} = 14\,\text{ohm}\,, \tag{3.24}$$

$$k_{eq} = \frac{r_w^2}{S_\gamma\left(\dfrac{S_\gamma}{S_\delta} + \dfrac{I_\gamma}{I_\delta}\right)} = \frac{0.6 \times 0.6 \times 10^{-4}}{0.046\left(\dfrac{0.046}{0.007} + \dfrac{0.05}{0.07}\right)} = 1.06 \times 10^{-4}\,\text{kgm}^2/\text{s} \tag{3.25}$$

and

$$cG = \frac{r_w}{S_\delta\left(\dfrac{S_\gamma}{S_\delta} + \dfrac{I_\gamma}{I_\delta}\right)} = \frac{0.6 \times 10^{-2}}{0.007\left(\dfrac{0.046}{0.007} + \dfrac{0.05}{0.07}\right)} = 0.12\,\text{kgm}^2/\text{Cs}\,. \tag{3.26}$$

3.4 Conclusion

We presented a theoretical description of motion of a toy train for a constant applied voltage V. The equations of motion for the speed of the train v

and the current i in the circuit are decoupled linear second-order equations. For $V(t)$, these equations are inhomogeneous equations, however, linear. We can construct exact analytical solution and analyze the time dependence of v and i and the influence of various parameters involved in the system. In [1] the effects of piecewise linear and sinusoidal forms of $V(t)$ and inductance have been investigated. Use of smartphone's light sensor to determine the angular velocity, linear velocity and centripetal acceleration of a light-emitting toy train moving on a circular track was reported [2]. A toy train system driven by intermittent wireless power supply has been proposed and studied [3].

3.5 Bibliography

[1] D.P. Wick and M.W. Ramsdell, Am. J. Phys. **72**, 863 (2004).
[2] S. Kapucu, The. Phys. Teach. **57**, 480 (2019).
[3] Y. Iyama, D. Sakashita and K. Hirahara, IEICE Tech. Rep. **17**, 9 (2017).

3.6 Exercises

3.1 Starting from Eqs.(3.4) arrive at the Eq. (3.5).

3.2 Decouple the Eqs. (3.5) and (3.9).

3.3 Solving the Eqs. (3.17) obtain the expressions for the parametes $f_{\rm k}$, $k_{\rm eq}$, cG and R.

3.4 The following table gives the average distances (d) moved by a toy train in 5 seconds and the measured steady current i for a range of values of applied voltage V. Compute the values of S_γ, I_γ, S_δ and I_δ by least-squares curve fitting and then find the values of $V_{\rm min}$, $i_{\rm c}$, $f_{\rm k}$, $k_{\rm eq}$, R and cG by assuming $r_{\rm w} = 0.7\,{\rm cm}$. Use the expressions obtained for $f_{\rm k}$, $k_{\rm eq}$, R and cG in the previous exercise.

V in volts	1	2	3	4	5
d in 5 seconds in m	0	0.015	0.029	0.05	0.07
i in ampere	0.065	0.07	0.076	0.08	0.084

6	7	8	9	10	11	12
0.085	0.105	0.125	0.14	0.16	0.18	0.195
0.09	0.095	0.101	0.106	0.11	0.116	0.12

3.5 For a toy train with an applied voltage $V > V_{\rm min}$ find an expression for $t_{\rm c}$ where for $t < t_{\rm c}$ the train is at rest.

Chapter 4

A Simple Electromagnetic Train

We describe a toy which is a simple and popular electromagnetic train. This toy consists of an AA battery, two strong neodymium spherical or disc magnets and a coil of copper wire. Figure 4.1 shows the train. At each end of the battery, magnets are attached. The orientation of the magnets is important. The free end of both magnets must have the same pole. That is, either south poles or north poles of both magnets should face each other. The battery with the magnets attached to both ends is the train. The track of the train is a coil of the copper wire made by winding the un-insulated wire around a wood dowel rod by hand. The track is like a solenoid. Insulated wire will be not useful since there will not be any flow of current. The diameter of the track is slightly bigger than the train and the turns in the solenoid to be smooth to allow the train to move smoothly. Winding the wire too closely will make us impossible to follow the train. Giving too much space between the successive turns will lead to a slow movement of the train. The adjacent turns should not touch otherwise the magnetic field created becomes weak.

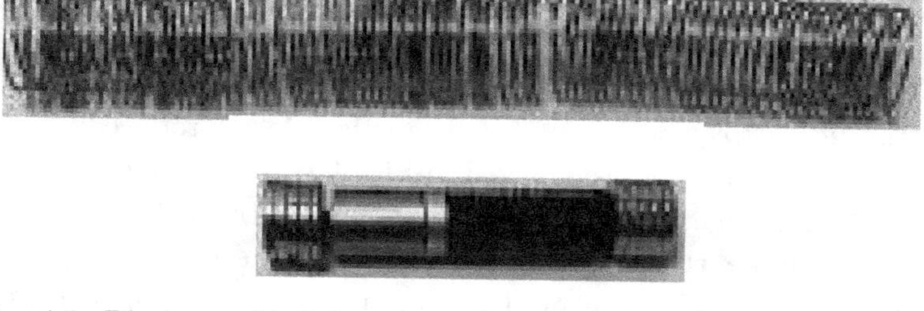

Fig. 4.1: Photograph of the setup of a simple electromagnetic train.

When the train is fully inserted into the one end of the track, it moves through the tunnel-like track in the forward direction and comes out via the other end. In case the train got pushed back out of the track, then reverse the end of the train and insert it into the track. Tracks with curves, spirals, ups and downs and ramps lead to increase in the entertainment [1-3].

Assume that north poles of the magnets are in touch with the ends of the battery. The orientation of the magnets are such that they try to push each other apart. However, they are already made apart by the battery. The repulsion of the magnets is overcome by the attraction of them to the metal ends of the battery. The magnets are made up of conducting materials boron and iron and so the neodymium magnets are capable to conduct electricity. When the train is fully placed inside the coil at one end, a local circuit is setup. The circuit starts from the positive terminal of the body, into the neodymium magnet, then to the copper wire of the coil and finally to the magnet attached to the other terminal of the battery. Suppose the positive terminal of the battery is in the left-side. A magnetic field is developed by the flowing current in the solenoid. This field interacts with the magnetic dipole moment associated with the magnets. The resulting force moves the train.

For the details of the laws of electromagnetism which govern the motion of the train refer to Sec. 1.10 of chapter 1 of part I. In the present chapter, first we consider the working principle of the electromagnetic train. We point out the various forces in action. We obtain the expressions for the forces due to the current in the solenoid and the eddy currents.

4.1 Principle of Working

Let us describe the physics behind the working of the electromagnetic train [3-6]. The train works under the two fundamental principles of magnetism and electromagnetism. The first is the magnetic field developed by the magnetic poles. The second is the magnetic field developed by the current carrying solenoid. *How are these two utilized in the train?*

A coil of wire wound in a helical shape is a *solenoid*. When a current flows through a solenoid a constant magnetic field is produced. The field lines of the resultant field is depicted in Fig. 4.2. Notice that the direction of the produced magnetic field is through the centre of the solenoid. In a magnetic dipole, the field lines originate from the north pole and ends on the south pole. The field produced in and around the solenoid sets up north and south poles as shown in Fig. 4.2.

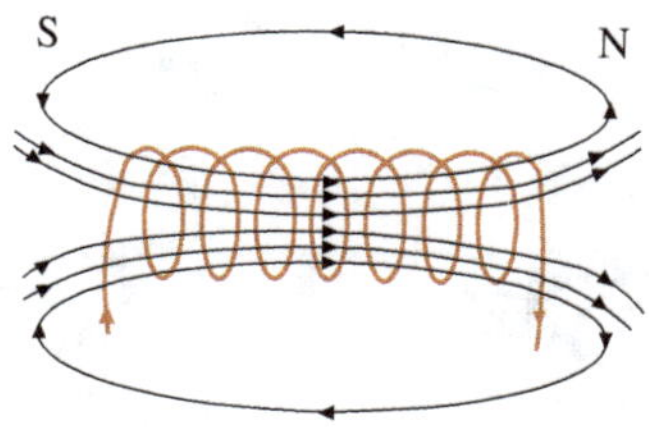

Fig. 4.2: The magnetic field lines of a solenoid.

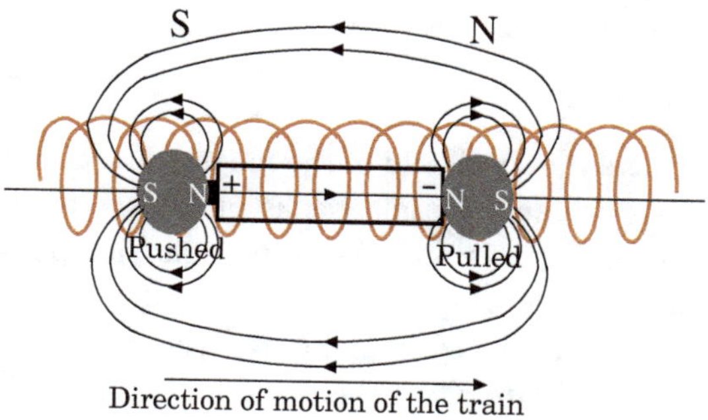

Fig. 4.3: The magnetic field lines of the train when south poles of the two magnets are facing outwards. The train moves from left to right.

In the train there are two magnets: one is attached to the positive terminal and the other to the negative terminal of the battery. These magnets can be cylindrical or spherical. Here onwards we choose spherical magnets. When the train is fully placed inside the solenoid-like track, a circuit is formed and the flowing current creates a magnetic field. In addition to this solenoid field, there are two other magnetic fields created by the two magnets attached to the battery. *What kind of interaction takes place between these three fields?*

Suppose the orientation of the poles of the magnets are as in Fig. 4.3. The south poles are facing outwards. The current flows from the left-side to the right-side and the south pole of the field of the solenoid is in the left-side of the train. The field lines of the three fields are shown in Fig. 4.3. The south pole setup by the solenoid and the south pole of the magnet attached to the positive terminal of the battery are in the left-side. Therefore, the solenoid field opposes this magnet and is pushed away (towards the right-side). On the other hand, in the right-side (front side) the north end of the

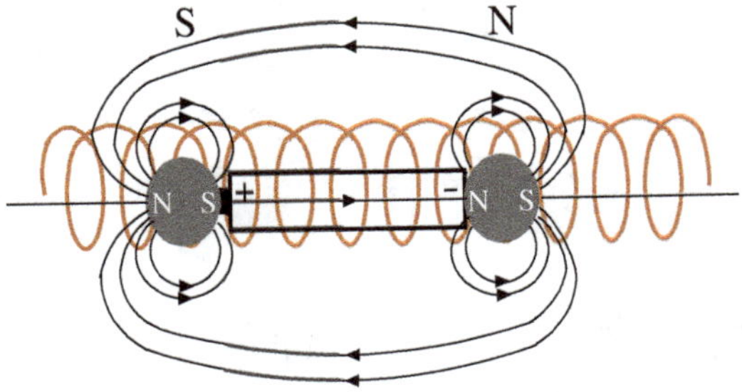

Fig. 4.4: The magnetic field lines of the train when opposite poles are facing outwards. There is no movement of the train.

solenoid field attracts the south end of the magnet attached to the negative terminal of the battery and this magnet is pulled right-side. Thus, there are *two forces* acting on the train: one in the right-side pulls the train and the second in the left-side pushes the train. That is, simultaneously the poles of the magnets are pushed in and out of the local circuit. The result is the *movement of the train* along the track.

As the train moves along the track, a new localized circuit is created thereby the movement of the train continues. Note that for the train to move, the poles of the ends of the train must be the same. In Fig. 4.3 the outer poles are set to the south. If the outer poles are the north by flipping the magnets, then the train can move but now in the other direction.

What will happen if the outer poles are as in Fig. 4.4 [4]? In this case there are also three magnetic fields. In the left- and as well as in the right-side of the train the outer poles of the magnet attached to the train and the solenoid field are opposite. On both sides the poles are attracted. The forces are thus in the opposite direction and *the train becomes stationary.*

Can a train move if a magnet is attached to one end only? The answer is yes. In the case of magnets in both ends, as in Fig. 4.3 there is a push in the left-side while there is a pull in the right-side and both in the same direction. This indicates that one magnet in one end is sufficient for the train to move, provided, the circuit is set into complete by a way.

4.2 Forces Opposing the Motion of the Train

There are two forces opposing the motion of the train. One is due to the sticking of the edge of the train in the spaces between the solenoid wires when the train moves. This force is normal to the edge of the train on the wire. This opposing force can be minimized by using a spherical shape magnet (in place of a cylindrical disk magnet) which will lead to a smooth motion through the track.

The second force is due to the eddy currents developed when the magnets pass through the coiled wire [5]. Such currents occur when the current flows in a closed loop. In the electromagnetic train setup, the eddy currents will flow in the portion of the coil between the magnets attached to the ends of the battery. This current is $I_{\mathrm{E}} = \varepsilon / R_{\mathrm{T}}$ where ε is the electromotive force (emf) over the closed loop under consideration and R_{T} is the resistance of the closed circuit. The magnetic force developed by the eddy currents on the train opposes its motion. This force is proportional to the velocity of the train and acts in the direction opposite to the direction of motion of the train. Therefore, the speed of the train will reach a terminal speed once it started moving [5].

Solved Problem 1:

Determine the magnetic field produced by a solenoid setup in the electromagnetic train.

Consider a solenoid with N turns in the length L. The number of turns per unit length is $n = N/L$. The radius of the solenoid is R. The magnetic field is created inside a solenoid when a current flows through it. The direction of the field depends on the direction of the flow of the current. Further, the created field depends on the number of turns per unit length. Outside the solenoid the field is zero or small.

P be the point at which the magnetic field is to be calculated. The point P is, say, at a distance x from the origin O_1 as shown in Fig. 4.5. The direction of the current is indicated by the symbols $\otimes$ and $\odot$. Here $r^2 = x^2 + R^2$. According to Biot-Savart law, the magnetic field due to a tiny segment of the coil, up at top of the loop is

$$\mathrm{d}\mathbf{B} = \frac{\mu_0 I}{4\pi} \frac{\mathrm{d}\mathbf{S} \times \widehat{\mathbf{r}}}{r^2} . \tag{4.1}$$

This field points downwards. The bottom segment of the loop produces a field that point downwards but of the same strength. Therefore, the vertical component cancels out. The horizontal component of $\mathrm{d}\mathbf{B}$ is [7]

$$\mathrm{d}B_1 = \frac{\mu_0 I}{4\pi} \frac{\mathrm{d}S}{(x^2+R^2)} \frac{R}{(x^2+R^2)^{1/2}} = \frac{\mu_0 I}{4\pi} \frac{R}{(x^2+R^2)^{3/2}} \mathrm{d}S . \tag{4.2}$$

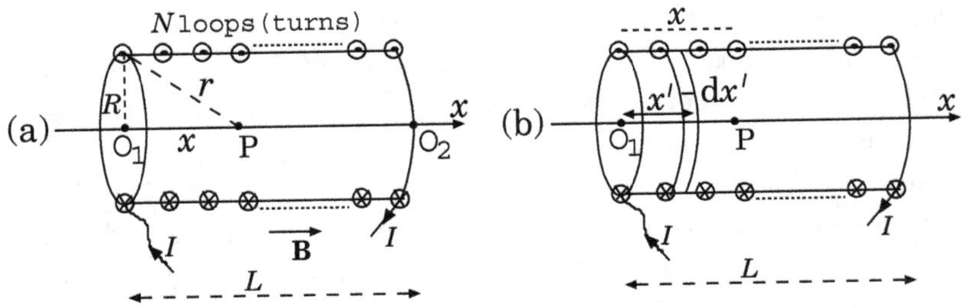

Fig. 4.5: (a) Geometry of a solenoid formed in the electromagnetic train setup. (b) A small volume element with width $\mathrm{d}x'$ at a distance x' from the origin O_1.

Then, the field over one loop along the axis of a single loop is obtained as

$$\begin{aligned} B_1 &= \frac{\mu_0 I}{4\pi} \frac{R}{(x^2+R^2)^{3/2}} \int_{\text{loop}} \mathrm{d}S \\ &= \frac{\mu_0 I}{4\pi} \frac{R}{(x^2+R^2)^{3/2}} 2\pi R \\ &= \frac{\mu_0 I}{2} \frac{R^2}{(x^2+R^2)^{3/2}} . \end{aligned} \tag{4.3}$$

Next, we determine the field due to all the loops. Consider Fig. 4.5b where the segment of the solenoid $\mathrm{d}x'$ is at a distance x' from the origin. The current in this section is

$$I_{\mathrm{T}} = \frac{IN}{L}\mathrm{d}x' . \tag{4.4}$$

The point P is at a distance $x - x'$ from this section. Then, the field due to this section of the solenoid is

$$\mathrm{d}B_1 = \frac{\mu_0 IN}{2L} \frac{R^2}{((x-x')^2+R^2)^{3/2}} \mathrm{d}x' . \tag{4.5}$$

The total field is

$$\begin{aligned} B_1 &= \frac{\mu_0 INR^2}{2L} \int_0^L \frac{1}{((x-x')^2+R^2)^{3/2}} \mathrm{d}x' \\ &= \frac{\mu_0 INR^2}{2L} \int_{x-L}^{x} \frac{1}{(y^2+R^2)^{3/2}} \mathrm{d}y \\ &= \frac{\mu_0 INR^2}{2L} \frac{1}{R^2} \left[\frac{y}{(y^2+R^2)^{1/2}} \right]_{x-L}^{x} \\ &= \frac{\mu_0 IN}{2L} \left[\frac{x}{(x^2+R^2)^{1/2}} - \frac{x-L}{((x-L)^2+R^2)^{1/2}} \right] . \end{aligned} \tag{4.6}$$

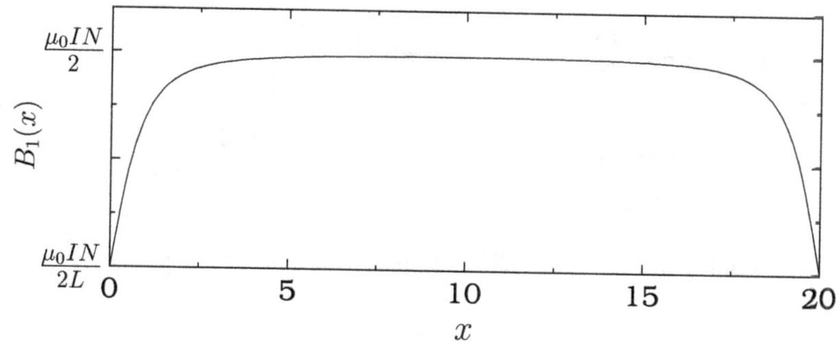

Fig. 4.6: $B_1(x)$ versus x for $R = 1$ units and $L = 20R$.

Note that

$$B_1(0) = B_1(L) = \frac{\mu_0 IN}{2\left(L^2 + R^2\right)^{1/2}} . \tag{4.7}$$

Figure 4.6 shows the variation of $B_1(x)$. The field becomes maximum at $x = L/2$.

4.3 Forces on the Train

There are two magnetic forces on the train. One is due to the current flowing in the solenoid and the other is due to the eddy currents. We calculate these forces [5].

4.3.1 *Force Due to the Current in the Solenoid*

In the case of a spherical magnet of radius a with magnetization $\mathbf{M}$ the field outside the magnet is simply that of a pure dipole [8] having the magnetic dipole moment $\mathbf{m} = (4/3)\pi a^3 \mathbf{M}$. The magnetic force on the dipole is $\mathbf{F} = \nabla(\mathbf{m} \cdot \mathbf{B})$. To bring the dipole from infinity to a point $\mathbf{r}$, we need to apply an opposite force. So, the work done is

$$U = -\int_{\infty}^{r} \mathbf{F} \cdot \mathrm{d}\mathbf{l} = -\int_{\infty}^{r} \nabla(\mathbf{m} \cdot \mathbf{B}) \cdot \mathrm{d}\mathbf{l} = -\mathbf{m} \cdot \mathbf{B}(\mathbf{r}) + \mathbf{m} \cdot \mathbf{B}(\infty) . \tag{4.8}$$

For $\mathbf{B}(\infty) = 0$, we have $U = -\mathbf{m} \cdot \mathbf{B}$. This is the potential energy associated with the permanent magnetic dipole of the magnetic moment $\mathbf{m}$ in $\mathbf{B}$ [9]. In the present problem there are two spherical magnetic dipoles - one centred at O_1 and another at O_2 (refer Fig. 4.7). We choose O_1 as the origin and the axes x, y and z as shown in Fig. 4.7. Within the interval

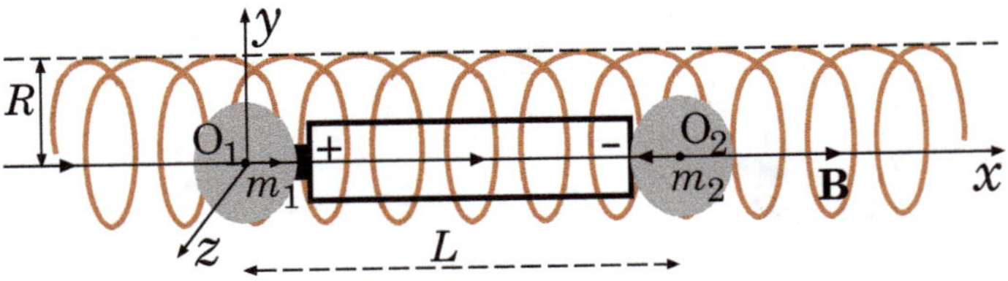

Fig. 4.7: Schematic representation of the train and the magnetic dipoles at O_1 and O_2.

L (the interval between O_1 and O_2), say, there are N turns of wire on the average.

The dipole moments of the magnets at O_1 and O_2 are $\mathbf{m}_1 = (m, 0, 0)$ and $\mathbf{m}_2 = (-m, 0, 0)$, respectively. Denote the field created by the current I flowing through the solenoid as $\mathbf{B} = (B_1, B_2, B_3)$, then for the dipole at O_1 the potential energy $U_1 = -mB_1(x)$. For the dipole at O_2, we have $U_2 = mB_1(x)$. The field $B_1(x)$ is given by Eq. (4.6). As $\mathbf{F} = -\nabla U$ the force on the dipole centred at O_1 with $x = 0$ is obtained as [5]

$$F(0) = -\frac{\mathrm{d}U_1}{\mathrm{d}x}\bigg|_{x=0} = m\frac{\mathrm{d}B_1}{\mathrm{d}x}\bigg|_{x=0} = \frac{m\mu_0 IN}{2L}\left[\frac{1}{R} - \frac{R^2}{(L^2+R^2)^{3/2}}\right]. \quad (4.9)$$

Next, the force on the dipole centred at O_2 at a distance L from O_1 is

$$F(L) = -m\frac{\mathrm{d}B_1(x)}{\mathrm{d}x}\bigg|_{x=L} = \frac{m\mu_0 IN}{2L}\left[\frac{1}{R} - \frac{R^2}{(L^2+R^2)^{3/2}}\right] = F(0). \quad (4.10)$$

The net force F is then

$$F = F(0) + F(L) = \frac{m\mu_0 IN}{L}\left[\frac{1}{R} - \frac{R^2}{(L^2+R^2)^{3/2}}\right]. \quad (4.11)$$

We can write

$$F = CI\,, \quad C = \frac{m\mu_0 N}{L}\left[\frac{1}{R} - \frac{R^2}{(L^2+R^2)^{3/2}}\right]. \quad (4.12)$$

4.3.2 *Force due to Eddy Currents*

When the conductor setup a closed loops of current to flow, then eddy currents are formed. In the train setup the battery is a conductor and there is a closed loop. So, in the portion of the coils between the magnets eddy currents are produced. As mentioned in Sec. 4.2 the eddy current is

$I_E = \varepsilon/R_T$. The magnetic force due to this current will oppose the motion of the train and hence the train will acquire a terminal velocity v. This velocity can be determined [5].

Let ε_1 and ε_2 are the emfs generated by the dipoles centred at O_1 and O_2, respectively. We find that $\varepsilon_1 = \varepsilon_2 = -vC/2$ (for a proof see the exercise 4.2 at the end of this chapter) where C is given in Eq. (4.12). Then,

$$I_E = \frac{\varepsilon}{R_T} = \frac{\varepsilon_1 + \varepsilon_2}{R_T} = \frac{2\varepsilon_1}{R_T} = -\frac{C}{R_T}v\,. \tag{4.13}$$

Here $R_T = R_{int} + R_{turns} + R_{CRM}$, where R_{int} is the internal resistance of the battery, R_{turns} is the resistance of the number of turns of the coil forming a local solenoid and R_{CRM} is the contact resistance of the magnets. To find R_{CRM} [5] measure the voltage at the terminal of the battery when the current is passing through the local solenoid and the train is kept stationary. Using Ohm's law R_{CRM} can be calculated. R_{CRM} calculated in this way will be slightly lower than its value when the train is in motion.

The total current is $I_T = I + I_E$, where I is the current flowing in the solenoid and I_E is the eddy current. The total emf is $F = C(I + I_E)$. The net force is $C(I + I_E) - Mg\mu_k = 0$, that is,

$$C\left(I - \frac{Cv}{R_T}\right) - Mg\mu_k = 0\,, \tag{4.14}$$

where Mg is the weight of the train and μ_k is the coefficient of the kinetic friction. From the above equation, we obtain the terminal velocity as

$$v = \frac{V}{C} - \frac{Mg\mu_k R_T}{C^2}\,, \tag{4.15}$$

where $V = IR_T$ is the open-circuit voltage of the battery.

4.4 Conclusion

The simple electromagnetic train considered in this chapter has no motors, gears and pistons. Even then, it is able to move through the track with considerable speed. Students can be asked to setup a certain number of tracks with different number of turns per cm and determine the number of turns that throw the train farthest out. In a closed track, the train can run as long as the battery has sufficient power. Performance of various brand batteries can be compared as demonstrated in [3].

4.5 Bibliography

[1] https://www.youtube.com/watch?v=J960J29OzAU.

[2] https://www.youtube.com/watch?v=9k7zywli4Vg.

[3] https://www.waynesthisandthat.com/How%20To%20Build%The%Simplest%20Electric%20Train.html.

[4] Zahra Moghimi, Young Scientist Research **2**, 10 (2018).

[5] C. Criado and N. Alamo, Am. J. Phys. **84**, 21 (2016).

[6] https://www.slideshare.net/DarshilKapadiya2/electromagnetic-train.

[7] https://spiff.rit.edu/classes/phys313/lectures/sol_f01_long.html.

[8] D.J. Griffiths, *Introduction to Electrodynamics* (Printice-Hall, New Jersey, 1999) 3rd edition, pp.265.

[9] Page 281 of the reference 8.

4.6 Exercises

4.1 The force on the train can be calculated using the Lorentz force. The total magnetic force $\mathbf{F}$ on a solenoid S due to the field $\mathbf{B}$ developed by the two magnets is given by $\mathbf{F} = I \int_S d\mathbf{X}' \times \mathbf{B}$. If the magnetic field of a dipole of moment $\mathbf{m}$ at a point $\mathbf{b}$ is $\mathbf{B} = (\mu_0/(4\pi b^3)) \left[3(\mathbf{m} \cdot \widehat{b})\widehat{b} - \mathbf{m}\right]$, then determine the x-component of $\mathbf{F}$ [5].

4.2 Determine the emfs ε_1 and ε_2 associated with the eddy currents developed in the local circuit formed in the electromagnetic train.

4.3 Given that $R = 1.1\,\text{cm}$, $L = 7\,\text{cm}$, $N = 25$, $M = 77.5\,\text{g}$, $V = 1.5\,\text{V}$, $m = 5.31\,\text{Am}^2$, $\mu_k = 0.45$, $\mu_0 = 4\pi \times 10^{-7}\,\text{NA}^{-1}$ and $g = 9.8\,\text{m/s}^2$ compute the terminal velocity.

4.4 For $I = 0.5\,\text{A}$, $R = 1.1\,\text{cm}$, $L = 7\,\text{cm}$, $N = 20$, $m = 5.31\,\text{Am}^2$ and $\mu_0 = 4\pi \times 10^{-7}\,\text{NA}^{-2}$ calculate the net force F on the train.

4.5 Calculate the magnetic field $B_1(0) = B_1(L)$ produced by a solenoid setup in the electromagnetic train for $\mu_0 = 4\pi \times 10^{-7}\,\text{NA}^{-1}$, $I = 0.5\,\text{A}$, $N = 20$, $L = 7\,\text{cm}$ and $R = 1.1\,\text{cm}$.

Chapter 5

Fun-Fly-Stick

The fun-fly-stick toy is a battery-operated Van de Graaff like electricity generator that repels and levitates objects. This is shown in Fig. 5.1. Van de Graaff is a device capable of generating very high voltages to accelerate particles to a high enough energy to produce new elements. A Van de Graaff generator has a hollow aluminium sphere seated on top of an insulated column containing a latex belt moving at a high speed. The belt would draw electrons from the belt roller in the base of the device. These electrons are then deposited on the outside of the aluminium sphere which become negatively charged. The fun-fly-stick appears as a mini Van de Graaff generator and functions in a similar way.

What are the parts inside the fun-fly-stick toy? Figure 5.2 shows the inside of it [1]. The names of various parts are specified. The toy has two small rollers – Teflon on the bottom and a metal (usually aluminium) on top. There is a rubber belt running over the two rollers. The belt is powered by a small motor. Near each of the two rollers, there is a copper comb. The combs are very close to the belt, however, they are not in touch with the belt directly. The Teflon roller is mounted on top of the axel of the motor. The motor is powered by two AA batteries. A cardboard control tube fits on top of the stick and is in tight contact with the exposed part of the upper comb. This tube acts as an electric charge accumulator. When you press the switch button you will hear a humming sound indicating that the stick is activated.

A fun-fly-stick package usually contains a fun-fly-stick and one or more *fun-flyers* (flying toys). The fun-flyers also called *mylars*, are objects of certain shape of thin aluminium strips. They are delicate and light-weighted. Some of the fun-flyers are shaped like butterfly, sphere and hour glass.

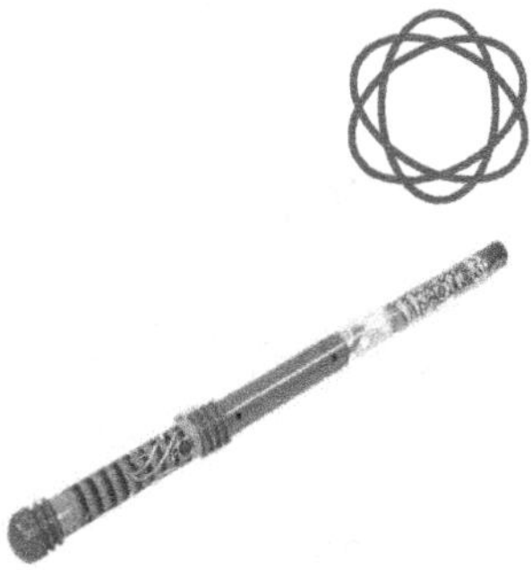

Fig. 5.1: A fun-fly-stick and a levitated object.

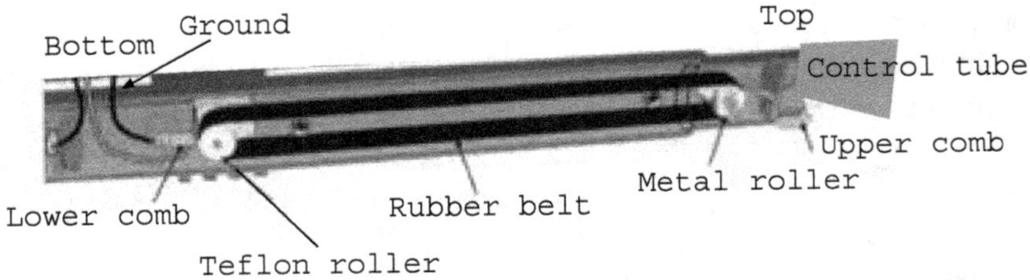

Fig. 5.2: Inside view of a fun-fly-stick.

Flyers with any desired pre-fixed shape can be made using thin aluminium foils.

To play with the fun-fly-stick, hold the sphere mylar lightly between your thumb and pointer finger of the left-hand. In the right-hand keep the stick, press the button of it with your thumb, wait for a few seconds and outstretch your arm. Now, extend your other arm and drop the sphere mylar onto the upper end of the stick. As the mylar strip is thin, it initially does not appear in spherical form but will be like tinsel ribbons. Allow the sphere to touch the end of the stick and then shake it off. The mylar shape will immediately expand to its pre-set sphere shape and repelled by the wand and begins to float. The levitated sphere can be moved about the room by moving the stick. Once it reached a desired height, release the switch button. The sphere starts dropping down. When it is close to the stick press the button again. The sphere starts moving up.

A flyer can be made to follow your hand. When the flyer is up to eye level, cup the left-hand with your fingers spread out. Bring it nearer to the sphere. When the palm of your hand is close enough to the sphere, then the sphere will be attracted by the palm like a magnet. Start moving your hand away, from the sphere. The sphere will follow. When it comes closer to your body, then use the stick to move it away. By a proper practice, we can hold the stick vertically and make the flyer to move left-side or right-side of the stick.

Another simple trick is to make the flyer to jump back and forth between your hand and the stick. Once the flyer started rising above the stick keep the palm of your hand some height above the stick. The rising flyer touches your hand and immediately dives down. Now, catch it with the stick again. It will rise up. The rise and dive down cycle will repeat.

In this chapter we explain the working of a fun-fly-stick and explore the physics behind certain tricks with it. The fun-fly-stick utilizes triboelectricity, electrostatic induction and corona discharge. Therefore, let us first introduce them.

5.1 Triboelectricity, Electrostatic Induction, Corona Discharge

The process of separation of charges by means of rubbing or contact between materials is known as *triboelectricity* [2]. Tribo means to rub. By this process a material can be made to have a net positive or negative charge. When we bring two different insulators in contact and separate, then one insulator will give away electrons to the other. The material which gains electrons becomes electrically negatively charged. For example, rubbing of a PVC pipe with a silk cloth makes the pipe a net negatively charged and the cloth will have a net positive charge. Our body undergoes discharge while getting into a car seat and crossing a woolen carpet. In winter sometimes, we get a shock while touching a door knob. This is because we gain a positive or negative charge when walking on the floor. We loose charges when we touch the door knob. Some of the materials that give up electrons by contact with another material are glass, nylon, hair, wool, silk, aluminium, paper, PVC pipe, a few to mention. For packaging and storing of electrostatic discharge sensitive devices, companies use materials which do not charge each other while in contact.

Charges can be separated without contact also by means of *electrostatic induction* [3]. It is a process of charging a material by bringing it nearer to an electrically charged object. This leads to redistribution of charges in the material causing one side of it having an excess of either positive or negative charges. An electrical conductor in a neutral state has an equal number of positive and negative charges and are intermingled in the conductor. Suppose an electrically charged material is brought near to the neutral conductor. The electrical charges on or near the surface of the electrically charged material attract the opposite charges in the neutral conductor. The result is the redistribution of charges in the conductor. The redistribution of charges persists as long as the electrically charged material is near the conductor. When the material is removed, then the thermal motion of electrons takes place in the conductor and the changes intermingle again.

The induction process can be witnessed in our every day life. For example, after combing our hair if we bring the comb near a piece of paper, the paper will be attracted to it. The reason is that while combing, it captures some electrons from our hair and becomes negatively charged. When it is brought near a piece of paper, an electrostatic induction process takes place which induces the paper to generate positive charges and hence the paper gets attracted. *Does the electrostatic induction work in noncounducting or dielectric materials? How?*

A *corona discharge* is an electrical discharge occuring on and around the surface of a charged conductor due to the ionization of the surrounding fluid such as air. A strong enough electric field of a charged conductor leads to a chain reaction: electrons in the air interact with neutral atoms or molecules and ionize them. The created and accelerated free electrons strike other atoms and ionize them, that is, create positive ions and free electrons. These opposite charges are accelerated by the electric field in opposite direction. In a positive corona, the electrons are attracted to the nearby positive electrode while the ions are repelled and the ions are attracted. Corona discharge is often observed as a bluish glow. In the air, coronas generate gases such as O_3, NO and NO_2. It can generate nitric acid if water vapour is present. These gases are toxic to humans and environments and degrade nearby materials.

5.2 Operation Inside the Stick

Let us bring out the physics of what is happening inside the stick and then explain the mechanism of floating of a fun-flyer [1].

When the power button in the fun-fly-stick is pressed, the rubber belt running over the Teflon and aluminium rollers moves. The roller and the belt are made of different materials. Teflon is able to capture negative charges when in contact with another material. Compared to rubber, Teflon has relatively very high tendency to capture electrons when both are in contact with each other. Due to the triboelectric effect, the Teflon gets electrons from the rubber belt. The result is that the belt becomes *positively charged*. The comb fixed close to the Teflon captures the excess electrons from the Teflon due to the electrostatic induction described in the previous section. The excess charges then flow to the ground through the person operating it via the metal rim of the button.

Actually, when the electric field of the negatively charged Teflon reaches the tips of the comb, because the like charges are repelled, the electrons in the tips of the comb move away and the tips become *positively charged*. The positive ions created due to corona discharge are attracted to the Teflon. Further, the electric field is strong enough that corona discharge occurs on a small scale. The free electrons created due to ionization of atoms in air are repelled by the Teflon roller and are attracted by the comb close to the Teflon roller.

The electrically positively charged belt travels to the end part of the stick where it contacts the second roller made of aluminium. Note that aluminium chooses to give up electrons, while rubber has the ability to capture electrons. Now, due to the triboelectric effect the rubber belt gains negative charges and the roller is now *positively charged*. The electric field of the aluminium roller causes electrostatic induction with the upper comb and corona discharge with the air. Negative charge is induced in the comb since the roller is positively charged. The free electrons and the positive ions in the air created by the corona discharge are attracted towards the roller and the comb, respectively. The electrons moving towards the roller see the belt on the way. The belt captures these electrons and move towards the Teflon roller and the cycle continues. In this way the level of charge in the stick is sustained.

At any instant of time, the belt carries both negative and positive charges. However, the upper belt moving away from the Teflon is positively charged while the lower part moving towards Teflon from the aluminium roller is negatively charged. The charge of the stick is the same as the charge of the belt moving from the Teflon roller to the aluminium roller and is in *positive charge*.

As one end of the comb near to the aluminium roller is negatively charged while the other end is in direct contact with the inside of the cardboard control tube. The control tube is an extension of the comb and is mounted on the top of the stick. Cardboard has a high electrical resistivity but is a conductor. Moreover, its discharge process is much slower than a metal, consequently, shocking sensation by the operator is eliminated.

Note that inside the stick, charges are separated in the conducting components. The charged components have an electric field around them. The largest charged component is cardboard tube (positively charged) and has a strong positive electric field around it. The second largest concentration of charge is with the two rollers. The weakest electric field is over the belt (*why*?).

5.3 Floating of a Fun-Flyer

Now, we describe the mechanism behind the floating of a fun-flyer [1,3,4].

Once the fun-fly-stick got charged, it can be used to float a light-weighted conductive mylar strips. When the mylar is dropped to the control tube its strips gains the positive charges *instantly* on contact or through induction. Due to the accumulation of charges in the control tube, the resultant electric field will be strong. Therefore, if the flyer is smaller in size, then it gets charged even without touching the control tube. Because the like charges repel each other, the mylar strips are repelled from the control tube. The excess free positive charges are able to move within the strips. Once they reach the surface of the strips they cannnot go farther. The excess positive charges are now on the entire surface of the strips. Due to the repulsion of like charges, every part of the strip is repelled from every other part. This causes the expansion of the mylar strips to the pre-fixed shape. The repelling process between the positive charges of the control tube of the stick and the mylar strips results in the mylar strips to have the *levitation effect*. The mylar strips float on the electric field.

We know that gravity is stronger than the electric field. To defy the gravity, the flyers are to be light-weighted. In order to gain charges from the control tube of the stick, the flyer is chosen as a conductive object. For a better effect, the flyer is chosen with a shape consisting of thin conductive ribbon tied together on both ends that opens up into a floating globe. When the shape floats, the electric field of the positive charge exists around it.

What will happen if we touch a floating fun-flyer by a finger? The shape collapses and falls down. This is because our body is a conductor. When we touch the shape (which is positively charged) the negative charges (electrons) from our body get on to the shape resulting in loss of charge and hence the collapse. This property of the flyer can be used to make it to jump between the stick and the palm of the other stretched hand kept above the stick. When the repelled flyer moves above the stick and touches the palm of the hand, then it instantly discharges. Then, it begins to fall below. However, when it arrives at the control tube of the stick, it is charged and begins to float above. This process repeats and creates jumping of the fun-flyer between the stick and the palm of the hand.

Solved Problem 1:

What will take place if a floating fun-flyer is brought near a wall or a furniture?

An electric field approaching a neutral object creates an electrostatic induction in conductors while polarization in insulators. Hence, as a fun-flyer is brought near a wall or any other object, its electric field makes redistribution of the charges in that object. Essentially, the positive electric field around the floating fun-flyer induces negative charge in neutrally charged objects. This causes the fun-flyer to attract to those neutral objects. If it touches the objects, then a discharge takes place and flyer loses its charge.

5.4 Some Other Fun Activities with the Fun-Fly-Stick

In this section we enumerate some of the fascinating activities with the fun-fly-stick [1,5-8].

1. Cover the control tube of the stick by a plastic cover in order to avoid it being damaged by water. Bring the control tube of the activated stick near a thin stream of water from a tap. *The water stream bend towards the stick.* This is because the water molecules are polar. Molecules having dipoles are termed as polar molecules. The electric field of positively charged stick makes the negative side of the hydrogen-oxygen dipole to rotate and attracted to the positively charged stick. This leads to the bending of the water towards the control tube.
2. Prepare a few thin light-weighted cups (or pie pans) made of a conducting material. Keep one within another and so on. Hold the stick vertically with control tube pointing upward. Place the set (stack) of cups on the control tube. Press the power button. *The cups bounce out of the control tube one after another. Why?* This is caused by the electrostaic forces of repulsion between the cups. Because of the cups' individual weights they got repelled one by one from the stack.
3. On a table place an empty soft drink metal cane. Activate the fun-fly-stick and hold the control tube, just away and parallel to the cane and say, left-side of it. As the cane is a conductor, the electrons in it

are attracted to the positively charged control tube. When the cane is close to the stick, quickly shift the stick to the other side of the cane. Now, the cane rolls back towards the stick. By shifting the stick left- and right-sides of the cane it can be made to *roll back and forth*.

4. Keep some rice krispies on a table. Activate the stick and bring it towards the rice krispies. They are attracted to the tube, gain positive charges and then repelled creating *dancing cereal*.

5.5 Conclusion

A physics teacher can make use of the fun-fly-stick as a magic stick and entertain the students by performing various tricks. The tricks can serve as practical demonstration of the following concepts about static electricity:

1. A charged object attracts a neutral object.
2. Like charges are repelled and unlike charges are attracted.
3. Electrically neutral objects may be positively charged or negatively charged.
4. Capture of charges can happen either in contact or through induction.
5. Charging of conductors and also insulators can be achieved by means of conduction and also by indution.

Practically, it is difficult to create strong positively charged objects in a lab for an electrostatic demonstration. The fun-fly-stick becomes a simple source of positive charges.

5.6 Bibliography

[1] https://www.myweb.wit.edu/santosr/Fu%20Fly%20Stick%20.pdf.
[2] L.M. Surhone, M.T. Timpledon and S.F. Marseken, *Triboelectric Effect* (VDM Publishing, Riga, 2010).
[3] https://www.real-world-physics-problems.com/fly-stick.html.
[4] https://asc-mag-media.s3.amazonaws.com/datasheet/11-0058DS.pdf.
[5] https://www.stevespanglerscience.com/lab/experiments/fun-fly-stick-static-generator/.
[6] https://www.firebox.com/pdfs/Magic_Wand_Manual.pdf.
[7] https://www.arborsci.com/blogs/cool/fun-fly-stick-electrostatic-exploration-2.
[8] http://www.funflystick.com/pdf/FunFlyStickEducationalWeb.pdf.

5.7 Exercises

5.1 What will happen if there is an oil between rubber belt and the rollers?

5.2 Let us isolate your friend from the ground. For this your friend can stand on an insulator. Ask your friend to have a flyer on the palm of a stretched hand with the palm facing up. You press the button of the fun-fly-stick and ask your friend to hold the control tube by the other hand. Explain your observation.

5.3 The charge of the control tube of the activated fun-fly-stick can be either positively or negatively charged and this depends on the setup inside the stick. Describe a way of identifying the charge on the stick without using any electronic device. Make use of the fact: "Repulsion always proves the charge on an object".

5.4 Against a white board or wall, hold a paper sheet. Press the power button of the fun-fly-stick and rub the control tube on the paper. The paper will stick to the board or wall for some time. Why? Do the above on different surfaces. Give an account of your observation.

5.5 Glue a piece of yarn to a light-weight ball and wrap the ball with a thin aluminium foil. Using a bar or retort stand suspend the ball between two metal cans kept on an insulator (polystyrene cups can be used). Make sure that the ball is mid-way between the bottom ends of the cans. Hold your finger to one of the canes and activate the fun-fly-stick. What does happen? What does happen when you remove the hand? Why?

Chapter 6

Plasma Ball

Plasma, the fourth state of matter different from solids, liquids and gases, is a hot ionized gas containing almost equal number of electrons and positively charged ions. A *plasma ball* is a sealed clear glass globe filled with a mixture of noble (inert) gases such as neon, argon, xenon and krypton at a pressure close to atmospheric pressure and a high voltage electrode at its centre (see Fig. 6.1). It is also called *plasma globe* or *plasma lamp*. It is popular for displaying bright electric luminescent colours. In a plasma ball, a highly energized gas of plasma is formed by supplying voltage.

This formed plasma appears as a beautiful, colourful and decorative filamentary arcs [1,2]. The plasma filaments extend from the electrode at the centre of the ball to the outer glass insulator and move around inside the ball. The colours of the filaments depend on the gases present inside the ball. It would be amazing to watch the ball light-up with electricity.

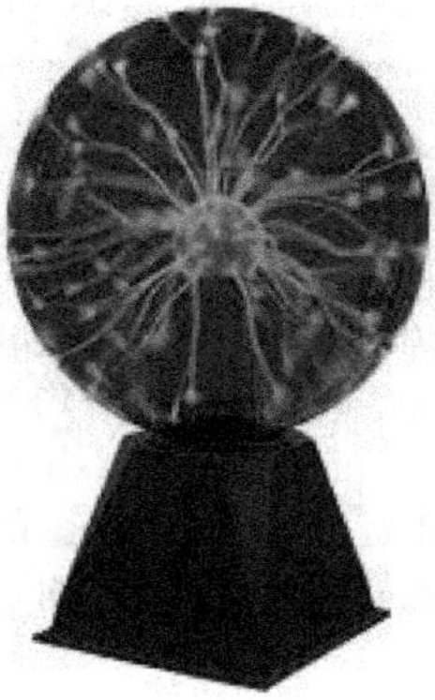

Fig. 6.1: A plasma ball.

The plasma ball was invented by Nikola Tesla when he experimentally investigating high-frequency electric currents in a vacuum tube. In 1894, Tesla got a patent for a plasma lamp. William (Bill) Parker, a student of the Massachusetts Institute of Technology, designed the modern plasma balls in 1971. James Falk produced plasma globes and marketed to science museums in 1970s and 1980s.

What can we do with a plasma ball? One can do a lot of tricks that surprise others. It can also be used in several experiments. Plasma balls are primarily used as toys for their lighting effects and the several tricks performed on them by moving our hands around them. For example, switch-on the ball and place a finger on the glass. A colourful strand of light (filament) is drawn from the central electrode to the finger. This happens due to the conductive character of our body. When the glass of the plasma ball is touched, a discharge path is created which has low resistance than the surrounding glass and gases. The charges in the ball go to the hand, then to the body and finally to the floor. The plasma balls can be used for high voltage experiments. For example, when a wire coil is kept on the ball, a capacitive coupling occurs. It can be able to transfer enough voltage to the coil thereby to produce a small arc or energize high voltage load.

In this chapter we present a description of a typical plasma globe and the physics of various tricks with it.

6.1 Description

A plasma globe [3] has a Tesla coil-like circuit which generates a high voltage (2.5kV) alternating current. There is a drive circuit in which current from a lower voltage dc supply powers a high-frequency circuit. Its output is stepped by a high voltage transformer. The transformer is connected to an electrode which is usually in the shape of a ball. These are kept at the central part of the globe. Inside the globe, there are noble gases, usually, neon and argon. The transformer produces radio-frequency energy and is transmitted through the electrode to the gases inside the globe. Stray electrons in the gases are radially accelerated by the high voltage produced electric field. The accelerated electrons crash into the electrons of the gas atoms leading to ionizing of them. These atoms become plasma. As the electrons fall to lower energy orbits they emit light and form like filamentary arcs or tendrils.

The filamentary arcs start from the electrode at the central place and extend to the outer glass. The current then passes into the surrounding air. The electrons are then attracted to the neutrally charged earth. When you touch the globe with your hand an easier path is created by you for the electrons to flow. Therefore, the filaments move toward the point of touch and the electrons travel to the earth by passing through your body.

Some globes have provision to control or vary the high voltage power supplied to the central electrode by means of a control knob. One can set a low voltage creating only one tendril. In this case, this single plasma channel transmit, the lowest light energy through the glass of the globe to outside world. By slowly increasing the power to the electrode, one can create two tendrils then three and so on. The filament is brighter because of more current is flowing through it while it is thinner because of the magnetic field around it creates a force that compresses the size of the channel.

Devices like cell phones, laptop computers, digital audio players can be affected by the rf field produced by the plasma globe. Therefore, a plasma globe has to be kept away from such electronic devices. Bringing a magnet and plasma globe closer would produce a potential giving shocks and burns.

6.2 Some Fun Tricks With a Plasma Globe

Let us enumerate some of the fun tricks with the plasma globe and the physics in them [4-6].

1. Lighting a Fluorescent Light Tube

The oscillating electrons in the plasma globe produce an electromagnetic field. Because of this, when a flourescent light tube is brought nearer to the globe (without touching it) the electrons inside the tube undergo oscillations and lead to an electric current causing the tube to light-up. Alternately, when you hold the tube and make it to touch the plasma globe, then the electrons passes through the tube, causing it to light-up, through you and into the ground.

2. Lighting a Plasma Globe Without It Turned On

First, switch-on the plasma globe and place your hand on its top. Then, switch-off it, immediately withdraw your hand and place again on to the top. You will observe the electric tentacles flash to your hand. Take out your hand and clap several times near the globe. This act induces a few more flashes of electricity.

3. Creating a Spark With Two Coins

Place a coin on the top of a plasma globe. Carefully touch the coin with another coin. Because the coins are conductors the electrons easily flow through them. When much electrons are build up in the first coin, they jump to the second coin and create a spark.

4. Convection

The plasma threads are hot enough that they can easily rise because of their buoyancy in the other gases present in the plasma globe. In view of this, it is practically difficult to find a horizontal filament to remain for more than a second. However, the buoyancy is able to stabilize the vertical filaments at the top, practically, a single one.

For two more activities see the exercises 6.3 and 6.4 at the end of this chapter.

6.3 Conclusion

Nowadays, plasma balls are available in the market with different shapes and with noble gases such as neon, xenon and krypton. USB plasma balls powered from the USB port of computer are also available. In addition to the various fun tricks listed earlier, the plasma balls are used in several experiments. The electric field produced by the plasma ball extends outside the ball and we can easily detect it. We can use this electric field to light up gas. That is, atomic ionization can be realized with only about 20 V.

6.4 Bibliography

[1] http://www.lightninglamps.com/plasma-ball-work/.
[2] http://www.wonderopolis.org/wonder/how-does-aplasma-ball-work/.
[3] https://en.wikipedia.org/wiki/Plasma_globe.
[4] https://www.real-world-physics-problems.com/plasma-ball.html.
[5] https://sciencestruck.com/what-is-plasma-ball-how-does-it-work.
[6] https://www.uclan.ac.uk/research/explore/groups/assets/Plasma_-Ball_ activity_1.pdf.

6.5 Exercises

6.1 Using an oscilloscope measure the plasma globe voltage V as a function of distance d between the globe and the probe of the oscilloscope. (The oscilloscope to be kept far away from the globe.) Draw a graph between V and d and then determine the field strength (E).

6.2 Why is the air around a plasma not ionized by the electric field present outside the globe (produced by the applied high voltage)? Propose a simple way of ionizing the air around the globe and test it.

6.3 Take an inexpensive calculator and bring it slowly towards the plasma globe. Give a qualitative explanation of observation made.

6.4 What do happens in the following experiments [6]? Give appropriate explanations.
(a) Stand on a chair and place your one hand on the plasma globe. Hold a flourescent light tube in your other hand and ask your friend to stand on the ground and touch the light tube.
(b) Do (a) by standing yourself also on the ground.

6.5 The electric field produced by the Tesla coil-like circuit reaches outside the globe. Propose a way of identifying its presence outside the globe. Does the field spatially decrease radially or circumferentially? Examine its spatial variation by means of a simple experiment with a neon bulb or light emitting diode (LED).

Part VIII: Miscellaneous Toys

Chapter 1

Some Other Toys

So far we have presented in detail various aspects of many interesting toys that can be useful for educational purposes. We covered history, construction, working principle and theoretical model of them and pointed out the understanding of certain concepts in physics through them. There are certain toys with the physics behind their working is simple to explain and the theoretical descriptions do not involve long derivation. For certain toys the working principle is simple, however, developing appropriate modelling is either difficult or the development of a model is in the preliminary stage. In some cases the forces involved and the factors influencing are not strictly deterministic. Examples of such kinds of toys are balancing toys, gyro-ring, floating ping-pong ball, helicopter toys, pinwheel, tornado bottle, just to mention a few. In this last chapter, we present the physics of such toys.

1.1 Balancing Toys

Balancing toys are a good choice for demonstrating the stability of equilibrium points. The role of the position of centre of mass (CoM) on the stability of an equilibrium point of balancing toys is easy to illustrate. The point at which the mass of an object is concentrated is the CoM and is also called centre of gravity (CoG). If the object is made to support at its CoM, then the net torque acting on the object is zero and it will remain in static equilibrium. Take a one foot scale. Keep your one finger vertically pointing up and place the scale horizontally at the tip of the finger. When the half-foot mark is exactly at the tip of the finger, then the scale is balanced. When you place a small coin or a small weight, say, on the edge of the scale, the balance is lost. Now, the CoM lies between the half-foot mark

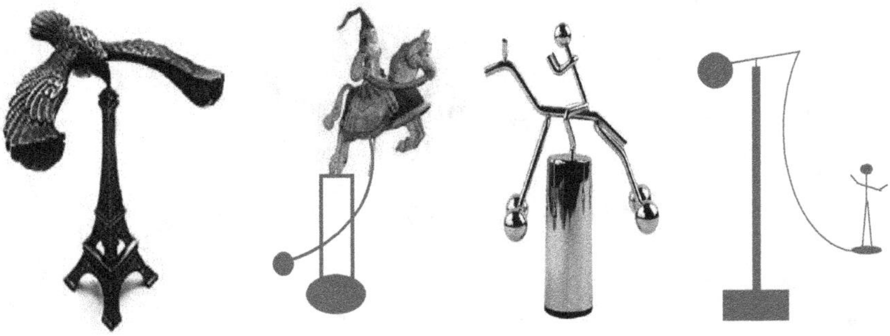

Fig. 1.1: A few balancing toys.

and the edge where the coin is placed. By means of various trials one can find out practically the new balancing point.

During 1859-1860 Michael Faraday used a roly-poly balancing toy to discuss the CoG and equilibrium points in his lectures presented at the Royal Institution of Great Britain [1,2]. In 1860 David Wells in his book used the horse and rider balancing toy in a section on CoG and stability of an equilibrium point [1,3]. Standing or balancing toys at the sharp edge or at the finger tip needs an explanation behind the balancing [1,4-10].

Figure 1.1 shows some of the balancing toys. A simple and popular balancing toy is the *balancing bird* shown in Fig. 1.1 (the first toy from left-side). The balancing point of this toy is its beak and is the point of CoM. The tip of the beak of the bird is kept in contact with the support (or tip of a finger or a tip of a pencil). The bird toy will not fall down if we tilt it in any direction. It will make a damped oscillation and return to its stable equilibrium state.

How do we explain the balancing of this toy? When viewed from the above with the tail towards us, we may think that most part of the bird is behind the balance point (also called *pivot point*). If we carefully look at the bird, we can notice that the tips of the wings have extra weights and the tips of the wings are just ahead of the beak. These two aspects can be clearly seen in Fig. 1.1. The additional weights and their locations are such that the mass of the toy is perfectly balanced at the pivot point and the CoM of the extra weights are at a height below the height of the beak. The wings of the bird are extended in front of the beak so that the weight of the bird in front of the beak and the weight of the bird behind the beak are balanced at the beak.

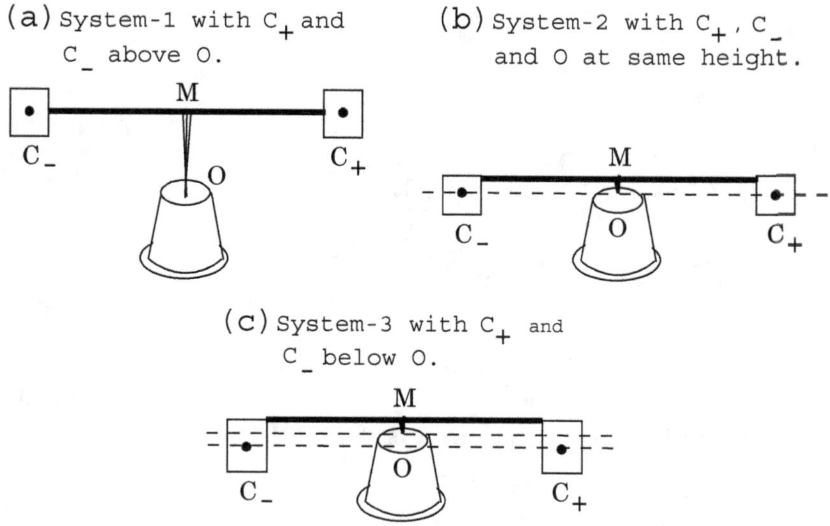

Fig. 1.2: A three model systems similar to the balancing bird toy. These systems are at equilibrium state. The systems are in contact with the top of the support at the point O. (Reproduced with permission from Institute of Physics Publishing Ltd: J. Fort, J.E. Llebot, J. Saurina and J.J. Sunol, Phys. Edu. **33**, 98 (1998).)

We discuss the stability of the balancing bird toy [6]. In the equilibrium state the heights of the bottom of the tip of the wings and the tip of the beak are not at same height from the bottom of the support. The tip of the beak is relatively at a higher hight than the bottom of the tips of the wings. *What is the significance of this?*

To explain the stability of the balancing toy consider the three systems designated as systems 1, 2 and 3. These systems are similar to the balancing bird toy, but in the systems-1, 2 and 3 the CoM of the two end masses are above, same and below the point of support of the toy, respectively. These three systems are depicted in Fig. 1.2 [6]. In all these systems a thin horizontal rod connects two equal masses (cubes) and their CoM are denoted as C_- and C_+ and O is the contact point of the system on the top of the supporter as in Fig. 1.2. m is the value of the two end masses. A second (vertical) rod starts from M and its other end is made in contact with the top of the support. For simplicity assume that the masses of the two rods are negligible.

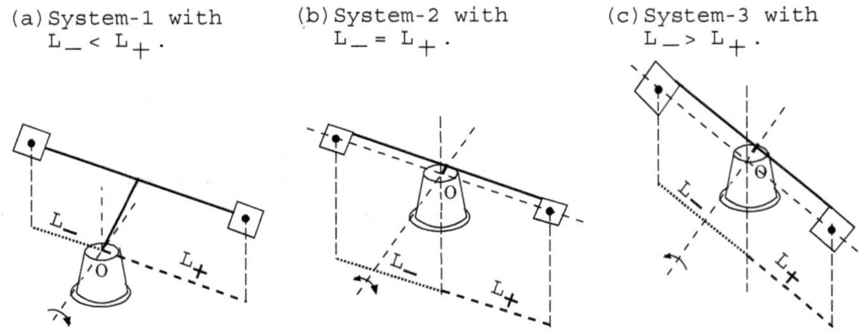

Fig. 1.3: Systems 1-3 disturbed from the equilibrium point. (Reproduced with permission from Institute of Physics Publishing Ltd: J. Fort, J.E. Llebot, J. Saurina and J.J. Sunol, Phys. Edu. **33**, 98 (1998).)

In system-1 (Fig. 1.2a) at the equilibrium position the CoM of the two end masses are above the contact point O. The torques exerted by the left- and right-side masses are $\tau_- = mgL_-$ and $\tau_+ = mgL_+$, respectively, where L_- (L_+) is the horizontal distance of the CoM of the left-side (right-side) mass attached to the horizontal rod. Here τ_+ makes the system to undergo rotation in the clockwise direction while τ_- leads to rotation in the anticlockwise direction. As $L_- = L_+$, we have $\tau_+ = \tau_-$ and hence these two effects are balanced at the equilibrium point as long as the system is undisturbed. This is not the case in the other two systems.

Suppose the toys are disturbed from the equilibrium position in a direction. Let the system-1 be, for example, tilted as in Fig. 1.3a. The distances L_+ and L_- are marked. Now, find the stability of the equilibrium point. In Fig. 1.3a $L_+ > L_-$ and $\tau_+ > \tau_-$. Then, $\Delta\tau = \tau_+ - \tau_- = mg(L_+ - L_-) > 0$. Consequently, this system rotates away from the equilibrium point. The equilibrium point becomes *unstable*. Next, consider the system-2 in Fig. 1.3b where the points of CoM of the two end masses and the point O are at same height from the bottom surface and in the disturbed case as from Fig. 1.3b $L_+ = L_-$. Hence, $\Delta\tau = \tau_+ - \tau_- = mg(L_+ - L_-) = 0$.

We obtain an interesting result for the system-3 in Fig. 1.3c. As $L_+ < L_-$ we have $\Delta\tau = \tau_+ - \tau_- = mg(L_+ - L_-) < 0$. As a result the system starts rotating towards equilibrium position. After making a few damped oscillations the toy settles into the equilibrium state and is thus a stable state. The balancing bird toy can be treated as a system similar to the

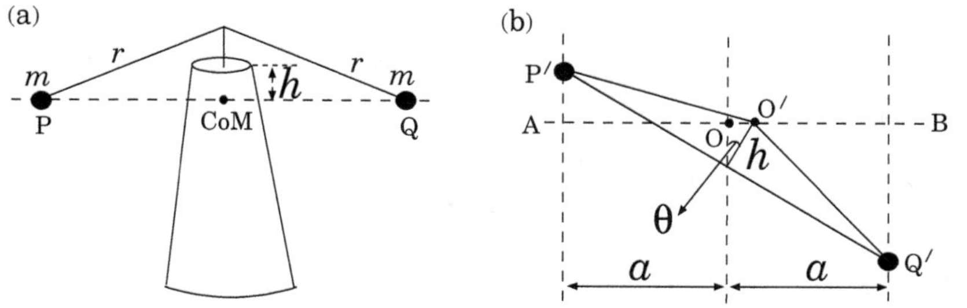

Fig. 1.4: (a) A Model of a balancing toy. (b) Representation of parameters of a blalancing toy tilted by an angle θ.

system-3. In the system-3 and in the balancing bird toy, the CoM of the two end masses (the outer end part of the wings in the balancing bird toy) are below the point O of the support at the stable equilibrium state. The toy considered is stable against a disturbance from the equilibrium state in any direction as long as the CoM points C_- and C_+ of the two masses at the ends of the wings are below the contact point O.

Solved Problem 1:

Consider a model of the balancing toy as shown in Fig. 1.4a. The rigid wires carrying the two identical masses m have negligible mass. Show that a small tilt of the mass arm in the vertical direction leads to the oscillation of the toy in the vertical plance with period $T = 2\pi r/\sqrt{gh}$.

Let θ be the angle of tilt. Then, the vertical distance h will be tilted by an angle with respect to the horizontal line and the CoM will shift to a new position, say, O′ from the initial position, say, O as shown in Fig. 1.4b.

As the CoM has shifted by a distance $\mathrm{OO}' = h\sin\theta$, we write

$$\mathrm{O'A} = \mathrm{OA} + \mathrm{OO'} = a + h\sin\theta\,, \tag{1.1a}$$

$$\mathrm{O'}B = \mathrm{OB} - \mathrm{OO'} = a - h\sin\theta\,. \tag{1.1b}$$

So, the restoring torque is

$$\begin{aligned}\tau &= mg\mathrm{O'B} - mg\mathrm{O'A} \\ &= mg(a - h\sin\theta) - mg(a + h\sin\theta) \\ &= -2mgh\sin\theta\,. \end{aligned} \tag{1.2}$$

As the wires have negligible mass, we get the moment of inertia of the two masses as $I = 2mr^2$ and the torques as

$$\tau = I\ddot{\theta}\,. \tag{1.3}$$

Fig. 1.5: A gyro-ring toy.

Equating the Eqs. (1.2) and (1.3) and for small θ we obtain

$$I\ddot{\theta} = -2mgh\sin\theta \approx -2mgh\theta\,. \tag{1.4}$$

That is,

$$\ddot{\theta} = -\frac{gh}{r^2}\theta = -\omega^2\theta, \quad \omega^2 = \frac{gh}{r^2}\,. \tag{1.5}$$

It is the equation of motion of a simple pendulum and the period of oscillation is $T = 2\pi/\omega = 2\pi r/\sqrt{gh}$.

1.2 Gyro-Ring

A mesmerizing active play toy for kids and adults, popularly known as *gyro-ring*, shown in Fig. 1.5, has a large ring about 30-cm-diameter made from a metal rod of about 5-6-mm-diameter and usually five small washers of slightly unusual design with holes twice the diameter of the metal rod are strung on to the ring. The washers are able to freely spin. This toy is also called *chatter ring* or *wobble ring* or *jitter ring*. It was first played in New Zealand [11].

To play with this simple toy [12,13], hold the ring in your non-dominant hand. Spin-up the five washers with the other hand. One appropriate way is to place the washers on your index finger midway up the ring and using your thumb flick them to the side. Alternately, with the palm of your hand hit the washers. Once the washers started rotating with high vibration, rotate the ring smoothly towards yourself repeatedly. Essentially, you are pulling the ring through the washers. As long as the ring is in motion, the washers spin around it. In a typical case the spinning and wobbling of the washers are too fast (even spin to 1000 rpm) for our eye to follow them. The faster you move the ring, the higher the washers move towards your hands. If you pull the ring slowly towards you, then the washers go downward. The goal of the play is not to permit the washers to go to the top or not to get into the bottom of the ring. That is, the play is to rotate the ring properly thereby keeping the washers in the midway between the top and bottom of the ring. The spinning washers have to be kept almost at the same height. The fast spinning of the washers makes a vibrating type feeling and produces a very distinctive sound. For video demonstrations of action of the gyro-ring see [14,15].

A practiced player can do a variety of tricks with the gyro-ring. One can *throw and catch* it once the washers started spinning. The washers should not cease its motion. In the *flip* play, toss the ring in a flip, catch it and repeat. In another trick called *sideways hold* turn the ring on its side, change your holding angle and play. For many other interesting tricks, one may refer to the Ref. [12].

What does keep the washers spinning? The source is the friction between the ring and the washers. Let us point out the significance of this friction [16]. Due to this, friction the washers roll along the ring without slipping. As they are spinning the gyrocopic precession comes into picture, makes them in contact with the ring and lean to one side. The friction force acts at the contact point. The spinning washers try to progress downward in order to get energy from gravity to keep spinning. The washers are prevented from the act of moving downward by the upward component of the friction force. The washers are spinning due to the component of the friction force along the direction of their spinning. As long as the ring is properly pulled towards the body of the player, the contact force keeps the washers in action at a certain height above the bottom of the ring.

Apparatus used to explain the motion of the gyro-ring [17] and measure the friction between the ring and the washers [18] are reported.

Solved Problem 2:

It is found that the ratio of torque free precession to spin of the disk is equal to $\dot{\phi}/\omega_{\text{disk}} = 2/\cos\theta$ where θ is the tilt angle. In an experiment with a gyro-ring, it is found that the rotational frequency of the ring is 0.44 Hz and the precession frequency of the disk is 21 Hz. The ratio of the inner radius of the ring to that of the disk is 22. Find the tilt angle.

We have

$$\frac{\omega_{\text{disk}}}{\omega_{\text{ring}}} = \frac{\text{inner radius of the ring}}{\text{inner radius of the disk}} = 22\,. \tag{1.6}$$

Then,

$$\frac{f_{\text{disk}}}{f_{\text{ring}}} = \frac{\omega_{\text{disk}}}{\omega_{\text{ring}}} = 22\,. \tag{1.7}$$

This equation gives $\omega_{\text{disk}} = 22 \times 0.44 \times 2\pi$. Using this we find

$$\frac{\dot{\phi}}{\omega_{\text{disk}}} = \frac{2\pi \times 21}{2\pi \times 22 \times 0.44} = 2.16942\,. \tag{1.8}$$

As $\dot{\phi}/\omega_{\text{disk}} = 2/\cos\theta$ we obtain

$$\theta = \cos^{-1}\left(\frac{2}{2.16942}\right) = \cos^{-1} 0.922 = 23^\circ\,. \tag{1.9}$$

1.3 Toy Guns

Toy guns appear like real guns, however, they are usually for casual play by children. There is a variety of toy guns [19-22]. Rubber band guns fire rubber bands. Spud guns fire a piece of potato. Water guns spray jets of water. Pop guns fire a small cork of foam out of a barrel using air pressure.

A cap gun, cap pistol or cap rifle is a toy pistol with a hammer action detonates a small piece of explosive cap [20]. A cap gun produces a loud sound like a gunshot and a gust of smoke when a small paper cap containing an explosive material is exploded. In a cap pistol, there is a lever to open the loading chamber. In the chamber there will be a pin where the paper cap gun roll can be fitted and then closed. When the trigger is pulled the cap strip advances forward, the hammer is pulled back to the point and released thereby strikes the cap. The sudden shock makes the cap to explode but harmlessly and producing the noise and smoke.

An air gun [21] is used to fire metallic projectiles (pellets or spherical shots) making use of compressed air or gases by mechanically pressurizing. In modern air guns, the power sources are spring-piston, pneumatic or bottled compressed gas. The part of the air gun containing a power source is called a *power plant*. In the spring piston, the elastic energy stored in a spring is used to propel projectiles. Pneumatic utilizes the stored pressurized gas. External sources of pressurized gas is used in the compressed gas type of air gun.

A spring gun has a spring-loaded piston pump setup in a compression chamber. This is separated from a gun barrel. The spring is a grease-lubricated steel coil spring. The user first to make the gun ready for shooting, manually cock it by flexing a lever attached to the pump setup. This causes pulling the pump piston rearwards and the spring is compressed until the back of the piston captures the sear.

When the trigger is pulled, the sear is disengaged causing the spring to decompress and leading to the release of the stored elastic potential energy. The piston is pushed forward, thereby the air in the pump is compressed. The pump outlet is just behind the pellet kept in the barrel chamber. When the air pressure is sufficient to overcome the static friction holding the pellet, then the pellet is driven forward due to the expansion of the pressurized air. This happens in a fraction of a second. In this process, an adiabatic heating of the air takes place and then as the air expands cooling occurs. Recoil effect occurs due to the forward pushing of the piston by the spring. Further, the spring undergoes transverse vibrations and torquing. These vibrations can be dampened by close-fitting spring guides and adding viscous silicone grease to the spring. Modern high-powered spring guns achieve muzzle velocities about the speed of sound. For details of explanation of working of gas spring gun, pneumatic air gun and water gun refer [20-22].

Solved Problem 3:

A ball of mass 10 g is shoot straight-up by a toy gun which uses a spring to shoot. The value of the spring constant is $k = 400\,\mathrm{N/m}$. The spring is compressed 5 cm shorter than its rest state length for shooting. Assume that the force is linear. Calculate the initial velocity of the ball and the maximum height the ball can reach from the rest position of the spring.

Assume that inside the gun there is no air resistance and the ball not touches the inside of the gun. The initial energy E_0 before shooting the ball is the potential energy due to the compression of the spring and the gravitational potential energy. This energy is

$$\begin{aligned} E_0 &= \frac{1}{2}ky^2 + mgy \\ &= \frac{1}{2} \times 400\,\frac{\mathrm{N}}{\mathrm{m}} \times (-0.05\,\mathrm{m})^2 + 0.01\,\mathrm{kg} \times 9.8\,\frac{\mathrm{m}}{\mathrm{s}^2} \times -0.05\,\mathrm{m} \\ &= 0.5\,\mathrm{Nm} - 0.005\,\mathrm{Nm}\,. \end{aligned} \tag{1.10}$$

As the gravitational potential energy is very small compared to the potential energy, we write $E_0 = 0.5\,\mathrm{J}$.

When the balls is shoot, its kinetic energy $mv^2/2$ is the potential energy E_0. This gives

$$v^2 = \frac{2E_0}{m} = 2 \times 0.5\,\mathrm{J} \times \frac{1}{0.01\,\mathrm{kg}} = 100\frac{\mathrm{kg\,m}^2}{\mathrm{kg\,s}^2} = 100\,\mathrm{m}^2/\mathrm{s}^2\,. \tag{1.11}$$

That is, $v = 10\,\mathrm{m/s}$.

Next, at the maximum height h_{max} the velocity of the ball is zero, the spring is at rest and the total energy is $E(h_{\mathrm{max}}) = mgh_{\mathrm{max}}$. According to

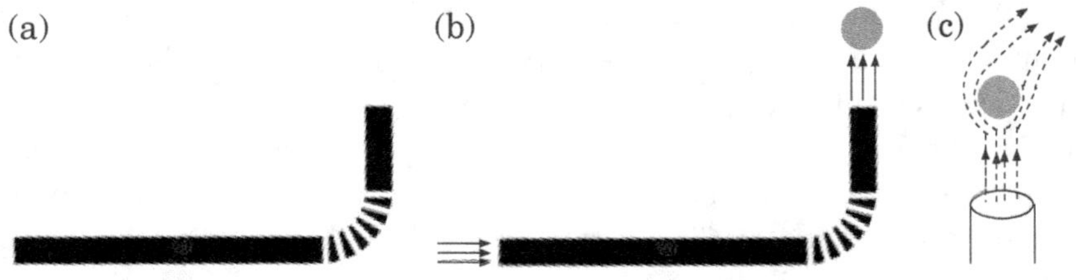

Fig. 1.6: (a) A drinking straw. (b) Floating of a ping-pong ball by blowing air into the straw. (c) Flow of air around the ball.

the conservation of energy this energy should be equal to E_0. This gives

$$h_{\text{max}} = \frac{E_0}{mg} = \frac{0.5\,\text{J}}{0.01\,\text{kg} \times 9.8\,\text{m/s}^2} = 5.1\,\text{m}\,. \tag{1.12}$$

1.4 Floating a Ping-Pong Ball

Blowing through an empty drinking straw or using a hair dryer, one can float and spin a ping-pong ball in air. For interesting demonstration of it see the videos [23-25]. Take a drinking straw and bend part of it point up as shown in Fig. 1.6a. Over the short end of the straw, keep a light-weight ping-pong ball and hold it with your fingers. Blow air into the straw and slowly release the ball. The ball will be suspended in the air.

Instead of a drinking straw, a hair dryer (blower) or a vacuum cleaner can be used. Switch-on a hair dryer and keep its air coming part vertically so that the invisible air flows vertically upward. Above the air stream, place a ping-pong ball. When the ball is placed at a certain height, it is balanced and stably levitated on the stream of air. In this process the ball initially moves up and down, spins and finally settles down near the centre of the air stream. This is dramatic to watch. When we try to pull the ball out of the air stream, we can easily realize a force pulling it back in. We can also notice that the ball deflected the air stream outward.

1.4.1 *Mechanism of Floating of the Ball*

What is actually happening? When the ball is placed over the air blown from the drinking straw or the hair dryer, the ball flies above with the air flowing up from, say, the hair dryer. As the ball reaches a certain height, at which there is a balance between the force of gravity that pulls the ball down and the force of air that pushes the ball up, the ball is suspended in the air column at that point of balance [26]. The air moving upward hits the bottom of the ball and its speed is slowed down. This creates a region of higher pressure under the ball and keeps the ball against the gravitational pull. Alternately, as shown in Fig. 1.6c [27], the air flows along the curved (arc) path around the ball. The air pressure on the ball decreases. According to Bernoulli's theorem, a moving fluid has a lower pressure. So, the pressure on the surface of the floating ball decreases due to the stream of air flowing around the surface of the ball. This permits the atmospheric pressure of the calm air on the other side of the ball to push it back to the air stream.

One can observe that the floating ball stays at the centre of the flowing air stream. *How do we account this?* In the centre of the air stream, the air moves much faster than that at the edge. This is because the air at the sides has friction with the stationary air in the room and is thus slowed. When the ball starts to move out of the centre, the relatively speedy air passing the ball become on the side back towards the centre. As the slow-down air on the outside creates higher pressure and the speedy air creates lower pressure the ball is recentred.

1.4.2 *Minimum Volumetric Flow Rate for Stable Floating*

It is possible to determine the minimum volumetric flow rate Q of air through the drinking straw for a stable float of the ball [28]. Denote m as the mass of the ball, R as its radius, r as the radius of the straw, g as the gravitational constant, the cross-sectional area of the ball as $A = \pi R^2$, v as the velocity with which air travels over the surface of the ball and ρ as the density of air. Q is equal to $\pi R^2 v$. Assuming the displacement of the ball due to gravity is smaller than one radius we have

$$mg < \pi R^2 v \rho v = Q \rho v = \frac{\rho Q^2}{\pi R^2}. \tag{1.13}$$

That is,

$$Q > R\left(\frac{\pi m g}{\rho}\right)^{1/2} \quad \text{and} \quad v > \frac{1}{R}\left(\frac{mg}{\pi \rho}\right)^{1/2}. \tag{1.14}$$

Solved Problem 4:

What will happen if the stably floating ball is slightly deflected?

In the neighbourhood of the surface of the ball, the air flows along curved paths. The radius of the curvature can be roughly the radius R of the ball. This curvature leads to a pressure gradient $\rho v^2/R$. That is, $\nabla P = -(\rho v^2/R)\mathbf{e}_r$. This pressure gradient points radially inwards to the balance point at which the net force on the ball is zero and the ball is stably floating. Assume that the ball is deflected from the stable equilibrium position normally by a distance z. The total volume of the ball displaced is $\approx Az = \pi R^2 z$. Then, the restoring force is given by [28]

$$\mathbf{F} = -\frac{\rho v^2}{R}\pi R^2 \mathbf{z} = -\pi \rho v^2 R \mathbf{z}. \tag{1.15}$$

We rewrite the above equation as

$$\mathbf{F} = -m\omega^2 \mathbf{z}, \quad \omega = v\left(\frac{\pi \rho R}{m}\right)^{1/2}. \tag{1.16}$$

F given by the above equation can be regarded as the force of harmonic oscillator with ω being the natural frequency of oscillation. The ball will undergo an oscillation. With $Q = \pi r^2 v$ the frequency of oscillation, $\nu = \omega/(2\pi)$, is

$$\nu = \frac{Q}{A}\sqrt{\frac{\rho R}{4\pi m}}. \tag{1.17}$$

Observe that ν is directly proportional to Q. This indicates that doubling of Q, that is, blowing air through the drinking straw twice as hard will lead to oscillations that are twice as fast [28].

Suppose $x = 0$ is the base of the straw at which initially a ping-pong ball is kept at rest. $x = x^*$ is the position at which the ball is made to float stably by blowing air. *Is it possible to drive the ball from $x = 0$ to $x = x^*$ by a constant blowing?* Through a model equation it has been found that a constant blowing does not make the ball to do so [29]. This is becasue the equilibrium point x^* is essentially an elliptic type point. If an equilibrium point is an elliptic (stable) point, then the trajectories form elliptical orbits around x^* and no trajectory ends on it. The blowing must be time-dependent to bring the ball to x^*.

1.5 Simple Helicopter Toys

A plastic helicopter toy [30] with two wings (propellers) and a cylindrical stem can be bought from toys stores. A typical helicopter toy is shown in Fig. 1.7. This type of toy had originated around 320 A.D. in Jin dynasty

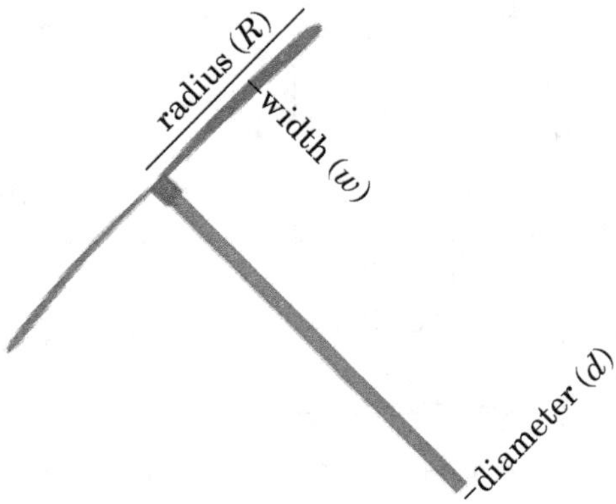

Fig. 1.7: A plastic helicopter toy.

China [31]. In the early period feathers and bamboo were used as wings and sticks were used as a stem.

How do we make a helicopter toy to fly? Hold the stem of the toy between your hands with wings above. Keep the left-hand stationary and fastly move the right-hand away to launch it. The toy flies to a height with counterclockwise spin.

Let us obtain a formula for the lift of the toy [30]. Denote D_{L} (in meter) as the distance the stem of the toy is rotated between the hands up and the duration of rotating the stem as t_{r} (in sec.). r is the radius of the stem. At the time of release, the rotational speed ω of the toy is given by

$$\omega = 2\pi \times \text{number of revolutions} \times \frac{1}{t_{\mathrm{r}}} = 2\pi \frac{D_{\mathrm{L}}}{2\pi r} \frac{1}{t_{\mathrm{r}}} = \frac{D_{\mathrm{L}}}{r t_{\mathrm{r}}}. \tag{1.18}$$

With R as defined in Fig. 1.7 the average speed v of the wing is

$$v = \omega \frac{R}{2} = \frac{D_{\mathrm{L}} R}{2 r t_{\mathrm{r}}}. \tag{1.19}$$

Here v is specified at the midpoint between the centre and the edge of the wing.

The lift force F_{L} depends on the density of air $\left(\rho \text{ in kg/m}^3\right)$, planform area of the wings ($A = 2Rw$ where w is the width of the wing defined in Fig. 1.7), speed v of the wing and the lift coefficient C_{L}. This coefficient is a function of the shape of the wing and the angle of attack. F_{L} is given by

$$F_{\mathrm{L}} = \frac{1}{2} C_{\mathrm{L}} \rho A v^2. \tag{1.20}$$

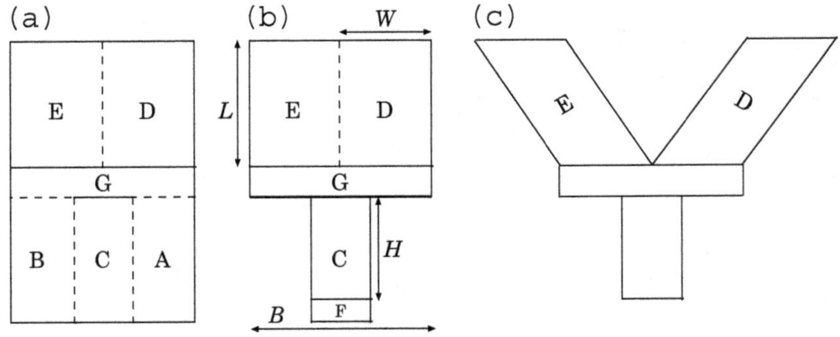

Fig. 1.8: Construction of a paper roto-copter toy. For details see the text.

For a successful flight $F_L > W$, weight of the toy. If the mass of the toy is 10 g, then $W = mg = 0.098\,\mathrm{N}$. For $R = 0.1\,\mathrm{m}$, $w = 0.02\,\mathrm{m}$ and assuming that the shape of the wing is roughly rectangular we have $A = 2Rw = 0.004\,\mathrm{m}^2$. For the choice $D_L = 0.1\,\mathrm{m}$, $r = 0.003\,\mathrm{m}$ and $t_r = 0.2\,\mathrm{s}$ we find $\omega = 167\,\mathrm{rad/s}$ and $v = 8\,\mathrm{m/s}$. Assume that $C_L = 1$ and $\rho = 1\,\mathrm{kg/m}^3$. We obtain $F_L = 0.128\,\mathrm{N}$ and is sufficiently higher than $W = 0.098\,\mathrm{N}$.

According to conservation of momentum the net force in the toy has to be equal to ma, that is m, where m is the mass of the toy and v is the total acceleration. Moreover, the net torque in the toy must be equal to $I\dot{\omega}$ where I is the total torque in the toy and $\dot{\omega}$ is the angular acceleration. We write [38]

$$m\dot{v} = mg - k_1 v^2 \omega, \tag{1.21a}$$

$$I\dot{\omega} = k_2\omega - k_3\omega. \tag{1.21b}$$

The solution of these equations should give the terminal linear and angular velocities as

$$v_t = \left(\frac{k_3 g m}{k_1 k_2}\right)^{1/3}, \quad \omega_t = \left(\frac{g m k_2^2}{k_1 k_3^2}\right)^{1/3} \tag{1.22}$$

In the above two equations k_1, k_2 and k_3 are appropriate constants.

Helicopter toys similar to the one shown in Fig. 1.7 are also called *roto-copter*. There are many variants of roto-copter toys. A simple version of roto-copter made of an ordinary paper with two or more wings (or blades) can be prepared at home. Prepare a paper template as shown in Fig. 1.8a [32-34]. Cut along the dotted lines. Fold the portions A and B on to the backside of the portion C. Now, we have the toy shown in Fig. 1.8b. Next, fold the part F to the backside of C. Bend the wings D and E on the opposite sides and realize the toy as shown in Fig. 1.8c. The wings are tilted at an angle to the ground. The toy is the two wings or two blades roto-copter. It resembles like the winged seeds of the maple tree. One can easily prepare a three wings roto-copter toy.

To make the roto-copter toy to fly, turn on fan in a room and keep it at maximum speed. Release the toy at a corner of the room. In case it is not flying but falls down rapidly, then remove some of its weight. During its fall to the ground it spins. *Why does the toy spin?* [32-34]. When it falls down there are two forces acting on it in vertical direction: one is its weight $F_{\rm W}$ and the other is the resistive force (air resistance by the wings) $F_{\rm D}$. Air uplifts it against the wings thereby bending them up since the wings are very flexible. Hence, as the copter flies, the wings take curved shapes and the parts of the wings further from the body bending up more. The air is flowing past the wings and got deflected by them. The wings are exerting a force on the air as the air got deflected by the wings. This is in accordance with Newton's third law. So, there are forces acting on different sides of the copter and more over in opposite direction thereby making it to spin. *What is the direction of spin?* When the wing E (D) is bend towards you and the wing D (E) is bend away from you, then the toy spins in clockwise (counterclockwise) direction.

For the study of different kinds of helicopter toys, one may refer to the Refs. [35-39].

1.6 Pinwheel

A *pinwheel* is a stick with a twisted paper which turn as it catches the wind. A typical pinwheel is shown in Fig. 1.9a and has four twisted blades. The pinwheel was earlier called as *whirligig* and was invented by a woman in 19th century. The pinwheel is similar to a wind turbine and is easy to make at home. The blades have curved surfaces and are connected to a shaft. The rotation of the pinwheel is due to the hitting of the wind on the curved surfaces of the blades. The wind hits the curved surfaces of the blades which are in the form of like cogs and as a result the wind force against the blades adds up in the curves. This causes the pinwheel to spin according to the Newton's third law of motion.

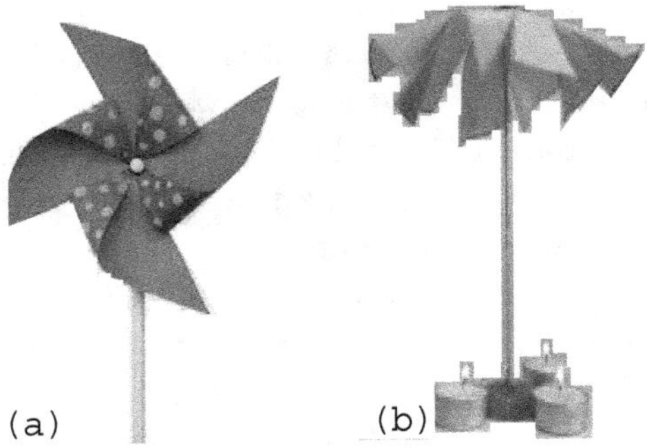

Fig. 1.9: (a) A pinwheel. (b) A pinwheel subjected to hot air developed by candles.

The following are certain interesting aspects of the pinwheel [40-43]:

1. As the shape of a pinwheel is circular and it has blades on all sides whatever be the direction of wind flow, the wind can make contact with the blades and develop rotational motion.
2. A typical pinwheel has rotational symmetry. The centre of the pinwheel is the centre of rotation.
3. The cups of the blades are such that the oncoming air is captured and the blades are pushed in that direction. Thus, if the wind is incident straight at the blades, then the pinwheel spins anticlockwise.
4. The pinwheel spins when air is blown into the cups of the blades. Keeping the front side of the pinwheel facing right to us and blowing air on the bottom half (into the cups) will make the pinwheel to spin quickly anticlockwise. Blowing on the top half, that is against the backs of the cup of the blades, makes the pinwheel to spin slowly in clockwise.
5. The faster the speed of the wind, the stronger the force and higher the speed of rotation.
6. It illustrates how a wheel can be made to spin by a straight movement of the wind.
7. A paper pinwheel with the arrangement shown in Fig. 1.9b spins round. Here, the hot air rises from the candles strikes on the blades of the pinwheel and makes it to spin. This is due to convection process and Newton's third law of motion. It can be used to demonstrate a convection phenomenon.

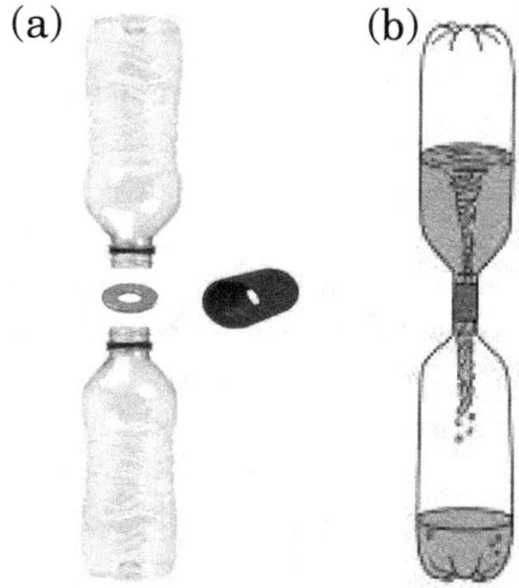

Fig. 1.10: (a) Making of a tornado bottle setup with two water bottles and a washer or a plastic connector. (b) Occurrence of a tornado.

1.7 Tornado Bottle

We are familiar with several examples of vortices including hurricanes, whirlpools, sunspots and spiral galaxies. Tornado is a kind of black cloud vortex with funnel-shaped and spiraling winds. It is easy to create vortex using simple materials. The required materials are two empty two-litre plastic colourless bottles without any label on them, a plastic connector or an appropriate metal washer and a thick electrical tape and food colouring (optional).

Begin by filling one of the two bottles about 2/3 full of water. Add a few drops of food colouring for clear visibility of the vortex to be setup. Place a washer on its mouth and then the second empty bottle over the washer (refer Fig. 1.10a). Tape the joining and mouths of the two bottles together using the electrical tape several times. Make sure that the water will not leak out and the bottles are secured. Turn the setup of the two bottles over. Spin the setup in a circular motion quickly for a few seconds and keep it on a table. The water in the top bottle drip down into the lower bottle through the washer. We can clearly observe the formation of a mini tornado in the water (see Fig. 1.10b). If unable to create a tornado, then you need to try it several times in order to obtain it working properly. One can observe the tornado with one bottle alone. *What is the advantage of using two bottles here?*

The forces of action involved in the tornado formation are the centripetal force and the gravitational force. Let us briefly point out the physics behind the setting of the tornado [44,45]. First, assume that the setup is not rotated initially. When the water filled bottle is kept top of the setup a thin layer of water bulge across the hole of the washer is developed by the surface tension. The weight of the water pushes out the bulge into a large drop of water. This drop drips into the lower bottle. Many drops are followed fastly. Consequently, the air pressure in the lower bottle increases leading to air bubbles getting into the upper bottle. At the same time as the water level and air pressure in the upper bottle decreases the water weight's pressure at the washer also decreases. This continues until the water in the upper bottle completely flows into the lower bottle. There is no formation of a tornado.

Next, consider the case of the setup rotated along a circular path. As the bottles rotating the water also rotating. When the setup is set into vertical with the water filled bottle as the upper one, then two forces come into act. The *gravitational force* pulls the water towards the washer hole. On the other hand, the water is pulled to the middle of the bottle by the *centripetal force*. A vortex is then formed as the water gets into the lower bottle. The water is moving in smaller circles. The speed of the water particles in the upper bottle increases as they are moving toward the washer hole. The centripetal force then increases. As a result, the slope of the water becomes steeper at the bottom of the upper bottle. The hole in the washer allows the air from the bottom bottle to get into the other bottle easily and water drains to the bottom bottle rapidly.

From the above, we note that the vortex in the *tornado bottle* setup is developed by the horizontal rotation of the water, gravity and the centripetal force. A tornado can be made to rotate either in clockwise or counter clockwise direction by initially rotating water in the respective direction.

1.8 Magnetic Levitation of Iron Balls

Earlier we considered the toy Levitron where a magnetized spinning top is levitated in mid air. A simple setup was proposed by Sasaki, Yagi and Murakami to realize levitation of iron balls under moderate magnetic field of a Fe-Nd-B ring [46]. The setup consisted of an Fe-Nd-B ring magnet with 5 cm of outer diameter, 3 cm of inner diameter and 4 cm of height, transparent plastic box and some iron balls of about 1 cm diameter and

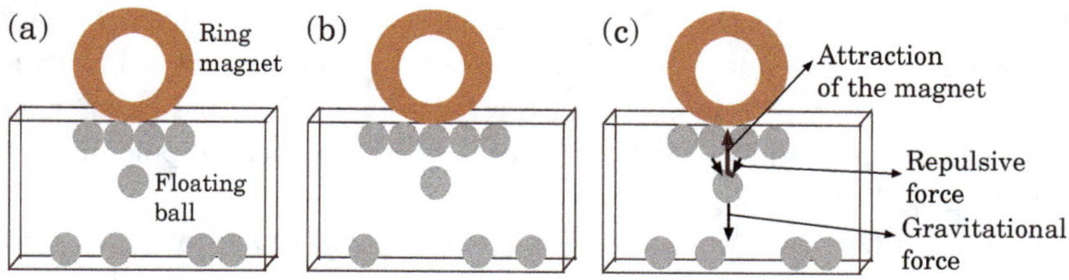

Fig. 1.11: (a) A configuration of floating an iron ball with four balls in the first row touching the box. (b) Another configuration of floating an iron ball with five balls in the first row. (c) Representation of forces acting on the floating iron ball in the second row [46].

4 g in weight. The size of the box used was 6 cm length, 1.25 cm of width, 5 cm of height and 1 mm thickness. The magnet was magnetized along its central axis. Using a glue tape the ring magnet was fitted on the plastic box with its axis of pole tilted 90° from a horizontal line and normal to the wide walls of the box.

The iron balls were made to bounce by shaking the box. After shaking the box the magnet attracted some iron balls. In one configuration, as shown in Fig. 1.11a, four balls were in the first row and these were in contact with the top of the box. Interestingly, in the second row there was an iron ball floating in mid air without in contact with the balls in the first row and the sides of the box. Figures 1.11a and 1.11b show two different equilibrium positions for stable floating of an iron ball. The equilibrium position could be altered by rotating the ring magnet.

Detailed experiments of measuring the field distribution around the floating iron ball(s) revealed the mechanism of the levitation [46-49]. The forces involved in the floating of an iron ball were the magnetic force and gravity. In the stable floating configuration, the field lines were more or less aligned parallel to the horizontal direction and all the balls in the first row were magnetized along the same horizontal direction. This resulted in repulsive forces between the iron balls. Though the balls in the first row were attracted by the ring magnet, the repulsive forces of the first row balls on the second row ball prevented it from lifting and hence it was suspended in mid air. (*How long such a ball can suspend in mid air?*) Therefore, to levitate an iron ball in the above setup a repulsive force was required and was provided by the first row iron balls. This is illustrated in Fig. 1.11c.

With reference to the configuration shown in Figs. 1.11a and 1.11c, the field lines are almost parallel to the horizontal line in the neighbourhood of the centre of the magnet. That is, the middle two iron balls in the first row are magnetized along the same horizontal direction. Let m_1 and m_2 are the magnetic dipole moments of the two balls and r_1 and r_2 are their heights from the ground. Denote M and h as the magnetic dipole moment and the height of the ring magnet.

The force between two magnetic dipoles $\mathbf{m}_1$ and $\mathbf{m}_2$ separated by $\mathbf{r}$ is given by

$$\begin{aligned}\mathbf{F}(\mathbf{r},\mathbf{m}_1,\mathbf{m}_2) = \frac{3\mu_0}{4\pi r^5}\Bigg[&(\mathbf{m}_1\cdot\mathbf{r})\,\mathbf{m}_2 + (\mathbf{m}_2\cdot\mathbf{r})\,\mathbf{m}_1 + (\mathbf{m}_1\cdot\mathbf{m}_2)\,\mathbf{r}\\ &-\frac{5\,(\mathbf{m}_1\cdot\mathbf{r})\,(\mathbf{m}_2\cdot\mathbf{r})\,\mathbf{r}}{r^2}\Bigg]. \end{aligned} \tag{1.23}$$

In our case, $\mathbf{m}_1$, $\mathbf{m}_2$ and $\mathbf{r}$ are in the z-direction. So, $\mathbf{m}_1 = m_1\mathbf{k}$, $\mathbf{m}_2 = m_2\mathbf{k}$, $\mathbf{r} = \mathbf{r}_{21}\cdot z\mathbf{k}$ and $r^5 = z^5$. Then, Eq. (1.23) gives the force on $\mathbf{m}_2$ as

$$\begin{aligned}\mathbf{F}_{21} &= \frac{3\mu_0}{4\pi z^5}\left[m_1m_2z\mathbf{k} + m_1m_2z\mathbf{k} + m_1m_2z\mathbf{k} - \frac{5m_1m_2z^3}{z^2}\mathbf{k}\right]\\ &= -\frac{3\mu_0m_1m_2}{2\pi z^4}\mathbf{k}\\ &= -\frac{3\mu_0m_1m_2}{2\pi(r_2-r_1)^4}\mathbf{k}. \end{aligned} \tag{1.24}$$

Therefore, the force on $\mathbf{m}_2$ due to $\mathbf{m}_1$ is in the $-\mathbf{k}$ direction. $\mathbf{m}_2$ repulses $\mathbf{m}_1$ with the force $\mathbf{F}_{21}$.

As $\mathbf{M}$, the magnetization of the magnet, is in the $-\mathbf{k}$ direction, the force on $\mathbf{m}_2$ due to $\mathbf{M}$ is

$$\mathbf{F}_{\mathrm{Mm}_2} = -\frac{3\mu_0}{2\pi z^4}\,(\mathbf{M}\cdot\mathbf{m}_2)\,\mathbf{k}. \tag{1.25}$$

As $\mathbf{M}\cdot\mathbf{m}_2 = -Mm_2$ we get

$$\begin{aligned}\mathbf{F}_{\mathrm{Mm}_2} &= \frac{3\mu_0}{2\pi z^4}Mm_2\mathbf{k}\\ &= \frac{3\mu_0}{2\pi(h-r_2)^4}Mm_2\mathbf{k}. \end{aligned} \tag{1.26}$$

As it is in the $+\mathbf{k}$ direction, $\mathbf{F}_{\mathrm{Mm}_2}$ is an attractive force.

Using Eqs. (1.24) and (1.26), we can write the balancing condition in the vertical direction as

$$\text{upward force} = \text{downward force} \implies F_{\mathrm{M2}} = F_{21} + mg, \tag{1.27}$$

where m is the mass of the floating ball and g is the acceleration of gravity. *What about the force balance along the horizontal direction?* Change in the balance conditions and the position of the magnet and the iron balls can be utilized to realize different floating configurations.

It is to be noted that the Earnshaw's theorem forbids a three-dimensional stable levitation like the one realized above. The theorem states that no fixed configuration magnet can be made stationary and stable by three-dimensional static magnet/or gravitational forces without any contact physically [49]. In the iron ball levitation, the ball is found to be not touching the sides of the box. It has been argued that possibly the walls of the plastic box have a small diamagnetic susceptibility and exerts a repulsive force to stabilize the levitation [49,50].

It is possible to realize a group of balls floating. For example, Murakami et al [48] considered a setup with 20 iron balls. In one peculiar configuration, they found 9 balls in the first row with each ball in contact with the nearest neighbours, in the second and third rows there were 6 and 2 balls, respectively, floating without touching the walls of the box as well as without touching any other ball. The levitated balls were found to move in a group in a direction along which the ring magnet made to move. The distribution of the balls was found to be symmetric with respect to the centre line of the ring magnet.

It may be possible to utilize the levitation of balls in mid air to transport objects without any physical contact. Particularly, it can be used in the production processes of semiconductor that are highly susceptible to contamination when physical contacts with the object got moved. The setup presented here can be treated as an educational toy (which can be easily setup) to demonstrate the levitation of objects and various experimental analysis can be performed to create different stable configuration and the associated field distribution.

Solved Problem 5:

If two magnets are pointed along the z-direction and separated by a distance z along the z-direction, then the magnetic field due to a magnet of dipole moment m_1 at a distance z is given by $\mathbf{B} = -\left(\mu_0/(4\pi)\right) \nabla \left(m_1/z^2\right)$. If this field acts on another magnet of magnetic moment m_2 pointing in the same z-direction, then the force between them is given by $\mathbf{F} = -\nabla\left(-\mathbf{B}\cdot\mathbf{m}_2\right)$. Determine the force between the two magnets.

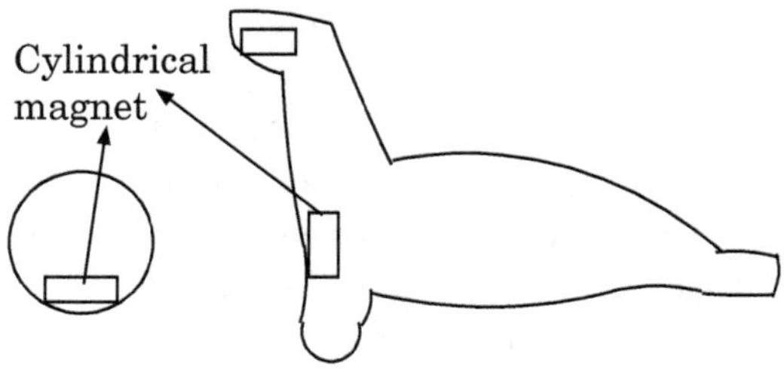

Fig. 1.12: The outline of the magnetic seal and ball toy.

We obtain

$$\begin{aligned}\mathbf{B} &= -\frac{\mu_0}{4\pi}\nabla\left(\frac{m_1}{z^2}\right)\\ &= -\frac{\mu_0 m_1}{4\pi}\left(\mathbf{i}\frac{\partial}{\partial x}+\mathbf{j}\frac{\partial}{\partial y}+\mathbf{k}\frac{\partial}{\partial z}\right)\frac{1}{z^2}\\ &= \frac{\mu_0 m_1}{2\pi z^3}\mathbf{k}\,. \end{aligned} \tag{1.28}$$

Then,

$$\begin{aligned}\mathbf{F}_{21} &= -\nabla(-\mathbf{B}\cdot\mathbf{m}_2)\\ &= \nabla\left(\frac{\mu_0 m_1\mathbf{k}\cdot m_2\mathbf{k}}{2\pi z^3}\right)\\ &= \frac{\mu_0 m_1 m_2}{2\pi}\nabla\frac{1}{z^3}\\ &= -\frac{3\mu_0 m_1 m_2}{2\pi z^4}\mathbf{k}\,. \end{aligned} \tag{1.29}$$

1.9 Magnetic Seal and Ball

One of the fascinating magnetic toys is the magnetic seal and ball. This toy consists of a plastic seal and a plastic ball both having permanent magnets inside and no battery. The sketch of the toy is shown in Fig. 1.12. Place the seal and the ball on a table both well separated. When the seal is brought near to the ball, the ball starts to spin. As the distance between them decreases, the ball spins faster.

The seal has two magnets, one in the head and another at the base. The purpose of the magnet in the head is to balance the ball on the head and is not responsible for the spinning of the ball placed on the table. The magnet at the base is cylindrical and its axis is horizontal pointing towards the ball.

The magnet in the ball is disk shape. Turner analysed the working of this toy [51] and here we summarize it. Studying with iron fillings and compass revealed that the disk magnet in the ball is perpendicularly magnetized to the plane of the disk. Further, the magnetic moment is in the vertical direction. In the seal, the magnet in the base is magnetized in the plane of the disk, while the magnetic moment points at an angle 45° to the vertical.

Let us say initially the magnet in the ball has the magnetic moment $\mathbf{M}$ vertically. $\mathbf{B}$ is the magnetic field of the base magnet in the seal. Its magnetic moment is about 45° to vertical. The magnetic moment of the magnet in the seal has two components: vertical and horizontal. The horizontal component of the base magnet in the seal exerts a torque $\boldsymbol{\tau} = \mathbf{M}\times\mathbf{B}$. On the other hand, the vertical component of the magnet gives a repulsive force $\mathbf{F}$. Both these torque and the repulsive force are exerted on the magnet in the ball. The torque causes the ball to tip slightly. The repulsive force provides a torque $\boldsymbol{\tau} = \mathbf{R}\times\mathbf{F}$ about the point P' where P' is the point of contact of the ball on the table and $\mathbf{R}$ is the distance between the centre of the magnet in the ball and the contact point. This torque enables the ball to spin about the vertical axis.

Essentially, the horizontal component of $\mathbf{M}$ of the magnet in the seal tips the ball while the spinning of the ball is caused by the vertical component of $\mathbf{M}$. In this action, actually the ball slowly moves away from the seal. In order to make the ball to spin continuously the seal has to be continuously moved towards the ball. If the magnet in the ball is tipped to the other side, then the spinning of the ball will take place in the opposite direction.

Where does the energy needed to sustain the motion come from? What will happen if the magnet at the base of the seal kept vertical?

1.10 Simple DC Motors

An electric motor is essentially an electrical machine that converts electrical energy into mechanical energy. A DC motor works under the principle that when a current carrying conductor is kept in a magnetic field, it experiences a mechanical force.

With a magnet and a 1.5 V battery as the prime parts, simple toys/devices called *homopolar motors* or *DC motors* can be easily setup in a classroom/home to demonstrate in a simple manner the effect of magnetic force [52-58]. Such devices are very easy to build, low in cost and moreover fascinate and surprise the students. Let us mention two such motors.

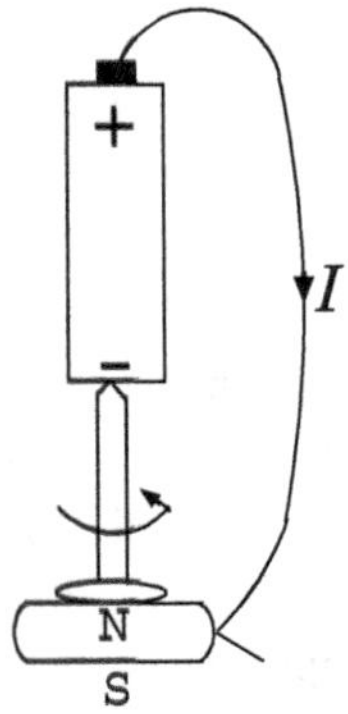

Fig. 1.13: The setup of a simple DC motor with a 1.5 V battery, a small magnet and a nail.

The components required are a 1.5 V battery, a small cylindrical magnet, a steel nail or a screw and a short piece of fluxible copper wire. Place the head of the nail on the top of the magnet. The head of the nail sticks to the magnet as it is attracted by the magnet. Connect the sharp tip of the nail to the negative terminal of the battery in a vertical position. As the nail got magnetized, the nail hangs from the bottom of the battery as shown in Fig. 1.13. Next, connect one end of the wire to the positive terminal of the battery. When the other end of the wire is touched to the side of the magnet (tangentially or perpendicularly to the curved surface of the magnet) a closed circuit is setup and current passes through the magnet. The magnet and the nail start to rotate fast around the vertical axis. The rotation will not take place if the wire is made to touch the centre of the free bottom side of the magnet. The current passing through the magnet feels a force due to the magnetic field through which it flows. The rotation of the magnet is not due to this force [57] since this force is internal one occurring within the magnet. The external circuit is responsible for the rotation of the magnet.

The field $\mathbf{B}$ of the magnet points parallel to its axis. As pointed out in [58] each radial current element, say, $i\mathrm{d}\mathbf{l}$ with current intensity i experiences a Lorentz force $\mathrm{d}\mathbf{F}$ where $\mathrm{d}\mathbf{F} = i\mathrm{d}\mathbf{l} \times \mathbf{B}$. This force acts in the direction orthogonal to the field and the current. This force produces a twisting force, torque, on the magnet leading to spinning of it about its axis in a counter-clockwise direction when viewed from the top.

Fig. 1.14: The setup of a simple motor with a 1.5 V battery, a small magnet, a copper coil, two safety pins and a rubber band.

There is an another simple DC motor [59-61]. To build it we need a 1.5 V battery, two safety pins, a rubber band, a small strong magnet and one foot of insulated copper wire and a small piece of sand paper. Create a small coil using the wire and straighten the two ends of the wire as shown in Fig. 1.14a. Remove the insulation coating on the two straight ends of the wire. Setup the motor as shown in Fig. 1.14b. Use the rubber band to attach the top of the safety pins on either terminal of the battery. Using a sticky putty or a piece of sticky tap, fit the battery on a floor or a table or a piece of flat wooden piece so that the setup will not move and rigidly fixed. Insert the two ends of the wire into the hole of the two safety pins so that the coil hangs between the pins as shown in Fig. 1.14b. Place the magnet on the top of the battery. Initially give a little spin to the coil. The coil will spin as long as the battery is not weak. *Why does the wire set into a coil?*

In the setup current flows from the positive terminal of the battery, through the safety pin attached to it, the coil and then to the negative terminal via the second safety pin. In this way a current carrying closed loop is formed. The loop creates a magnetic field. The direction of this field can be determined using the *right-hand rule*. There is another magnetic field created by the magnet attached to the battery. If the coil is perfectly perpendicular to the magnet, then the two fields are antiparallel. Further, if the fields produced by the magnet and the coil are equal in strength and are antiparallel, then the fields are cancelled and the coil got locked and stationary. But the fields are not perpendicular to each other and not equal in strength. The field of the coil can be increased by increasing its turns. The magnet attracts one side of the coil and repel the other side of the coil. That is, an electromagnetic interaction is developed causing the coil to spin. In this way the electrical energy is changed into mechanical energy.

Fig. 1.15: A kaleidoscope toy and some of the images viewed in it.

Does the coil spin in both directions? Why? How do you determine the field produced by the coil? By bringing another small magnet the speed of the coil can be increased or decreased by changing the pole of the second magnet.

Based on the principles of working of the two simple DC motors presented in this section, students may setup a variety of DC motors with a very few materials.

1.11 Kaleidoscope

Among the fascinating toys, the kaleidoscope is a unique delightful device-like one mesmerizing people of all ages by its creation of infinite variety of rich pictures through the interplay of mirrors. A single plane mirror by means of reflection gives an identical image. However, the use of multiple mirrors magically generates marvelous simple to complex patterns. A kaleidoscope is a simple optical device cum toy consisting of two or more reflecting surfaces (reflectors) tilted by an angle to each other such that objects or part of them on one end of the mirrors could be seen as a regular symmetrical pattern due to repeated reflection if seen from the other end. Commercially available kaleidoscopes are small in size, cylindrical in shape, contain three reflecting mirrors and have no great external appearance but provide unimaginable mesmerizing secret images which even been changed quickly. Figure 1.15 shows a kaleidoscope and some of the images seen in it.

Generating images with multiple mirrors have been described more than 500 years ago by Giambattista Della Porta (1535-1615), Athanasius Kircher

(1601-1690) and Richard Bradley (1688-1732) [62]. The study of theory of polarization of light by multiple reflections by the Scottish physicist Sir David Brewster (1781-1868) led to the invention of kaleidoscope in 1816. When he looked certain things at the end of two mirrors, he observed patterns and colours in different arrangements. These observations led him to the invention of kaleidoscpe [63]. He derived this name from the Greek words *kalos* (beautiful), *eidos* (aspect) and *skopein* (to see/I see). In 1817 he filed a patent (number 4136) for his kaleidoscope which was then granted. Even after 200 years of its invention, the kaleidoscope is commercially available throughout the world in a variety of versions.

To build a simple kaleidoscope of your own, see [64]. The body of the kaleidoscope is often made from tin metal or cardboard tube within which reflectors (2 or 3) are kept. The reflectors are usually plane mirrors. In the case of three mirrors, the mirrors are arranged to form an equilateral triangle. The reflectors have equal width and length. They are held at certain angles such that the surfaces of the reflectors face inwards. At one end, an aperture is held acting as a viewing eye-hole. At the other end, a transluent object cell is fixed. This contains coloured pieces which can move when the cell or the whole device is rotated. Light is crucial to produce fascinating patterns in the kaleidoscopes [63]. Certain kaleidoscopes have oil filled wands with glitter and small foil shapes. This gives rise sparks of light when the kaleidoscope is rotated by 180°.

The two basic principles behind the function of a kaleidoscope are the laws of reflection and the well-known fact that white light contains VIBGYOR (Violet, Indigo, Blue, Green, Yellow, Orange, Red). A light incident on a surface is reflected. The direction of the reflection and the angle of reflection depend on the direction of the incidence and the angle of incident. The law reflection is that the angle of incident is equal to the angle of reflection. The light entering into the kaleidoscope passes through the traiangular tunnel, the walls of which are the reflecting mirrors. Occurrence of multiple reflections of light and absorption of some of the colours of light by the coloured pieces placed at the end of the tunnel creates innovative beautiful patterns. The angle between the mirrors essentially determines the number of reflections and the complexity of the patterns.

Nowadays, computer graphics and new technologies allow great exploration of the kaleidoscope phenomenon. The beautiful patterns generated by a kaleidoscope are used by fashion designers. For creation of uniform distribution of light in laptop screens, kaleidoscopes can be utilized. The mirror symmetry in a kaleidoscope paves ways for new design for optical tools and technologies. Students may be asked to predict and analyse the effect of varying the number of mirrors, angle between the mirrors, length of the mirrors, width of the mirrors and the position of the eye-hole.

1.12 Conclusion

Certain other interesting educational toys for which studies have been reported include wobbling christmas tree toy [65-67] which exhibits unusual oscillatory and phase transitions, Jacob's ladder [68], kalliroscope displaying spatially extended flow structures developed by heat from fingers tips [69,70], tumbling toy [71,72], thermal swing working under the principle of conversion of heat energy into kinetic energy [72,73], ooze tube [74,75], water rocket[76] and the horizontal flying carousel swing [77].

Some other toys useful to explain certain principles of physics are [78] a clown on a monocycle (balancing toy), drunk falling backwards (due to shift in the CoG) the boy on a trapeze (mechanical bank toy), a climbing monkey (a pulley system), Newtonian nut cracker, propeller on a screw, the cracking frog (wherein an application of a bending force to a flexible steel plate in it producing a cracking noise), pendulum cart (to and fro motion), toy rocket (working with jet propulsion) and the bird in a shell (the working of it is an analogy of a first-order phase transition).

In recent years, many novel materials and their surprising properties have been realized. These have been used to develop new kinds of toys. Most of these toys are not yet manufactured in bulk and not yet reached to toy shops world wide. Lots of such fascinating toys and short explanation of physics of their functioning are described in physicsfunshop.com. Some of these toys are tapered 4-mirror kaleidoscope, the Hoverpen, earth mass black hole, corner cube prism, hydro-gyro, 2d-magnetic levitation, bismuth diamagnetic trap levitation, the spin geared kinetic toy, diffraction spectrum handle, ferro-fluid interactive lava lamp, buda ball, vortex dome, swing thing, magnetic top and snake and chaotic magnet pendulum.

1.13 Bibliography

[1] R.C. Turner, Phys. Edu. **30**, 542 (1992).
[2] M. Faraday, *A Course of Six Lectures on the Various Forces of Matter and Their Relations to Each Other* (Harper & Brothers, New York, 1868).

[3] D. Wells, *Natural Philosophy* (Ivision, Phinney & Co, New York, 1860).
[4] http://cdn.teachersource.com/downloads/lesson_pdf/CTR-200.pdf.
[5] R.C. Turner, Am. J. Phys. **55**, 84 (1987).
[6] J. Fort, J.E. Llebot, J. Saurina and J.J. Sunol, Phys. Edu. **33**, 98 (1998).
[7] T.B. Greenslade, The Phys. Teach. **19**, 554 (1981).
[8] T.B. Greenslade, The Phys. Teach. **59**, 540 (2021).
[9] R.H. Romer, The Phys. Teach. **43**, 68 (2005).
[10] L. Minkin and A.C. Zable, The Phys. Teach. **60**, 549 (2022).
[11] https://en.wikipedia.org/wiki/Chatter_ring.
[12] http://www.jugglingworld.biz/tricks/prop-manipulation-tricks/jitter-ring-tricks/.
[13] H.R. Crane, The Phys. Teach. **30**, 306 (1992).
[14] https://www.youtube.com/watch?v=DP86wo.a8Ue0.
[15] https://www.youtube.com/watch?v=wWpVYd8Qo-U.
[16] https://www.real-world-physics-problems.com/gyro-rng.html.
[17] https://www.rose-hulman.edu/~moloney/AppComp/2001Entries/e0-4a/chatteringtoy.htm.
[18] G. Flores and H. Fearn, Gyro-ring; https://physics.fullerton.edu/~heidi/PROJECTS/gyro-ring.doc.
[19] https:/en-wikipedia.org/wiki/Toy_gun.
[20] https://en.wikipedia.org/wiki/Cap_gun.
[21] https://en.wikipedia.org/wiki/Air_gun.
[22] http://www.madehow.com/Volume-6/Water-Gun.html.
[23] https://www.youtube.com/watch?v=V2deFgyj3XQ.
[24] https://www.youtube.com/watch?v=FZeZcu_gN9o.
[25] https://www.youtube.com/watch?v=ofmi7Fbe8jl.
[26] https://www.physicscentral.com/experiment/physicsathome/ping-pong-physics.cfm.
[27] https://www.exploratorium.edu/snacks/balancing-ball.
[28] https://physics.stakeexchange.com/questions/69731/floating-a-ping-pong-ball-in-the-air-just-using-a-pen.
[29] P.K. Newton and Y. Ma, Am. J. Phys. **89**, 134 (2021).
[30] S. Shakerin, The Phys. Teach. **51**, 310 (2013).
[31] https://en.wikipedia.org/wiki/Bamboo-copter.
[32] https://www.exploratorium.edu/science_explorer/roto-copter.html.
[33] D.H. Annis, The Phys. Teach. **59**, 320 (2005).
[34] https://www.essexmums.com/educationalactivities/paper-roto-copt-er/.

[35] G.E.P. Box, Qual. Engin. **4**, 453 (1992).
[36] Y.K. Ng, S.Y. Mak and C.M. Chung, The Phys. Teach. **40**, 181 (2002).
[37] D.H. Annis, Am. Statis. **59**, 320 (2005).
[38] M.M. Williams, The Review: J. Undergraduate Student Res. **8**, 33 (2006).
[39] M. Liebl, The Phys. Teach. **48**, 458 (2010).
[40] https://www.scientificamerican.com/article/strong-wind-science-the-power-of-a-pinwheel/.
[41] https://www.sciencebuddies.org/stem-activities/strong-wind-science-the-power-of-a-pinwheel.
[42] https://www.science-sparks.com/wind-experiment-make-a-pinwheel/.
[43] https://schoolscienceexperiments.com/2nd-grade-science-projects-pin-wheel-experiment/.
[44] http://www.exploratorium.edu/snacks/.
[45] https://aamboceanservice.blob.core.windows.net/oceanservice-prod/education/for_fun/TornadoBottle.pdf.
[46] S. Sasaki, I. Yagi and M. Murakami, J. Appl. Phys. **95**, 2090 (2004).
[47] M. Sakai, Y. Takabayashi, K. Sugawara, T. Shibayama, Y. Shimomura, T. Orikasa, K. Kamishima and N. Hirastuka, J. Appl. Phys. **97**, 083908 (2005).
[48] M. Murakami, Y. Nishimura, T. Hirooka, S. Sasaki and I. Yagi, J. Appl. Phys. **97**, 083911 (2005).
[49] Y. Sakurai, J. Appl. Phys. **104**, 044503 (2008).
[50] A.L. Geim, M.D. Simon, M.I. Boamfa and L.O. Heflinger, Nature **400**, 323 (1999).
[51] R.C. Turner, The Phys. Teach. **25**, 568 (1987).
[52] C. Chia Verina, The Phys. Teach. **42**, 553 (2004).
[53] D. Featonby, Phys. Edu. **40**, 505 (2005).
[54] D. Featonby, Phys. Edu. **41**, 292 (2006).
[55] S.M. Stewart, Rev. Bras. Ensinode Fisica **29**, 275 (2007).
[56] H.K. Wong, The Phys. Teach. **47**, 124 (2009).
[57] H.K. Wong, The Phys. Teach. **47**, 463 (2009).
[58] A.K.T. Assis and J.P.M.C. Chaib, Am. J. Phys. **80**, 990 (2012).
[59] J. Yap and D. MacIsaac, Phys. Edu. **41**, 427 (2006).
[60] https://www.education.com/science-fair/article/no-frills-motor/.
[61] https://www.spsnational.org/the-sps-observer/summer/2017/simple-motor.

[62] K.D. Graf and B.R. Hodgson, *Popularizing Geometrical Concepts: the Case of the Kaleidoscope* the paper presented at the ICMI Symposium on Popularization of Mathematics, University of Leeds, 17-22, September 1989.
[63] D. Correia, Bullettin of the Scientific Instrument Society **131**, 1 (2016).
[64] C.R. Conwell and J.S. Paschal, The Science Teacher, November 1986, pages 41-44.
[65] J. Sivardiere, Am. J. Phys. **51**, 1016 (1983).
[66] R.B. Prigo, Am. J. Phys. **52**, 225 (1984).
[67] B. Rodewald and H.J. Schlichting, Am. J. Phys. **53**, 1172 (1985).
[68] https://en.wikipedia.org/wiki/Jacob%27s_ladder_(toy).
[69] D. Borrero-Echeverry, C.J. Crowley and T.P. Riddick, Phys. Fluids **30**, 087103 (2018).
[70] https://www.paulmatisse.com.
[71] J. Guemez, C. Fiolhais and M. Fiolhais, Phys. Edu. **44**, 53 (2009).
[72] R.I. Leine, D.H. Van Campen and C.H. Glocker, J. Sound Vib. Control **9**, 25 (2003).
[73] P.W. Hewson, Thys. Phys. Teach. **13**, 350 (1975).
[74] C.M. Saviz and S. Shakerin, The Phys. Teach. **52**, 332 (2014).
[75] S. Shakerin, The Phys. Teach. **56**, 248 (2018).
[76] A. Romanelli, H. Bove and F.G. Madina, Am. J. Phys. **81**, 762 (2013).
[77] A. Ferstl and E.R. Duden, The Phys. Teach. **60**, 428 (2022).
[78] H. Levinstein, The Phys. Teach. **20**, 358 (1982).

1.14 Exercises

1.1 When the ball in the floating of a ping-pong ball in air being in a balance state, move the blower/hair dryer towards a wall. The height of the levitated ball increases greately. Why does the ball move to a greater height when it is near the wall?

1.2 Why does air in the floating ping-pong ball follow the surface of the ball?

1.3 A ping-pong ball of mass $m = 15\,\text{g}$ is set to float in air by means of a jet of water coming out of a nozzle. Assume that the water strikes the ball with a speed $v_\text{i} = 0.3\,\text{m/sec}$ and after collision the water falls dead. Calculate the rate of flow of water in the nozzle.

1.4 Why does not the roto-copter toy move sideways through the air even though part of the thrust being along horizontal?

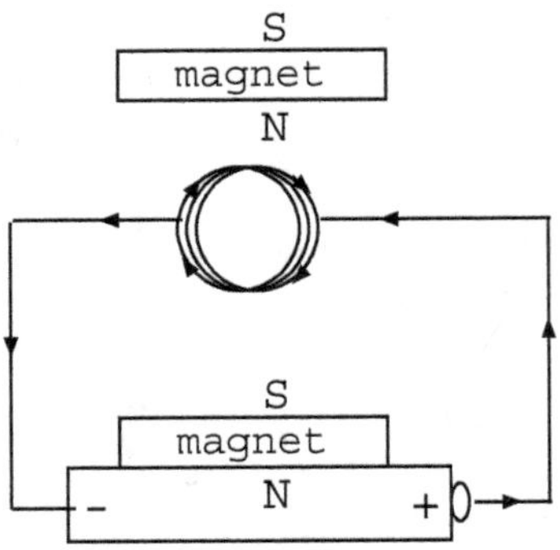

Fig. 1.16: The modified setup of the DC motor shown in Fig. 1.14 with two magnets.

1.5 Consider a roto-copter toy with two wings in the equilibrium case wherein the toy attains its terminal velocity, that is a constant velocity. Obtain an expression for the terminal velicity [37].

1.6 What will happen if the front side of a pinwheel facing your left-side (a) blown into the side of the blades and (b) blown on the bottom half (against the back of the cups)?

1.7 Consider the DC motor setup made with a 1.5 V battery supplying 5 A current, a copper coil of resistance 0.3 ohm with 5 turns with radius 0.008 m and a small magnet. Determine the field produced by the coil.

1.8 Consider the DC motor setup shown in Fig. 1.16. Ignoring the fields developed by the coil and the battery sketch the field lines of the setup [59].

1.9 Does the formation of images in a kaleidoscope due to total internal reflection?

1.10 Place two mirrors face to face. Tape over one edge to form a hinge. They form a corner when the mirrors are opened. The number of images reflected is determined by the angle formed. On a paper using a protractor draw lines along say, 60° on either sides of the protractor. Keep the hinged mirrors on the 60° angle lines. Place a small object between the mirrors. Position yourself behind the object in line with the corners formed by the mirrors. How many images formed? Note down the number of images formed for various angles. Write a note on the observations made [64].

Index

www.ingramcontent.com/pod-product-compliance
Lightning Source LLC
LaVergne TN
LVHW010327220926
848484LV00004BB/852